Hydrodynamic Fluctuations
in Fluids and Fluid Mixtures

Cover design by Alfredo García (Draft, S.A. – Madrid)

Hydrodynamic Fluctuations
in Fluids and Fluid Mixtures

by

José M. Ortiz de Zárate
Departamento de Física Aplicada I
Universidad Complutense, Madrid, Spain

Jan V. Sengers
Institute for Physical Science and Technology
and Burgers Program for Fluid Dynamics
University of Maryland, College Park
Maryland, USA

ELSEVIER
Amsterdam – Boston – Heidelberg – London – New York – Oxford
Paris – San Diego – San Francisco – Singapore – Sydney – Tokyo

Elsevier
Radarweg 29, PO Box 211, 1000 AE Amsterdam, The Netherlands
The Boulevard, Langford Lane, Kidlington, Oxford OX5 1GB, UK

First edition 2006

Library of Congress Cataloging-in-Publication Data
A catalog record for this book is available from the Library of Congress

British Library Cataloguing in Publication Data
A catalogue record for this book is available from the British Library

ISBN-13: 978-0-444-51515-5
ISBN-10: 0-444-51515-1

For information on all Elsevier publications
visit out website at books.elsevier.com

Transferred to digital printing in 2007.

06 07 08 09 10   10 9 8 7 6 5 4 3 2 1

Preface

Our aim in this book is to present a systematic guide how to use linearized fluctuating hydrodynamics for interpreting fluctuations in fluids and fluid mixtures, that are either in thermodynamic equilibrium or in non-convective, non-turbulent nonequilibrium states. The topics covered in this book summarize research pursued during the last two decades with a large number of collaborators including G. Ahlers, J.R. Dorfman, J.A. Fornés, R.W. Gammon, B.M. Law, W.B. Li, J.C. Nieuwoudt, T.R. Kirkpatrick, L. Muñoz Redondo, J. Oh, F. Peluso, R. Pérez Cordón, R. Schmitz, P.N. Segrè, and K. J. Zhang. We also acknowledge valuable discussions with M.A. Anisimov, E.G.D. Cohen, M.E. Fisher and H. van Beijeren. We are especially indebted to J.R. Dorfman, T.R. Kirkpatrick and J.M. Rubí for constructive comments on several subjects described in this book. We have greatly appreciated the stimulating environment of our colleagues at the Institute for Physical Science and Technology at the University of Maryland, where a significant part of this book was written. We are indebted to Alfredo García for the cover design. J.M.O.Z. wishes to thank his family and friends for support, encouragement and, especially, patience during the (longer than expected) time it has taken to complete this book. J.V.S. wants to express his gratitude to his wife, Anneke Levelt Sengers, for her life-long supportive partnership in the pursuit of research in science and engineering.

This book was completed during 2005, which has been declared World Year of Physics on the occasion of the centennial of Einstein's "miraculously" productive 1905 year. As is well known, one of the papers published by Einstein in 1905 dealt with the fluctuating motion usually referred to as Brownian motion. The ideas contained in the 1905 paper were later revisited by Einstein himself, and lead to the formulation of the Einstein hypothesis, Eq. (3.38) in our book. The Einstein hypothesis is at the basis of the theory of fluctuations in fluids, and this monograph is our small contribution to the World Year of Physics celebration.

Madrid and College Park, November 2005.

Contents

Chapter 1

Introduction

The purpose of this book is to present a systematic treatment of hydrodynamic fluctuations in fluids and fluid mixtures. The theory of fluctuations in fluids that are in thermodynamic equilibrium is well developed (Boon and Yip, 1980). Specifically, the intensity of density fluctuations is proportional to the isothermal compressibility and the intensity of concentration fluctuations in mixtures is proportional to the osmotic compressibility. Moreover, the decay of thermally excited fluctuations is governed by Onsager's regression hypothesis which says that the decay rates of the fluctuations is determined by the transport coefficients that appear in the linear relations between fluxes and gradients in nonequilibrium thermodynamics. A theoretical frame work for dealing with fluctuations in fluids in thermodynamic equilibrium states is provided by Landau's fluctuating hydrodynamics (Landau and Lifshitz, 1958, 1959; Fox and Uhlenbeck, 1970a). Such fluctuations can be investigated experimentally by light scattering and neutron scattering (Berne and Pecora, 1976).

An interesting issue is the nature of thermal fluctuations in fluids and fluid mixtures that are in thermodynamic nonequilibrium states. While it is well known that nonequilibrium fluids can exhibit large fluctuations associated with convection patterns or turbulence, a rather new subject is that of fluctuations in fluids that are subjected to a temperature gradient or to shear in the absence of convective patterns or turbulent flows. About 50 years ago a microscopic picture of nonequilibrium phenomena in fluids was proposed by Bogoliubov (1946, 1962). It was based on a postulate that a fluid away from equilibrium would proceed to a thermodynamic equilibrium state in two distinct stages: first a microscopic kinetic stage with a time scale of the order of the time between molecular collisions, which for dense fluids or liquids is of the same order as the duration of the molecular collisions, after which local equilibrium is established; second a macroscopic hydrodynamic stage during which the fluid evolves in accordance with the hydrodynamic equations. Implicit in this postulate was the idea that no long-ranged dynamic correlations would be present in a fluid of molecules with short-ranged forces, unless the system would be near an incipient thermodynamic or hydrodynamic instability.

The subsequent history of nonequilibrium statistical physics has revealed a basic flaw in this picture. In evaluating the randomizing nature of molecular collisions one may distinguish between quantities like mass, momentum and energy, that are conserved in molecular collisions, and non-conserved physical quantities. However, it turns out that the slow hydrodynamic modes associated with the conserved quantities and the fast modes associated with non-conserved quantities are not independent, but can interact to cause a coupling between modes resulting in long-ranged (mesoscopic) dynamic correlations. The classical picture of short-ranged dynamic correlations first appeared to be inadequate in fluids near the critical point when experiments revealed a divergent thermal conductivity which could not be explained by the Van Hove theory of critical slowing down of the fluctuations that was based on strictly thermodynamic considerations (Michels et al., 1962; Sengers, 1966; Sengers and Keyes, 1971); this observation led to the development of the mode-coupling theory of critical dynamics (Fixman, 1967; Kadanoff and Swift, 1968; Kawasaki, 1970). Some time later it turned out that the same mode-coupling theory could also account for the presence of long-time tails in the Green-Kubo correlation functions for the transport coefficients that were originally noticed in computer simulations of molecular dynamics (Pomeau and Résibois, 1975; Dorfman, 1975; Ernst et al., 1976a). The presence of mesoscopic dynamic correlations also caused the appearance of a divergence in the virial expansion for the transport coefficients of moderately dense gases (Dorfman and Cohen, 1967; Brush, 1972).

Around 1980 it became evident that the mode-coupling theory would also predict the existence of long-ranged fluctuations in fluids that are kept in stationary nonequilibrium states (Kirkpatrick et al., 1982a,b; Fox, 1982; Tremblay, 1984). Specifically, when a fluid is subjected to a stationary temperature gradient, the temperature gradient causes a coupling between the component of the velocity fluctuations parallel to the gradient and the temperature fluctuations, leading to an algebraic divergence of the fluctuations in the limit of small wave numbers (Kirkpatrick et al., 1982b; Ronis and Procaccia, 1982). An algebraic dependence of the nonequilibrium fluctuations as a function of the wave number is now believed to be a general feature of fluctuations in fluids in stationary nonequilibrium states (Grinstein, 1991; Dorfman et al., 1994) and is a manifestation of a general principle of generic scale invariance in nonequilibrium statistical mechanics (Kirkpatrick et al., 2002; Belitz et al., 2005). The ultimate divergence of the intensity of nonequilibrium fluctuations for small wave numbers, *i.e.*, for large wavelengths, in the presence of temperature or concentration gradients will be prevented by gravity and finite-size effects. This is the reason that gravity and finite-size effects can play an important role in the wave-number dependence of fluctuations in fluids in nonequilibrium states.

In this book we shall elucidate how fluctuating hydrodynamics can be used to deal with both equilibrium and nonequilibrium fluctuations in fluids and fluids mixtures. Fluctuating hydrodynamics is a stochastic fluid dynamics approach in which it is assumed that the fluctuations can be described by the usual hydrodynamic equations but supplemented with random noise terms whose correlation functions are determined by a fluctuation-dissipation theorem. This approach can be extended to fluctuations in fluids in nonequilibrium states by assuming that the noise correlations (in contrast to the correlation functions of the hydrodynamic variables) satisfy local thermal equilibrium, that is, they are assumed to be given by the same fluctuation-dissipation theorem as for fluids in thermodynamic equilibrium, but with local values of the temperature, density and concentration.

We shall find that hydrodynamic correlation functions in fluids in nonequilibrium states always extend over a spatial range that is longer than what one would expect for the corresponding local-equilibrium expressions for these correlation functions, even when the system is far away from any hydrodynamic instability. The long-ranged nature of the nonequilibrium fluctuations causes a volume element inside the fluid to receive information from neighboring volume elements. Thus nonequilibrium fluctuations even far away from any hydrodynamic instability already foreshadow to some extent the ultimate appearance of complex spatio-temporal behavior of fluids far from equilibrium like the onset of convection or the onset of turbulence. In this book we shall demonstrate these features by presenting a detailed analysis of fluctuations in fluid layers subjected to a temperature gradient. However, the fluctuating-dynamics approach elucidated in this book can be used to deal with a wide variety of nonequilibrium fluctuations in fluids and fluid mixtures.

We shall proceed as follows. In Chapter 2 we present a review of some of the basic concepts of nonequilibrium thermodynamics and a derivation of the hydrodynamic equations for fluids and fluid mixtures. Chapter 3 presents the theory of fluctuations in fluids that are in thermodynamic equilibrium. The major part of the book, namely Chapters 4 through 10, deal with thermal fluctuations in fluids and fluid mixtures in the presence of a stationary temperature gradient, to which we refer as the Rayleigh-Bénard problem. The Rayleigh-Bénard problem has been historically the most thoroughly studied nonequilibrium system, both from a theoretical and an experimental point of view. It may be considered as a kind of paradigm for nonequilibrium systems. Ideas and insights obtained from a careful study of the Rayleigh-Bénard problem can be extended and used to interpret other nonequilibrium systems. We start our study of the Rayleigh-Bénard problem in Chapters 4 and 5, where we consider "bulk" thermal nonequilibrium fluctuations in fluids and fluid mixtures, namely, fluctuations with wavelengths much smaller than any finite height of the fluid layer. In Chapter 6 we develop a general procedure for incorporating finite-size effects in fluctuating hydrodynamics. In Chapter 7 we apply the procedure to a study of thermal nonequilibrium fluctuations in one-component fluid layers. In Chapter 8 we consider the nature of such nonequilibrium fluctuations close to the onset of Rayleigh-Bénard convection. Chapter 9 deals with thermal nonequilibrium fluctuations in binary-fluid layers. In this book we consider fluctuations at hydrodynamic space and time scales that are experimentally accessible with light scattering or shadowgraph techniques. In Chapter 10 we present a review of the experimental attempts to measure nonequilibrium fluctuations. In Chapter 11 we give a brief discussion of some other types of nonequilibrium fluctuations, namely nonequilibrium fluctuations in fluids subjected to shear, nonequilibrium interface fluctuations, nonequilibrium fluctuations in liquid crystals and in mixtures with chemical reactions. This chapter confirms the general theme of the book about the long-ranged nature of nonequilibrium fluctuations as summarized in the Epilogue.

One of the problems we encountered when preparing this volume is a lack of uniform notation among the various communities (such as statistical physicists, physical chemists, fluid dynamists, mechanical and chemical engineers) to whom this book is addressed. Even more striking is the presence in the literature of different definitions for some of the thermophysical properties and dimensionless numbers that enter in the description of fluids. In this book, we tried to follow the definitions and nomenclature recommendations of both IUPAP (Cohen and Giacomo, 1987) and IUPAC (Mills et al., 1988). Regarding the description of thermal diffusion, where the coexistence of various definitions and sign conventions is particularly confounding, we have adopted the recommendations contained in the book edited by Köhler and Wiegand (2002). For the benefit of the reader, we have included at the end of the book a *List of symbols and corresponding SI units*, where the nomenclature used is summarized, and reference is made to the page number where the property is first introduced. In addition, we have prepared a *List of abbreviations* which the reader can find at the end of the volume, prior to the *Subject index*.

Chapter 2

Nonequilibrium thermodynamics

In this book we shall consider thermodynamic fluctuations, *i.e.*, spontaneous variations around a mean value of various thermodynamic quantities in fluids. The nature of such fluctuations will be characterized by correlation functions of the fluctuating thermodynamic quantities. To specify the spatial and time dependence of these correlation functions we need to employ basic equations from nonequilibrium thermodynamics and from fluid dynamics. For this reason, we review in this chapter those concepts that will be amply used throughout this book. For a more detailed introduction to nonequilibrium thermodynamics and fluid dynamics the reader can consult many monographs on these subjects. For instance, introductions to nonequilibrium thermodynamics can be found in the classical text of de Groot and Mazur (1962), in a book of Haase (1969) and in recent books of Demirel (2002) and Öttinger (2005). Introductions to fluid dynamics are provided by books of Landau and Lifshitz (1959), of Batchelor (1967), of Tritton (1988), and of Bird et al. (2002).

2.1 Local thermodynamic properties

2.1.1 Equilibrium thermodynamics

The reader of this book is assumed to be familiar with equilibrium thermodynamics of fluids. A system or a subsystem is said to be in thermodynamic equilibrium if its properties would not change when the system or the subsystem would be isolated from its environment. As an example we consider here a one-component fluid in the absence of external forces that is in a homogeneous thermodynamic equilibrium state, *i.e.*, in an equilibrium state with only one phase being present. The most relevant thermodynamic quantities to be considered are the temperature T, the pressure p, the chemical potential, the volume V, the (internal) energy U, and the entropy S. Examples of derived thermodynamic properties are the isochoric heat capacity C_V and the isobaric heat capacity C_p. In classical thermodynamics the thermodynamic properties can be defined regardless of the microscopic structure of the system. For a system consisting of molecules, like a fluid, a thermodynamic description can be justified on the basis of equilibrium statistical physics. For thermodynamics to be valid the number of molecules needs to be sufficiently large. How large is a fundamental question in statistical mechanics, a treatment of which falls outside the scope of the present monograph. For all applications in this book, the systems will be sufficiently macroscopic for the concept of thermodynamic averages to be valid.

One distinguishes between intensive and extensive thermodynamic properties. In thermodynamic equilibrium intensive thermodynamic properties, like temperature, pressure and chemical potential, have uniform values throughout the system independent of the size of the system. The values of extensive thermodynamic properties are proportional to the size of the system. More precisely, at a given temperature and pressure extensive thermodynamic properties are proportional to the total mass of the system. Hence, extensive thermodynamic properties, like V, U, S, C_V or C_p, can be made independent of the size of the system by dividing them with the total mass of the system. Extensive thermodynamic properties per unit mass are referred to as specific thermodynamic properties and are designated in this book by lower case symbols. Thus v is the specific volume, u the specific energy, s the specific entropy, c_V the isochoric specific heat capacity, and c_p the isobaric specific heat capacity. Instead of specific volume we prefer to use in practice the (mass) density $\rho = v^{-1}$ as a more convenient thermodynamic variable. A fluid is called to be in *heterogeneous* thermodynamic equilibrium when more than one phase are present. In that case one can also define extensive thermodynamic properties and corresponding specific thermodynamic properties for each individual phase. In heterogeneous thermodynamic equilibrium, intensive thermodynamic properties have the same uniform value in all phases, but the values of the specific thermodynamic properties of the individual phases will in general be different.

The thermodynamic energy U as function of the entropy S and the volume V is an example of a characteristic function from which all other thermodynamic properties can be obtained by differentiation. Other examples of characteristic functions are the enthalpy H as a function of S and p, the Helmholtz energy A as a function of T and V, and the Gibbs energy G as a function of T and p. The relationship between pressure p, temperature T, and mass density ρ is called an equation of state. In differential form:

$$\mathrm{d}\rho = \rho\left[\varkappa_T\mathrm{d}p - \alpha_p\mathrm{d}T\right], \tag{2.1}$$

where $\varkappa_T$ is the isothermal compressibility and α_p the thermal expansion coefficient. In this book use will be made of a variety of thermodynamic relations for one-component fluids and for binary fluid mixtures. Derivations of the relevant thermodynamic relations can be found in many texts on equilibrium thermodynamics.

2.1.2 Continuum hypothesis and the assumption of local equilibrium

Although a fluid consists of molecules, statistical mechanics confirms that at a macroscopic level the state of the fluid can be described by a restricted number of thermodynamic variables. To extend these concepts to systems that are not in thermodynamic equilibrium one adopts the principle of local thermodynamic states. Specifically, it is assumed that one can imagine subsystems that are infinitely small relative to the macroscopic level, but that still contain a sufficiently large number of molecules, so that one can define at any given time for these small subsystems local values of the thermodynamic properties by the methods of equilibrium statistical physics. Furthermore, one adopts a continuum hypothesis by considering the system as a continuum of local thermodynamic states with thermodynamic properties that now depend on the position $\mathbf{r}$ and the time t, such as $\rho(\mathbf{r}, t)$, $T(\mathbf{r}, t)$, $p(\mathbf{r}, t)$, etc. Thus the continuum hypothesis allows us to replace the thermodynamic quantities by corresponding thermodynamic fields that are continuous functions of space and time. In addition to the thermodynamic fields, one introduces a local center-of-mass velocity $\mathbf{v}(\mathbf{r}, t)$, also called barycentric velocity, as a relevant field.

The assumption of local equilibrium implies that the local thermodynamic properties defined for the infinitesimal subsystems, as well as their derivatives, at any given time satisfy the same thermodynamic relations as those for systems in thermodynamic equilibrium. For example, the local mass density $\rho(\mathbf{r}, t)$ is assumed to be a functional of the local pressure $p(\mathbf{r}, t)$ and the local temperature $T(\mathbf{r}, t)$, independent of the local velocity $\mathbf{v}(\mathbf{r}, t)$, with $\rho[p(\mathbf{r}, t), T(\mathbf{r}, t)]$ satisfying the same equation of state as in thermodynamic equilibrium. By combining the continuum hypothesis with the local equilibrium assumption we then deduce from Eq. (2.1) a corresponding expression for the time derivative of the mass density $\rho(\mathbf{r}, t)$ at a given position $\mathbf{r}$:

$$\frac{\partial}{\partial t}\,\rho(\mathbf{r}, t) = \rho(\mathbf{r}, t)\left[\varkappa_T\,\frac{\partial}{\partial t}\,p(\mathbf{r}, t) - \alpha_p\,\frac{\partial}{\partial t}\,T(\mathbf{r}, t)\right]. \tag{2.2}$$

and for the spatial derivative at a given time t:

$$\boldsymbol{\nabla}\rho(\mathbf{r}, t) = \rho(\mathbf{r}, t)\left[\varkappa_T\,\boldsymbol{\nabla}p(\mathbf{r}, t) - \alpha_p\,\boldsymbol{\nabla}T(\mathbf{r}, t)\right]. \tag{2.3}$$

In the differential equations above, the thermodynamic coefficients $\varkappa_T$ and α_p are now, in principle, spatiotemporal fields. However, in practical applications the dependence of these thermodynamic coefficients on $\mathbf{r}$ and t is often neglected. The local-equilibrium assumption means that any classical thermodynamic relation can be adapted to nonequilibrium thermodynamics by replacing the thermodynamic quantities by fields, as we have done here for the equation of state.

The continuum hypothesis asserts that the local states of a nonequilibrium fluid can be described in terms of thermodynamics fields, obtained as averages over small volume elements, that depend on the position $\mathbf{r}$ and the time t. Since for a molecular fluid one can always define local averages of the conserved quantities, one does not need strictly speaking the assumption of local thermodynamic equilibrium in formulating equations for the conserved quantities. In practice, however, one assumes that the local values of the conserved quantities can be identified with the local equilibrium values. The local-equilibrium hypothesis implies that the thermodynamic fields satisfy at any given $\mathbf{r}$ and t the same relations as the equilibrium thermodynamic properties. These assumptions will be valid when the spatial and temporal dependence of the thermodynamic fields are negligible at molecular length and time scales, so that one can identify small volume elements that are homogeneous while containing so many molecules that statistical averages can be performed locally. Thus the assumption of local equilibrium will be valid when the molecular length and time scales are small compared to the hydrodynamic length and time scales. This assumption will not be valid in complex systems, such as glasses, polymer blends, colloidal systems, etc. Attempts to formulate thermodynamics "beyond" local equilibrium have been made (Villar and Rubí, 2001), but such complex systems will not be considered in this book. However, while the local-equilibrium hypothesis is valid for the thermodynamic properties in simple fluids, it is important to note that even in simple fluids the concept of local equilibrium, as formulated here, is no longer valid for the fluctuations of the thermodynamic properties. As already mentioned in Chapter 1, a major theme of this book is that the fluctuations of the (local) thermodynamic properties in nonequilibrium states are very different from those predicted by statistical mechanics for thermodynamic equilibrium.

2.2   Balance laws

In the previous section we discussed that the state of the fluid is specified by thermodynamic fields and by a fluid velocity that are functions of the position $\mathbf{r}$ and the time t. In this section we consider the rate of change of the (local) mass density $\rho(\mathbf{r}, t)$, the (local) momentum density $\rho(\mathbf{r}, t)\mathbf{v}(\mathbf{r}, t)$ and the (local) energy density $\rho(\mathbf{r}, t)u(\mathbf{r}, t)$ as they follow from the principles of conservation of mass, conservation of momentum and conservation of energy.

2.2.1 Mass balance

Let us consider a volume element ΔV inside a continuum where a velocity field $\mathbf{v}(\mathbf{r}, t)$ exists. Since there is not a net creation or destruction of mass inside, the temporal variation of mass contained in ΔV will be due only to the movement of fluid across the boundary S. The mass carried by the fluid crossing a surface element d$\mathbf{S}$ in unit time is: $-\rho\mathbf{v}\cdot$d$\mathbf{S}$. The minus sign arises because of the mathematical convention that d$\mathbf{S}$ on a closed surface points to the outside, so that velocity parallel to the director means decrement of the mass inside. Then, the balance of mass can be expressed as:

$$\frac{d}{dt} \int_{\Delta V} \rho(\mathbf{r}, t) d\mathbf{r} = - \int_{S} \rho(\mathbf{r}, t) \, \mathbf{v}(\mathbf{r}, t) \cdot d\mathbf{S}, \tag{2.4}$$

which, after application of Gauss' theorem, implies:

$$\frac{\partial}{\partial t} \rho(\mathbf{r}, t) = -\boldsymbol{\nabla} \cdot (\rho(\mathbf{r}, t) \mathbf{v}(\mathbf{r}, t)) . \tag{2.5}$$

Equation (2.5) represents the conservation of mass for a pure one-component fluid. It has the form of a time derivative of a field (mass density in this case) that equals the divergence of a flux (mass density times velocity in this case). Notice that the field under the time derivative on the left-hand side (LHS) has a tensorial character 0, while the field on the right-hand side (RHS), under the divergence, has a tensorial character equal to 1. This structure is typical for conserved fields. As will be discussed in more detail below, the balance of momentum, which is a vector field with tensorial character 1, involves the divergence of a second-order tensor. The structure of Eq. (2.5) also shows that mass (or mass density) is a *locally* conserved field. However, fields like the momentum density and energy density are only conserved *globally*, and in their corresponding balance laws there appear extra terms on the RHS which cannot be expressed as the divergence of something with a larger tensorial character. These extra terms are called source terms and they will be discussed in detail below.

Equation (2.5) may be rewritten as:

$$\left(\frac{\partial}{\partial t} + \mathbf{v}(\mathbf{r}, t) \cdot \boldsymbol{\nabla}\right) \rho(\mathbf{r}, t) = -\rho(\mathbf{r}, t) \, \boldsymbol{\nabla} \cdot \mathbf{v}(\mathbf{r}, t), \tag{2.6}$$

which is usually abbreviated as:

$$\frac{d}{dt} \rho(\mathbf{r}, t) = -\rho(\mathbf{r}, t) \, \boldsymbol{\nabla} \cdot \mathbf{v}(\mathbf{r}, t). \tag{2.7}$$

The operator d/dt, defined by Eqs. (2.6) and (2.7) is called a *material or substantial time derivative*. It represents the time derivative of a quantity in a volume element that moves with the fluid. After a small time increment Δt, a volume element located at $\mathbf{r}$ at time t, will be located at $\mathbf{r} + \mathbf{v}(\mathbf{r}, t)\Delta t$ at $t + \Delta t$. Hence, a material derivative is defined as:

$$\frac{d}{dt}\,\rho(\mathbf{r}, t) = \underset{\Delta t \to 0}{\text{Lim}}\; \frac{\rho(\mathbf{r} + \mathbf{v}(\mathbf{r}, t)\Delta t, t + \Delta t) - \rho(\mathbf{r}, t)}{\Delta t}. \tag{2.8}$$

Expanding $\rho(\mathbf{r} + \mathbf{v}(\mathbf{r}, t)\Delta t, t + \Delta t)$ in a Taylor series and taking the limit $\Delta t \to 0$, we see that the LHS of Eq. (2.6) actually represents the mass density derivative as seen by a volume element moving with the fluid. The differential operator $\mathbf{v} \cdot \nabla$ acting upon ρ in the LHS of the mass balance (2.6) is usually referred to as *advection* or streaming term. We finalize our comments about the mass balance by remarking that Eq. (2.6) can also be deduced without using Gauss' theorem, but by considering a volume element that moves with the fluid, a procedure usually referred to as using Euler coordinates (Batchelor, 1967).

In the case of a multi-component fluid one may specify a set of partial densities $\rho_k(\mathbf{r}, t)$ that represent the mass of the various components k per unit volume. The total mass density of a multi-component fluid is obtained by summation over the partial densities of the components:

$$\rho(\mathbf{r}, t) = \sum_{k=1}^{N} \rho_k(\mathbf{r}, t). \tag{2.9}$$

In addition, we need to consider different velocity fields $\mathbf{v}_k(\mathbf{r}, t)$ for each component. The phenomenon that, in a mixture, different components move at different velocities is called diffusion. To characterize diffusion, a barycentric (center of mass) velocity is defined as:

$$\mathbf{v}(\mathbf{r}, t) = \frac{1}{\rho(\mathbf{r}, t)} \sum_{k=1}^{N} \rho_k(\mathbf{r}, t)\, \mathbf{v}_k(\mathbf{r}, t). \tag{2.10}$$

Then, the diffusion flow $\mathbf{J}_k(\mathbf{r}, t)$ of component k is defined as:

$$\mathbf{J}_k(\mathbf{r}, t) = \rho_k(\mathbf{r}, t)\left[\mathbf{v}_k(\mathbf{r}, t) - \mathbf{v}(\mathbf{r}, t)\right]. \tag{2.11}$$

The SI units of the diffusion flux are kg m^{-2} s^{-1}. If the fluid is at rest, the barycentric velocity $\mathbf{v}(\mathbf{r}, t) = 0$ and the diffusion flux represents the amount of mass of component k crossing the unit surface per unit of time. For multicomponent systems, the balance of mass has to be considered independently for each component. Applying the same arguments leading to Eq. (2.5), one obtains the set of differential equations:

$$\frac{\partial}{\partial t}\rho_k(\mathbf{r}, t) = -\nabla \cdot \left(\rho_k(\mathbf{r}, t)\mathbf{v}_k(\mathbf{r}, t)\right). \tag{2.12}$$

Summing over the components k, with the help of Eq. (2.10), we observe that for a multicomponent fluid Eq. (2.5) continues to hold with $\mathbf{v}(\mathbf{r}, t)$ representing now the center of mass velocity. With the help of Eq. (2.11), Eq. (2.12) takes the simpler form:

$$\frac{\partial}{\partial t} \rho_k(\mathbf{r}, t) = -\boldsymbol{\nabla} \cdot [\rho_k(\mathbf{r}, t) \, \mathbf{v}(\mathbf{r}, t) + \mathbf{J}_k(\mathbf{r}, t)], \qquad (2.13)$$

which for a multicomponent fluid, together with Eq. (2.5), is the form of the mass balance more often used in applications. It should be noted that Eq. (2.5) and the k equations (2.13) are not linearly independent, since a knowledge of the mass density $\rho(\mathbf{r}, t)$ and of the $k - 1$ partial densities completely determines the remaining partial density.

2.2.2 Pressure tensor

As is well known from Newton's second law of dynamics, the momentum balance of a body depends on the forces exerted on it. Therefore, before discussing momentum balance, we first consider the description of forces acting in a continuum

Let us again consider a fluid element with volume ΔV inside a continuum with boundary S. The forces that may act on it can be divided in two classes: volume forces and surface forces. Volume forces are acting at all points inside ΔV, so that the total force on the volume element depends on the amount of matter contained in it. Volume forces are usually described by a force density $\mathbf{f}(\mathbf{r}, t)$, which expresses the amount of force per unit volume. The integration over ΔV of the force density gives the total volumetric force exerted on the element:

$$\mathbf{F}_{\text{vol}}(t) = \int_{\Delta V} \mathbf{f}(\mathbf{r}, t) \, d\mathbf{r} \simeq \mathbf{f}(\mathbf{r}, t) \, \Delta V, \qquad (2.14)$$

where the last equality applies in the limit of small volume element, $\Delta V \to 0$. One typical example of a volume force is gravity. In that case the force density is given by: $\mathbf{f}(\mathbf{r}, t) = -\rho(\mathbf{r}, t)g\hat{\mathbf{z}}$, where g is the gravitational acceleration constant and $\hat{\mathbf{z}}$ is a unit vector in the vertical direction. By convention, when effects of gravity are considered, we shall always choose the vertical direction antiparallel to gravity. Other examples of volume forces are electric or magnetic forces, in the case of charged, conducting or polarizable fluids. They shall not be treated in detail here, but they will need to be considered in discussing electroconvection in Chapter 11. Volumetric forces result from the application of an external field; they arise from the presence of gravity, of an electric field, of a magnetic field, etc. We shall use here the concepts of volumetric forces and external sources as completely equivalent.

Surface forces in a fluid element element with volume ΔV are exerted on its boundaries and they arise from the contact with adjacent fluid elements. To describe the surface forces mathematically, we first subdivide the boundary of the fluid element in many differential surface elements, represented by corresponding differential vectors dS. To describe surface forces, we need to assign a differential force vector $d\mathbf{F}_{sur}$ to each one of the differential components dS of the boundary S of ΔV. If we divide a dS in two pieces, $d\mathbf{S} = d\mathbf{S}_1 + d\mathbf{S}_2$, the (differential) force acting on dS has to be the force acting on $d\mathbf{S}_1$ plus the force acting on $d\mathbf{S}_2$. The more general linear relation between the vectors $d\mathbf{F}_{sur}$ and dS is expressed as:

$$d\mathbf{F}_{sur} = -\mathsf{P}(\mathbf{r}, t) \cdot d\mathbf{S}, \tag{2.15}$$

where $\mathsf{P}(\mathbf{r}, t)$ is a second-rank tensor, which is called pressure tensor, and which is evaluated at the center $\mathbf{r}$ of the differential surface, *i.e.*, the point where the vectors $d\mathbf{F}_{sur}$ and dS are supposed to be applied. The overall minus sign arises because the orientation of dS is, by convention, from the inside to the outside of the volume element ΔV and pressure is positive when fluids are in a state of compression[‡]. The components of the second-order pressure tensor can be represented by a square matrix $P_{ij}(\mathbf{r}, t)$. A classical more detailed introduction of the pressure tensor uses the so-called tetrahedron of Cauchy as can be found, *e.g.*, in the book of Batchelor (1967).

We have so far discussed the representation of surface forces over a differential surface element dS of the boundary of ΔV. Now we proceed to calculate the total surface force exerted on ΔV, which is obtained by summing over surface elements. For this purpose it is helpful to consider the parallelepiped ΔV shown in Fig. 2.1. Its center is located at $\mathbf{r}$, its boundary surfaces are parallel to the reference planes and its (small) volume is $\Delta V = \Delta x \Delta y \Delta z$. We assume that each of the six faces of the parallelepiped is infinitesimally small, so that the surface force over it can be represented by Eq. (2.15), while acting on the center point of the face. For instance, the force over the surface ABCD will be:

$$\Delta \mathbf{F}_{sur}^{ABCD} = -\left[\mathsf{P}(\mathbf{r} + \tfrac{1}{2}\Delta x\, \hat{\mathbf{x}}, t) \cdot \hat{\mathbf{x}} \right] \Delta y \Delta z,$$

or, in components:

$$\Delta \mathbf{F}_{sur}^{ABCD} = -\begin{bmatrix} P_{xx}(\mathbf{r} + \tfrac{1}{2}\Delta x\, \hat{\mathbf{x}}, t) \\ P_{yx}(\mathbf{r} + \tfrac{1}{2}\Delta x\, \hat{\mathbf{x}}, t) \\ P_{zx}(\mathbf{r} + \tfrac{1}{2}\Delta x\, \hat{\mathbf{x}}, t) \end{bmatrix} \Delta y \Delta z. \tag{2.16}$$

[‡]Instead of a pressure tensor some books primarily devoted to the theory of elasticity, use a stress tensor, that is equal to the pressure tensor, except for a change of sign (Liu, 2002)

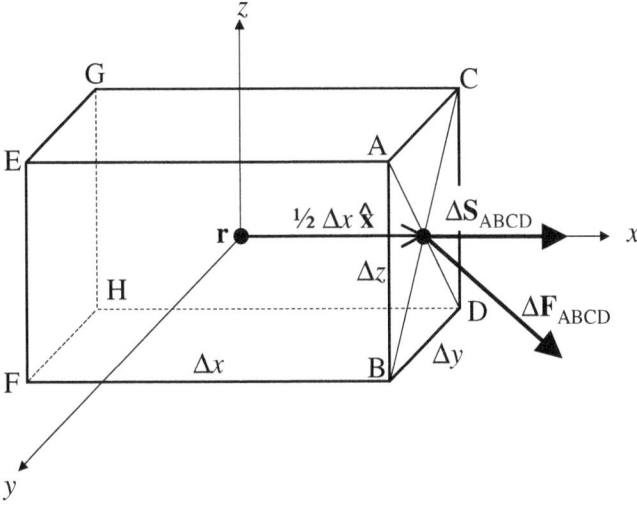

Figure 2.1: Calculation of the total surface force acting on a volume element.

Applying the same argument to the surface EFGH, we obtain:

$$\Delta \mathbf{F}_{\text{sur}}^{\text{EFGH}} = + \begin{bmatrix} P_{xx}(\mathbf{r} - \frac{1}{2}\Delta x \ \hat{\mathbf{x}}, t) \\ P_{yx}(\mathbf{r} - \frac{1}{2}\Delta x \ \hat{\mathbf{x}}, t) \\ P_{zx}(\mathbf{r} - \frac{1}{2}\Delta x \ \hat{\mathbf{x}}, t) \end{bmatrix} \Delta y \Delta z, \qquad (2.17)$$

where the global plus sign arises because the director of the surface EFGH is parallel to $-\hat{\mathbf{x}}$. Adding Eqs. (2.16) and (2.17), multiplying and dividing the resulting expression by Δx, we obtain:

$$\Delta \mathbf{F}_{\text{sur}}^{\text{ABCD}} + \Delta \mathbf{F}_{\text{sur}}^{\text{EFGH}} = - \begin{bmatrix} \dfrac{P_{xx}(\mathbf{r} + \frac{1}{2}\Delta x \ \hat{\mathbf{x}}, t) - P_{xx}(\mathbf{r} - \frac{1}{2}\Delta x \ \hat{\mathbf{x}}, t)}{\Delta x} \\[2ex] \dfrac{P_{yx}(\mathbf{r} + \frac{1}{2}\Delta x \ \hat{\mathbf{x}}, t) - P_{yx}(\mathbf{r} - \frac{1}{2}\Delta x \ \hat{\mathbf{x}}, t)}{\Delta x} \\[2ex] \dfrac{P_{zx}(\mathbf{r} + \frac{1}{2}\Delta x \ \hat{\mathbf{x}}, t) - P_{zx}(\mathbf{r} - \frac{1}{2}\Delta x \ \hat{\mathbf{x}}, t)}{\Delta x} \end{bmatrix} \Delta V$$

$$\simeq - \begin{bmatrix} \dfrac{\partial}{\partial x} P_{xx}(\mathbf{r}, t) \\[2ex] \dfrac{\partial}{\partial x} P_{yx}(\mathbf{r}, t) \\[2ex] \dfrac{\partial}{\partial x} P_{zx}(\mathbf{r}, t) \end{bmatrix} \Delta V,$$

where the last equality arises from the fact that Δx is small. Alternatively, it may be understood as the previous relation in the limit $\Delta V \to 0$. Applying a similar argument to the other four faces of the parallelepiped ΔV, we can express the total surface force $\Delta \mathbf{F}_{\text{sur}}$ over it as:

$$
\Delta \mathbf{F}_{\text{sur}} = - \begin{bmatrix} \dfrac{\partial}{\partial x} P_{xx}(\mathbf{r}, t) + \dfrac{\partial}{\partial y} P_{xy}(\mathbf{r}, t) + \dfrac{\partial}{\partial z} P_{xz}(\mathbf{r}, t) \\[2mm] \dfrac{\partial}{\partial x} P_{yx}(\mathbf{r}, t) + \dfrac{\partial}{\partial y} P_{yy}(\mathbf{r}, t) + \dfrac{\partial}{\partial z} P_{yz}(\mathbf{r}, t) \\[2mm] \dfrac{\partial}{\partial x} P_{zx}(\mathbf{r}, t) + \dfrac{\partial}{\partial y} P_{zy}(\mathbf{r}, t) + \dfrac{\partial}{\partial z} P_{zz}(\mathbf{r}, t) \end{bmatrix} \Delta V,
$$

or, in abbreviated form:

$$
\Delta \mathbf{F}_{\text{sur}} = - \left[\boldsymbol{\nabla} \cdot \mathsf{P}^{\mathsf{T}}(\mathbf{r}, t) \right] \Delta V, \tag{2.19}
$$

where $\mathbf{r}$ refers to the center of the volume element. We note that, although Eq. (2.19) has been obtained from a parallelepiped, it includes a limiting $\Delta V \to 0$ step in its deduction, so that the final result becomes independent of the shape of the volume element, as long as ΔV is small enough. Therefore Eq. (2.19) is a completely general representation of surface forces. Here we assume the pressure tensor to be symmetric: $P_{ij}(\mathbf{r}, t) = P_{ji}(\mathbf{r}, t)$, so that, in Eq. (2.19), P^{T} can be replaced by P. The symmetry assumption is strictly correct only if the torque that external forces can exert in a fluid element is of the order $(\Delta V)^{4/3}$. While this is the case for ordinary (isotropic) fluids; it is not true, for instance, in the case of polarizable liquids (which are anisotropic) in the presence of external electric fields. Again the interested reader should consult more specialized references for a detailed discussion of the conditions under which the pressure tensor is symmetric (Batchelor, 1967; Tritton, 1988).

The form of the pressure tensor becomes simple when a fluid is at rest ($\mathbf{v} = 0$). Since a fluid cannot withstand tangential forces, any shear will cause the fluid to move[‡]. Hence, in the case of a fluid at rest, the surface force $d\mathbf{F}_{\text{sur}}$ corresponding to any given $d\mathbf{S}$ must be normal to the surface, thus, (anti)parallel to $d\mathbf{S}$. Taking into account Eq. (2.15), we observe that this is only possible if the pressure tensor is proportional to the identity tensor everywhere in the fluid. Therefore we conclude that, for a fluid at rest:

$$
\mathsf{P}_{\text{rest}}(\mathbf{r}, t) = p(\mathbf{r}, t) \begin{pmatrix} 1 & 0 & 0 \\ 0 & 1 & 0 \\ 0 & 0 & 1 \end{pmatrix}, \tag{2.20}
$$

where the proportionality factor $p(\mathbf{r}, t)$ is to be identified with the hydrostatic pressure. In the case of a fluid in thermodynamic equilibrium that is

---

[‡] We are excluding here visco-elastic fluids, which can withstand to some extend shear stresses without moving

obviously at rest, the proportionality factor is equal to the thermodynamic pressure that will be independent of **r** and t if external forces are absent.

It is convenient to decompose the pressure tensor as:

$$\mathsf{P}(\mathbf{r}, t) = p(\mathbf{r}, t)\,\mathbb{I} - \Pi(\mathbf{r}, t), \tag{2.21}$$

where $p(\mathbf{r}, t)$ is the local (hydrostatic) pressure related to the local mass density $\rho(\mathbf{r}, t)$ and the local temperature $T(\mathbf{r}, t)$ through the local equilibrium equation of state, and where $\mathbb{I}$ represents the identity matrix. In Eq. (2.21), $\Pi(\mathbf{r}, t)$ is a (symmetric) tensor, usually called deviatoric stress tensor (Batchelor, 1967; Tritton, 1988). The physical interpretation of Eq. (2.21) is that the deviatoric stress tensor represents that part of the surface forces that causes movement of the fluid, while the hydrostatic pressure represents the part of the surface forces that can be withstood by the fluid without moving. We finally remark that, although for a fluid at rest the hydrostatic pressure equals one third of the trace of the stress tensor, this is not generally true when the fluid is moving.

2.2.3 Momentum balance

We can obtain the balance of momentum by applying Newton's second dynamic law to a volume element centered at the point **r**, and with mass $\rho(\mathbf{r}, t)\Delta V$. Representing the total surface force acting on ΔV by Eq. (2.19) and representing the total volume force by Eq. (2.14), we arrive at:

$$\rho(\mathbf{r}, t)\,\frac{d}{dt}\mathbf{v}(\mathbf{r}, t) = -\boldsymbol{\nabla} \cdot \mathsf{P}(\mathbf{r}, t) + \mathbf{f}(\mathbf{r}, t), \tag{2.22}$$

after having divided both sides of the equation by a factor ΔV, and using the symmetry of the pressure tensor. Note that in the LHS of Eq. (2.22) the material time derivative of the velocity field appears, as it should. We emphasize that Eq. (2.22) is assumed to be valid for both one-component and multi-component fluids.

Using the definition of the material time derivative (*c.f.*, Eqs. (2.6) and (2.7)) and expression (2.21) for the pressure tensor, we may rewrite Eq. (2.22) as:

$$\frac{\partial}{\partial t}(\rho\mathbf{v}) = -\boldsymbol{\nabla} \cdot [(\rho\mathbf{v})\mathbf{v} - \Pi] + \mathbf{f} - \boldsymbol{\nabla}p, \tag{2.23}$$

where, for simplicity, we have dropped the notation indicating the dependence of the local fields on **r** and t. In Eq. (2.23), $(\rho\mathbf{v})\mathbf{v}$, denotes the dyadic product (second-order symmetric tensor) defined by:

$$[(\rho\mathbf{v})\mathbf{v}]_{ij}(\mathbf{r}, t) = \rho(\mathbf{r}, t)v_i(\mathbf{r}, t)v_j(\mathbf{r}, t). \tag{2.24}$$

Equation (2.23) is the ordinary equation of motion of a fluid. It has the form of a balance law for the momentum density $\rho\mathbf{v}(\mathbf{r}, t)$. From the RHS of Eq. (2.23) we observe that the local change of momentum at a given point $\mathbf{r}$ has three causes:

1. $(\rho\mathbf{v})\mathbf{v}$: This term represents a flux (it appears under a divergence) of momentum. It is the momentum towed by the fluid as it moves, since $\rho\mathbf{v}$ represents the flux of mass towed by the fluid.

2. Π: This term also represents a flux of momentum, which comes from the adjacent volume elements through the surface forces. It vanishes for a fluid at rest.

3. $\mathbf{f} - \nabla p$: This term represents a local source of momentum. The presence of an external force or a pressure gradient tends to move the fluid creating a momentum.

Even though an external force tends in general to move a fluid, we see that a fluid still can remain at rest if the external force density $\mathbf{f}$ is balanced by a corresponding pressure gradient:

$$\mathbf{f}(\mathbf{r}, t) = \nabla p(\mathbf{r}, t). \tag{2.25}$$

The presence of an external force induces a nonuniform thermodynamic-pressure distribution. A nonuniform pressure distribution in an isothermal fluid is accompanied by a nonuniform density distribution in accordance with Eq. (2.3) (except in the case of an incompressible fluid: $\varkappa_T = 0$). When $\mathbf{v} = 0$, the mass balance Eq. (2.5) can only hold if the density field is stationary, *i.e.*, if it does not explicitly depend on time. Hence, an isothermal and quiescent state of a fluid is only possible with conservative external forces, *i.e.*, forces that do not depend on time and whose spatial distribution can be represented as the gradient of a potential so as to satisfy Eq. (2.25). Such isothermal states, under the influence of external forces are referred to as inhomogeneous equilibrium.

2.2.4 Energy balance

In this section we shall assume that only conservative external forces with density proportional to the mass density are present. For a multi-component fluid, this means volume forces that can be derived from a set of specific potential energies $\psi_k(\mathbf{r})$ which do not depend explicitly on time, so that:

$$\mathbf{f}(\mathbf{r}, t) = \sum_k \rho_k(\mathbf{r}, t)\, \mathbf{F}_k(\mathbf{r}) = -\sum_k \rho_k(\mathbf{r}, t)\, \nabla\psi_k(\mathbf{r}), \tag{2.26}$$

where the sum is over the different components of the fluid. A typical example of a conservative external force is gravity, for which $\psi_k(\mathbf{r}) = -gz$. A total potential energy $\psi(\mathbf{r}, t)$ per unit mass can be defined by:

$$\psi(\mathbf{r}, t) = \frac{1}{\rho(\mathbf{r}, t)} \sum_k \rho_k(\mathbf{r}, t) \, \psi_k(\mathbf{r}). \qquad (2.27)$$

Using the mass balance (2.13), we obtain the balance equation for the density of potential energy (energy per unit volume):

$$\frac{\partial}{\partial t} (\rho\psi) = -\boldsymbol{\nabla} \cdot \left[(\rho\psi)\mathbf{v} + \sum_k \psi_k \mathbf{J}_k \right] - \sum_k \mathbf{J}_k \cdot \mathbf{F}_k - \mathbf{f} \cdot \mathbf{v}. \qquad (2.28)$$

The flux of potential energy has two components: First $\rho\psi\mathbf{v}$, which is the potential-energy flux as a result of the movement of the fluid. Second $\sum \mathbf{J}_k \cdot \mathbf{F}_k$, which accounts for the differential energies carried by diffusion of the various components. In addition, there is a local source of energy, which arises from the work performed by the external force field to move the fluid.

Next, we consider the balance of kinetic-energy density defined as: $\frac{1}{2}\rho(\mathbf{r}, t)\mathbf{v}^2(\mathbf{r}, t)$. We do not consider any contribution to the kinetic energy from diffusion, usually referred to as *inertia* terms. Inertia terms may be neglected in most cases (de Groot and Mazur, 1962). Then, from Eq. (2.22), after some manipulation of derivatives, we obtain:

$$\frac{\partial}{\partial t} \left(\tfrac{1}{2}\rho\mathbf{v}^2 \right) = -\boldsymbol{\nabla} \cdot \left[\left(\tfrac{1}{2}\rho\mathbf{v}^2 \right) \mathbf{v} - \mathsf{P} \cdot \mathbf{v} \right] - \mathsf{P} : (\boldsymbol{\nabla}\mathbf{v}) + \mathbf{f} \cdot \mathbf{v}, \qquad (2.29)$$

where $\mathsf{P} : (\boldsymbol{\nabla}\mathbf{v})$ is the scalar quantity (invariant under orthonormal change of variables):

$$[\mathsf{P} : (\boldsymbol{\nabla}\mathbf{v})] (\mathbf{r}, t) = \sum_{i,j} P_{ij}(\mathbf{r}, t) \frac{\partial}{\partial x_j} v_i(\mathbf{r}, t). \qquad (2.30)$$

From Eq. (2.29) we see that kinetic energy is not a locally conserved variable; there is a local source of kinetic energy due to both an internal force ($\mathsf{P} : (\boldsymbol{\nabla}\mathbf{v})$) and an external force ($\mathbf{f} \cdot \mathbf{v}$). This reflects the fact that forces tend to move the fluid.

To obtain the balance of total mechanical energy (kinetic plus potential) we add Eqs. (2.28) and (2.29), so that

$$\frac{\partial}{\partial t} \left(\tfrac{1}{2}\rho\mathbf{v}^2 + \rho\psi \right) = -\boldsymbol{\nabla} \cdot \left[\left(\tfrac{1}{2}\rho\mathbf{v}^2 + \rho\psi \right) \mathbf{v} - \mathsf{P} \cdot \mathbf{v} + \sum_k \psi_k \mathbf{J}_k \right] \qquad (2.31)$$
$$- \mathsf{P} : (\boldsymbol{\nabla}\mathbf{v}) - \sum_k \mathbf{J}_k \cdot \mathbf{F}_k.$$

Note that Eq. (2.31) is at first sight a surprising result. Although there is some compensation in the source term, it turns out that mechanical energy is not a locally conserved variable. Why? While we are assuming that the external forces are conservative, the internal forces arising from the interaction between molecules are also conservative forces. We expect local conservation of energy, but Eq. (2.31) shows a net source (or sink) of mechanical energy. The solution to this apparent paradox is that the total energy of the fluid is not just its mechanical energy, but we need to consider also an internal thermodynamic energy $u(\mathbf{r}, t)$, which is a kind of energy that absorbs the source term in Eq. (2.31). Therefore, the total energy density $e(\mathbf{r}, t)$ will be the sum of three contributions, such that:

$$\rho(\mathbf{r}, t)\, e(\mathbf{r}, t) = \frac{1}{2}\rho(\mathbf{r}, t)\, \mathbf{v}^2(\mathbf{r}, t) + \rho(\mathbf{r}, t)\, \psi(\mathbf{r}, t) + \rho(\mathbf{r}, t)\, u(\mathbf{r}, t). \quad (2.32)$$

Having introduced the internal energy $u(\mathbf{r}, t)$, we need to establish the corresponding balance law. It requires a source term to compensate the source in Eq. (2.31). The flux term will have a contribution $\rho(\mathbf{r}, t)u(\mathbf{r}, t)\mathbf{v}(\mathbf{r}, t)$ due to the movement of the fluid. In addition, we need to consider the flow of internal energy from adjacent volume elements. This additional energy flow is called heat flow and it will be represented by $\mathbf{Q}(\mathbf{r}, t)$. Hence, the balance of internal energy can be written as:

$$\frac{\partial}{\partial t}\,(\rho u) = -\boldsymbol{\nabla} \cdot [\rho u \mathbf{v} + \mathbf{Q}] + \mathsf{P} : (\boldsymbol{\nabla}\mathbf{v}) + \sum_k \mathbf{J}_k \cdot \mathbf{F}_k. \quad (2.33)$$

Upon adding Eqs. (2.31) and (2.33), we see that the total energy density (kinetic plus potential plus internal), as given by Eq. (2.32), becomes a locally conserved quantity, *i.e.*, its time derivative equals the divergence of a vector field. The movement of the fluid causes a continuous transfer of energy from mechanical to internal.

Using the decomposition (2.21) of the pressure tensor as a the sum of an hydrostatic part and a deviatoric part, we can rewrite Eq. (2.33) as:

$$\frac{\partial}{\partial t}\,(\rho u) = -\boldsymbol{\nabla} \cdot [\rho u \mathbf{v} + \mathbf{Q}] - p\,\boldsymbol{\nabla} \cdot \mathbf{v} + \Pi : (\boldsymbol{\nabla}\mathbf{v}) + \sum_k \mathbf{J}_k \cdot \mathbf{F}_k. \quad (2.34)$$

At this point in the derivation one usually uses the assumption of the pressure tensor being symmetric, and that the double contracted product of a symmetric tensor and an antisymmetric tensor vanishes. If we split the tensor $\boldsymbol{\nabla}\mathbf{v}$ of velocity derivatives in its symmetric and antisymmetric part:

$$\boldsymbol{\nabla}\mathbf{v} = [\boldsymbol{\nabla}\mathbf{v}]^{(\mathrm{s})} + [\boldsymbol{\nabla}\mathbf{v}]^{(\mathrm{a})}, \quad (2.35)$$

with:

$$
\begin{aligned}
[\boldsymbol{\nabla}\mathbf{v}]_{ij}^{(s)} &= \tfrac{1}{2}\left\{\frac{\partial v_i}{\partial x_j} + \frac{\partial v_j}{\partial x_i}\right\}, \\
[\boldsymbol{\nabla}\mathbf{v}]_{ij}^{(a)} &= \tfrac{1}{2}\left\{\frac{\partial v_i}{\partial x_j} - \frac{\partial v_j}{\partial x_i}\right\},
\end{aligned}
\tag{2.36}
$$

then only the symmetric part $[\boldsymbol{\nabla}\mathbf{v}]^{(s)}$ actually contributes to the energy balance. This is a real simplification, since a symmetric second-rank tensor has only six independent components instead of nine. Thus, we finally write the balance of (internal) energy as:

$$
\frac{\partial}{\partial t}(\rho u) = -\boldsymbol{\nabla}\cdot[\rho u\mathbf{v} + \mathbf{Q}] - p\,\boldsymbol{\nabla}\cdot\mathbf{v} + \Pi : (\boldsymbol{\nabla}\mathbf{v})^{(s)} + \sum_k \mathbf{J}_k \cdot \mathbf{F}_k. \tag{2.37}
$$

The symmetric part $[\boldsymbol{\nabla}\mathbf{v}]^{(s)}$ in Eq. (2.36) equals the definition of the Cauchy's deformation tensor for elastic bodies (Liu, 2002) with the displacement field replaced by the velocity field. For this reason $[\boldsymbol{\nabla}\mathbf{v}]^{(s)}$ is often called Cauchy's velocity tensor. Physically, it represents deformations in the shape of the volume element ΔV due to the movement of the fluid, while the antisymmetric part represents rigid (infinitesimal) rotations of the volume element preserving the shape. A more detailed discussion of the physical interpretation of the symmetric and antisymmetric parts of the tensor $\boldsymbol{\nabla}\mathbf{v}$ of velocity derivatives may be found in more specialized monographs (Batchelor, 1967; Tritton, 1988; Liu, 2002).

2.2.5   Additional remarks

We have used conservation of mass, conservation of momentum and conservation of energy. In general one should also consider conservation of angular momentum. However, when the pressure tensor is symmetric (as we are assuming here), conservation of momentum is automatically satisfied and does not need to be considered independently. Specifically, when the pressure tensor is symmetric, the torque that external forces may exert over a volume element diminishes faster than ΔV for $\Delta V \to 0$ and the balance of momentum (2.23) also assures balance of torques (Batchelor, 1967).

When the pressure tensor is not symmetric, external forces can exert torques on fluid elements that are proportional to ΔV. For example, this situation occurs in the case of liquid crystals, where external electric fields may exert a torque in the fluid. In such cases, it is necessary to account for a balance of angular momentum to complete the hydrodynamic description. A detailed discussion of angular-momentum balance in liquid crystals can be found in the book by de Gennes and Prost (1993). See also the discussion in Sect. 11.3.

The goal of nonequilibrium thermodynamics is to calculate the spatiotemporal evolution of fields like mass density $\rho(\mathbf{r}, t)$, center-of-mass velocity $\mathbf{v}(\mathbf{r}, t)$, energy density $\rho u(\mathbf{r}, t)$, etc. In the previous subsections we have shown how this evolution is governed by balance laws. In formulating the balance laws, we needed to introduce new fields that do not exist in equilibrium thermodynamics: diffusion fluxes $\mathbf{J}_k(\mathbf{r}, t)$ in Eq. (2.13), a deviatoric stress tensor $\Pi(\mathbf{r}, t)$ in Eq. (2.21) and a heat flux $\mathbf{Q}(\mathbf{r}, t)$ in Eq. (2.33). All these fields appear in the corresponding balance laws under a divergence and represent an extra contribution to the total flux, in addition to the flux caused by the movement of the fluid. They are being referred to as dissipative fluxes since they will appear later in the dissipation function, Eq. (2.41). While a local equilibrium thermodynamic property like the pressure $p(\mathbf{r}, t)$ can be eliminated with the aid of a local-equilibrium equation of state, diffusion fluxes, heat flux and deviatoric stress do not exist in equilibrium, so that they cannot be eliminated from the balance laws. Therefore, to close the set of differential equations for the spatiotemporal evolution of density, velocity and energy, one needs to supplement the balance equations with phenomenological laws relating the fluxes to corresponding gradients and transport coefficients. How to do this systematically will be explained in the next section.

2.3 Entropy balance, dissipative fluxes and thermodynamic forces

The balance of entropy plays a very important role in (linear) nonequilibrium thermodynamics. We obtain it by applying the local-equilibrium assumption to the classical thermodynamics Gibbs-Duhem equation for the specific entropy s:

$$T ds = du + p dv - \sum_k \mu_k dc_k, \qquad (2.38)$$

where μ_k are the chemical potentials of the components of the mixture, c_k the corresponding concentrations in terms of mass fractions $c_k = \rho_k/\rho$, and $v = \rho^{-1}$ the specific volume. Just as we proceeded in Sect. 2.1.2 by applying the local-equilibrium assumption to Eq. (2.1) for the equation of state, we now apply the local-equilibrium assumption to Eq. (2.38) and obtain:

$$T(\mathbf{r}, t)\, \partial_t s(\mathbf{r}, t) = \partial_t u(\mathbf{r}, t) - \frac{p(\mathbf{r}, t)}{\rho^2(\mathbf{r}, t)} \partial_t \rho(\mathbf{r}, t) - \sum_k \mu_k(\mathbf{r}, t) \partial_t c_k(\mathbf{r}, t), \quad (2.39a)$$

$$T(\mathbf{r}, t)\, \boldsymbol{\nabla} s(\mathbf{r}, t) = \boldsymbol{\nabla} u(\mathbf{r}, t) - \frac{p(\mathbf{r}, t)}{\rho^2(\mathbf{r}, t)} \boldsymbol{\nabla} \rho(\mathbf{r}, t) - \sum_k \mu_k(\mathbf{r}, t) \boldsymbol{\nabla} c_k(\mathbf{r}, t). \quad (2.39b)$$

The chemical potentials $\mu_k(\mathbf{r}, t)$ in Eqs. (2.39) depend on the (local) partial density $\rho_k(\mathbf{r}, t)$ the (local) pressure $p(\mathbf{r}, t)$ and the (local) temperature $T(\mathbf{r}, t)$ by their equilibrium equations of state $\mu_k = \mu_k(\rho_k, p, T)$ which is now interpreted as a functional. The SI units of $\mu_k(\mathbf{r}, t)$ are J kg^{-1}, since we are using here chemical potentials defined per unit mass.

From the differential Eqs. (2.39) for the local specific entropy $s(\mathbf{r}, t)$, using the balance laws and the equation of state, one can obtain, after some manipulation of derivatives, the balance equation (de Groot and Mazur, 1962):

$$\frac{\partial}{\partial t}(\rho s) = -\boldsymbol{\nabla} \cdot \left[(\rho s)\,\mathbf{v} + \frac{1}{T}\left(\mathbf{Q} - \sum_k \mu_k \mathbf{J}_k \right) \right] + \dot{S}, \qquad (2.40)$$

with

$$\dot{S} = -\frac{1}{T^2}\,\mathbf{Q} \cdot \boldsymbol{\nabla}T + \frac{\Pi : (\boldsymbol{\nabla}\mathbf{v})^{(s)}}{T} - \frac{1}{T}\sum_k \mathbf{J}_k \cdot \left[T\,\boldsymbol{\nabla}\left(\frac{\mu_k}{T} \right) - \mathbf{F}_k \right]. \quad (2.41)$$

We observe that Eq. (2.40) has again the typical form, encountered previously, of a balance law with a flux and a source term $\dot{S}$ given by Eq. (2.41). Such a source term is usually referred to as *entropy-generation-rate* density. At first glance, the way the different contributions have been collected in the flux or source terms may seem somewhat arbitrary; for a detailed discussion of this point and for alternative expressions of the entropy balance we refer to de Groot and Mazur (1962).

Generically, the entropy source term is not zero and the entropy is a non-conserved variable in the spatiotemporal evolution of the fluid. In contrast to Eq. (2.31) for the mechanical energy, this is not a surprising result. From the second law of thermodynamics, we know that in irreversible processes entropy is not conserved, so we actually expect a non-zero entropy source term in the nonequilibrium (*i.e.*, irreversible or dissipative) evolution of the fields. The entropy source term, Eq. (2.41) contains the irreversible part of the evolution of the fluid.

The product $\Psi = T\dot{S}$ is usually referred to as *dissipation function*; it has units of energy density rate and for a one-component fluid can be easily obtained from Eq. (2.41). It is very interesting to analyze its structure: the dissipation function is a scalar quantity expressed as the sum of three terms. Each term is the contraction (or double contraction) of two tensors with the same rank. Notice that for each one of the three contractions, one of the contracted terms is precisely one of the three unknown fluxes we had to introduce in the formulation of the balance laws, as discussed at the end of Sect. 2.2.5. These fluxes are the heat flux $\mathbf{Q}$, the deviatoric stress tensor Π and the diffusion fluxes $\mathbf{J}_k$, referred to as *dissipative fluxes*. The vectors and tensor that are contracted with the dissipative fluxes are called *conjugate thermodynamic forces.* When the fluid is in thermodynamic equilibrium, the dissipative fluxes, as well as the corresponding forces, vanish: uniform stationary temperature, uniform stationary velocity and uniform chemical potentials (in the absence of external forces).

2.3.1 Phenomenological relations

We are considering here nonequilibrium thermodynamics of fluids subjected to external forces that do not create torque on volume elements. Nevertheless, the conclusions about the structure of dissipation elucidated above are completely general and they can be extended to *any* nonequilibrium system, such as elastic solids (Liu, 2002), liquid crystals (de Gennes and Prost, 1993), conducting fluids (Robinson and Stokes, 2002), plasmas (Krall and Trivelpiece, 1973; Choudhuri, 1998), etc. That is, any nonequilibrium system can be described in terms of fields; the spatiotemporal evolution of such fields satisfies appropriate balance laws; on establishing the balance laws some nonequilibrium dissipative fluxes need to be introduced; and, finally, the dissipation function of the system is a scalar obtained by contraction of the dissipative fluxes with some thermodynamic forces. This observation allows us to affirm that, in general, the dissipation Ψ of a nonequilibrium system can be expressed as:

$$\Psi = T\dot{S} = \sum_{\alpha} J_{\alpha}(\mathbf{r}, t)\, X_{\alpha}(\mathbf{r}, t), \tag{2.42}$$

where $J_{\alpha}(\mathbf{r}, t)$ are the dissipative fluxes and $X_{\alpha}(\mathbf{r}, t)$ the corresponding conjugate thermodynamic forces. The index α runs over the components of the various thermodynamic forces. Of course, the number of dissipative fluxes, their tensorial character, as well as the expression of the thermodynamic forces in terms of the nonequilibrium fields, do depend on the particular system under consideration. When the system is in equilibrium, all dissipative fluxes, as well as the thermodynamic forces, vanish. Notice that here we are defining the thermodynamic forces in terms of the dissipation function (Demirel, 2002), while other authors (de Groot and Mazur, 1962) define such forces in terms of the entropy-production rate $\dot{S}$.

Now we are in a position to understand the "closing" of the equations of nonequilibrium thermodynamics. In general one assumes that dissipative fluxes are (analytical) functionals of the thermodynamic forces (at least close to equilibrium, where the thermodynamic forces are small), so that: $J_{\alpha}(\mathbf{r}, t) = J_{\alpha}[X_{\beta}(\mathbf{r}, t)]$. In practice, a linear approximation is used for the relationship between the fluxes and the thermodynamic forces of the form:

$$J_{\alpha}(\mathbf{r}, t) = \sum_{\beta} M_{\alpha\beta}(\mathbf{r}, t) X_{\beta}(\mathbf{r}, t), \tag{2.43}$$

where the functional derivatives of the dissipative fluxes, $M_{\alpha\beta}(\mathbf{r}, t)$, are *phenomenological coefficients*, commonly referred to as *Onsager coefficients*. In principle, these coefficients depend on space and time through the local state variables, but in most practical applications any spatial and temporal dependence of the Onsager coefficients can be neglected. Equations (2.43), when introduced in the balance laws, close the equations of nonequilibrium thermodynamics yielding the hydrodynamic equations to be specified in Sect. 2.4.

Equations (2.43) are called *phenomenological relations*, because they were initially introduced on an experimental basis (Demirel, 2002). However, nonequilibrium statistical physics, provides a theoretical justification for states that are not very far from equilibrium (Boon and Yip, 1980; Kubo et al., 1991). We emphasize that according to Eqs. (2.43) a given dissipative flux $J_\alpha(\mathbf{r}, t)$ may depend on all thermodynamic forces, not just its conjugate one. This phenomenon is called *coupling* among thermodynamic forces and it is a generic feature of nonequilibrium thermodynamics. For example, as discussed in more detail in Sect. 2.4.2, the presence of a heat flux will cause a diffusion flux (Soret effect) and vice versa (Dufour effect). However, when two thermodynamic forces have different tensorial character they do not couple, or equivalently, the corresponding phenomenological cross-coefficients among their components vanish. The absence of coupling among the components of thermodynamic forces with different tensorial character is called the Curie, or sometimes Curie-Prigogine, principle (de Groot and Mazur, 1962; Demirel, 2002).

Substituting (2.43) into the expression (2.42) for the dissipation function, we conclude that:

$$\Psi(\mathbf{r}, t) = \sum_{\alpha\beta} M_{\alpha\beta}(\mathbf{r}, t) X_\alpha(\mathbf{r}, t) X_\beta(\mathbf{r}, t). \tag{2.44}$$

The matrix $\mathsf{M}(\mathbf{r}, t)$ formed with the coefficients $M_{\alpha\beta}(\mathbf{r}, t)$ is called *dissipation matrix*, or alternatively, matrix of phenomenological coefficients.

As a consequence of (2.44) the entropy-production rate $\dot{S}$ becomes a bilinear function of the thermodynamic forces. The second law of thermodynamics requires that $\dot{S}(\mathbf{r}, t) \geq 0$ everywhere in the fluid. This imposes further restrictions on the dissipation matrix M: all diagonal elements have to be positive $M_{\alpha\alpha}(\mathbf{r}, t) \geq 0$, while off-diagonal coefficients must satisfy conditions like, for instance,: $M_{\alpha\alpha} M_{\beta\beta} \geq \frac{1}{4}(M_{\alpha\beta} + M_{\beta\alpha})^2$ (de Groot and Mazur, 1962). We shall not further elaborate this point.

2.3.2 Onsager's reciprocal relations

As was originally shown by Onsager (1931a,b), time reversal invariance of the equations of motion of the molecules implies that the dissipation matrix must be symmetric:

$$M_{\alpha\beta}(\mathbf{r}, t) = M_{\beta\alpha}(\mathbf{r}, t). \tag{2.45}$$

This is one of the most important results from nonequilibrium statistical physics; its derivation is outside the scope of the present chapter and the interested reader is referred to the relevant literature (de Groot and Mazur, 1962; Kubo et al., 1991). Equations (2.45) are usually called Onsager's reciprocal relations. Equality of the cross-coefficients in the dissipation matrix does not always imply that both cross effects are equally important in practice, as we shall see later, when discussing Soret and Dufour effects in more detail.

Before closing this section, we emphasize that we are restricting ourselves to linear nonequilibrium thermodynamics, expected to be valid when the system is not too far from thermodynamic equilibrium. In linear nonequilibrium thermodynamics the dissipative fluxes depend only on the local (in space and time) values of the thermodynamic forces. During the past decades attempts have been made to develop Extended Irreversible Thermodynamics (Jou et al., 1993), which uses "phenomenological" relationships that are nonlocal in time, *i.e*, the local values of the forces depend on the fluxes and its time derivatives. However, the issues when one goes beyond linear nonequilibrium thermodynamics are far from settled (García-Colín, 1995).

2.4 Hydrodynamic equations

Having considered in the previous section the formal structure of the relationship between dissipative fluxes and thermodynamic forces, we now apply these relations to formulate the hydrodynamic equations for one-component and two-component fluids. The more general case of multicomponent fluids, that is, fluid mixtures with more than two components, will not be considered in this book.

2.4.1 Hydrodynamic equations for a one-component fluid

In the case of a one-component fluid there is no diffusion. Therefore all diffusive fluxes are zero, and from the entropy-production rate $\dot{S}$ given by Eq. (2.41) we obtain for the dissipation function the simpler expression:

$$\Psi = -\frac{1}{T}\,\mathbf{Q} \cdot \boldsymbol{\nabla}T + \Pi : (\boldsymbol{\nabla}\mathbf{v})^{(\mathrm{s})}. \tag{2.46}$$

We immediately recognize the presence of two dissipative fluxes in Eq. (2.46): a heat flux $\mathbf{Q}$ and a deviatoric stress tensor Π with their two corresponding conjugate thermodynamic forces: $-\left(\nabla T\right)/T$ and $\left(\nabla\mathbf{v}\right)^{(\mathrm{s})}$. Since the two pairs of fluxes and forces have in this case different tensorial character, they do not couple by virtue of the Curie principle. Therefore, specifying the more general (linear) phenomenological relations (2.43) for this case, we have:

$$Q_i(\mathbf{r},t) = -\sum_j M_{ij}^{(11)}(\mathbf{r},t) \; \frac{1}{T(\mathbf{r},t)} \; \frac{\partial}{\partial x_j} T(\mathbf{r},t) \tag{2.47}$$

for the heat flux, and

$$\Pi_{ij}(\mathbf{r},t) = \sum_{kl} M_{ijkl}^{(22)}(\mathbf{r},t) \; \frac{1}{2}\left\{ \frac{\partial}{\partial x_k} v_l(\mathbf{r},t) + \frac{\partial}{\partial x_l} v_k(\mathbf{r},t) \right\} \tag{2.48}$$

for the deviatoric stress tensor. Therefore the matrix M of phenomenological coefficients has, in this case, two parts: a second-rank tensor $\mathsf{M}^{(11)}$ relating the components of the heat flux with the components of the (inverse) temperature gradient; and a fourth-rank tensor $\mathsf{M}^{(22)}$ relating the components of the deviatoric stress tensor with the components of Cauchy's velocity tensor.

By virtue of Onsager's reciprocal relations, the matrix representing the tensor $\mathsf{M}^{(11)}$ is symmetric, so it can diagonalized. Isotropy means that there are no privileged directions, so we expect the diagonalized matrix to be proportional to the identity matrix. We further assume that in practical application the spatiotemporal dependence of the phenomenological coefficients can be neglected. Then, the phenomenological relation (2.47) for the heat flux reduces to:

$$\mathbf{Q}(\mathbf{r},t) = \frac{-K}{T(\mathbf{r},t)} \; \nabla T(\mathbf{r},t) \simeq -\lambda \, \nabla T(\mathbf{r},t), \tag{2.49}$$

where the last approximation assumes that the temperature gradients are not very large, so that the term T^{-1} can be treated as a constant independent of $\mathbf{r}$ and t. Equation (2.49) is the well-known Fourier's law for heat conduction with the coefficient λ being the thermal conductivity.

The second phenomenological equation (2.48) initially requires the knowledge of $3^4 = 81$ functions $M_{ijkl}^{(22)}(\mathbf{r},t)$. However, recalling that both the deviatoric stress tensor $\Pi_{\alpha\beta}$ and its conjugate force $\left(\nabla\mathbf{v}\right)^{(\mathrm{s})}$ are symmetric, this number reduces to $6 \times 6 = 36$ independent coefficients. Additionally, Onsager's reciprocal relations reduce the number of independent coefficients to 21 (the 6×6 matrix has to be symmetric). We are not giving more details, but it turns out that, because of other symmetries for an isotropic fluid, the number of independent components of $\mathsf{M}^{(22)}$ can be finally reduced to only two[‡]. The final form of the linear phenomenological relation (2.48) is usually

---

[‡]The arguments here are exactly the same employed to justify that an isotropic elastic solid has only two independent elastic modula (Liu, 2002)

written as:

$$\Pi_{ij}^{(\mathrm{s})} = \eta \left(\frac{\partial v_i}{\partial x_j} + \frac{\partial v_j}{\partial x_i} \right) + \left(\eta_{\mathrm{v}} - \frac{2}{3} \eta \right) \delta_{ij} \frac{\partial v_l}{\partial x_l}, \tag{2.50}$$

where η_{v} and η are the only two independent coefficients in the 4th order tensor $\mathsf{M}^{(22)}$ left after considering all symmetries, homogeneity and isotropy. The coefficient η is the shear viscosity and η_{v} is called bulk viscosity. Equation (2.50) is Newton's law for viscosity. Equation (2.50) implies that shearing force between adjacent fluid layers is proportional to the velocity gradient. The bulk viscosity η_{v} represents extra forces required to compress volume elements. There is an important class of fluids, for which the effective shear viscosity depends on the shear rate and which do not obey Newton's law (2.50). While this is an active field of research relevant to a variety of applications (Skelland, 1967; Bird et al., 1986; Denn, 2004), non-newtonian fluid behavior is not considered in this book.

After applying symmetries, isotropy and homogeneity, we have reduced the set of phenomenological coefficients to three scalar quantities, namely λ, η and η_{v}, which are generically referred to as *transport coefficients*. As one can see from Eq. (2.49), the transport coefficients are related to but are not identical with the Onsager coefficients.

Next, substituting Newton's viscosity law (2.50) into the momentum-balance equation (2.23), using the continuity equation (2.5) and, after some manipulation of the derivatives, one obtains the Navier-Stokes equation:

$$\rho \left[\frac{\partial \mathbf{v}}{\partial t} + (\mathbf{v} \cdot \boldsymbol{\nabla}) \, \mathbf{v} \right] = -\boldsymbol{\nabla} p + \eta \, \nabla^2 \mathbf{v} + (\tfrac{1}{3}\eta + \eta_{\mathrm{v}}) \, \boldsymbol{\nabla} (\boldsymbol{\nabla} \cdot \mathbf{v}) + \mathbf{f}, \tag{2.51}$$

which, together with the mass balance (2.6), is the second of the hydrodynamic equations for a one-component fluid. It is worth noticing the nonlinear character of the advection term $(\mathbf{v} \cdot \boldsymbol{\nabla}) \, \mathbf{v}$. The intrinsic nonlinearity of hydrodynamics makes it extremely difficult to solve its equations in a general case. The nonlinearity leads to such complicated phenomena as turbulence, which lies outside the scope of our present book.

To complete the set of hydrodynamic equations for a one-component fluid we consider the entropy balance, Eqs. (2.40) and (2.41); with the help of the mass balance (2.5) it may be rewritten as:

$$\rho T \left(\frac{\partial s}{\partial t} + \mathbf{v} \cdot \boldsymbol{\nabla} s \right) = -\boldsymbol{\nabla} \cdot \mathbf{Q} + \Pi : (\boldsymbol{\nabla} \mathbf{v})^{(\mathrm{s})}. \tag{2.52}$$

Equation (2.52) is an example of a heat equation. The second term in the RHS of Eq. (2.52) represents the viscous heating, and can be neglected for small and moderate gradients. Indeed, by inverting Newton's viscosity law (2.50), the components of the tensor $\boldsymbol{\nabla} \mathbf{v}^{(\mathrm{s})}$ may be expressed as a linear combination of the components of the tensor Π, so that the viscous heating is second order in the dissipative fluxes, while $\nabla \cdot \mathbf{Q}$ is first order. For practical reasons it is convenient to switch variables in Eq. (2.52), from entropy density $s(\mathbf{r}, t)$ to temperature $T(\mathbf{r}, t)$. This can be achieved by applying the local-equilibrium principle to the thermodynamic relationship:

$$T \, ds = c_V \, dT - \frac{T}{\rho^2} \left(\frac{\partial p}{\partial T} \right)_\rho d\rho, \tag{2.53}$$

where c_V is the specific heat capacity at constant volume. By further substituting Fourier's law (2.49) into the RHS of Eq. (2.52) and using the mass balance (2.5) one obtains:

$$\frac{\partial T}{\partial t} + \mathbf{v} \cdot \nabla T = \frac{\lambda}{\rho c_V} \nabla^2 T - \frac{T \alpha_p}{\rho c_V \varkappa_T} \nabla \cdot \mathbf{v}. \tag{2.54}$$

In deriving the heat equation (2.54) for the temperature, we adopted T and p as the independent variables. Alternatively we may use the $T ds$ equation with T and ρ as the independent variables, namely:

$$T ds = c_p \, dT + \frac{T}{\rho^2} \left(\frac{\partial \rho}{\partial T} \right)_p dp. \tag{2.55}$$

where now c_p is the specific heat capacity at constant pressure. The heat equation then becomes:

$$\frac{\partial T}{\partial t} + \mathbf{v} \cdot \nabla T = a_T \nabla^2 T + \frac{\alpha_p T}{\rho c_p} \left[\frac{\partial p}{\partial t} + \mathbf{v} \cdot \nabla p \right], \tag{2.56}$$

where $a_T = \lambda / \rho c_p$ is the thermal diffusivity, also called thermometric diffusivity. The mass balance Eq. (2.5), the Navier-Stokes Eq. (2.51), and the heat Eq. (2.54) or Eq. (2.56) constitute the complete set of hydrodynamic equations for a one-component fluid. These equations have to be solved for the fields $\mathbf{v}(\mathbf{r}, t)$, $T(\mathbf{r}, t)$ and $\rho(\mathbf{r}, t)$, with the pressure in Eq. (2.51) related to the density by the local-equilibrium equation of state $\rho = \rho[p(\mathbf{r}, t), T(\mathbf{r}, t)]$. Alternatively, if we use the heat equation (2.56), the fields to be solved for are $\mathbf{v}(\mathbf{r}, t)$, $T(\mathbf{r}, t)$ and $p(\mathbf{r}, t)$, while the density follows from the local equation of state. Depending on the problem under consideration, one or the other approach may be more convenient.

For an incompressible fluid the density is uniform under spatiotemporal evolution, so that the mass balance Eq. (2.5) reduces to:

$$\nabla \cdot \mathbf{v} = 0. \tag{2.57}$$

Notice that for an incompressible (divergence-free) fluid an uniform temperature solves the heat equation (2.54). Thus, for an incompressible isothermal fluid the Navier-Stokes Eq. (2.51) is the only relevant hydrodynamic equation. It is customary to eliminate then the pressure from the Navier-Stokes equation by applying a rotational to both sides of Eq. (2.51), so that it becomes a closed equation for the fluid velocity. The curl of the velocity field is called the vorticity, $\boldsymbol{\omega}$. In classical fluid dynamics the assumption of an incompressible isothermal fluid is very often adopted, so that fluids are studied in terms of the equation for the vorticity (Batchelor, 1967; Tritton, 1988). It is generally a good assumption for dense fluids and liquids. Compressibility effects are important in gas dynamics (Liepmann and Roshko, 1957; Shapiro, 1953). Compressibility effects are also important in fluids near the critical point (Onuki, 2002).

Recently, Brenner (2005a,b) has pointed out that in deriving the hydrodynamic equations, one should distinguish between a mass velocity and a volume velocity of the fluid, the difference being related to a diffusion flux. This observation has subsequently been supported by Öttinger (2005) and leads to some corrections to the hydrodynamic equations, ideas usually referred to as GENERIC. These correction terms do not affect the expressions for the hydrodynamic correlation functions to be presented in this book. Furthermore, it is worth mentioning that the hydrodynamic equations can also be derived from statistical mechanics starting from the Liouville equation. The transport coefficients then appear as integrals of time-correlation functions of molecular fluxes in an equilibrium ensemble, sometimes referred to as Kubo-Green formulas (Zwanzig, 1965; Steele, 1969; Kubo et al., 1991). One final remark is that it is quite simple to verify the Galilean invariance of the hydrodynamic equations. Relativistic hydrodynamics (Choudhuri, 1998) is outside the scope of the present volume.

2.4.2   Hydrodynamic equations for a binary mixture

In the case of a binary fluid mixture, we need to consider the complete entropy production, given by Eq. (2.41) but with the number of components reduced to two. Therefore we obtain for the dissipation in this case:

$$\Psi = -\frac{1}{T}\,\mathbf{Q}\cdot\boldsymbol{\nabla}T + \Pi : (\boldsymbol{\nabla}\mathbf{v})^{(s)} - \sum_{k=1}^{2}\mathbf{J}_k\cdot\left[T\,\boldsymbol{\nabla}\left(\frac{\mu_k}{T}\right) - \mathbf{F}_k\right]. \qquad (2.58)$$

We note that, in a binary mixture, the two diffusive fluxes are not independent, since Eqs. (2.10) and (2.11) imply: $\mathbf{J}_1 + \mathbf{J}_2 = 0$. Taking $\mathbf{J} = \mathbf{J}_1 = -\mathbf{J}_2$ we can rewrite Eq. (2.58) as:

$$\Psi = -\frac{1}{T}\,\mathbf{Q}\cdot\boldsymbol{\nabla}T + \Pi : (\boldsymbol{\nabla}\mathbf{v})^{(s)} - \mathbf{J}\cdot\left[T\,\boldsymbol{\nabla}\left(\frac{\mu}{T}\right) - (\mathbf{F}_1 - \mathbf{F}_2)\right], \qquad (2.59)$$

where $\mu = \mu_1 - \mu_2$, is the difference in chemical potentials between the two components. Next we neglect the presence of external forces. Note that, from the definition of $\mathbf{F}_k$, Eq. (2.26), when gravity is the only relevant external force, we have $\mathbf{F}_k = -g\hat{\mathbf{z}}$ for all k. Since in Eq. (2.59) external forces appear as a difference between the two components, the gravity effect will cancel. Of course, this may not be true when there are other external forces, as electromagnetic forces for a charged fluid or an electrolyte solution. Such cases are usually dealt with by defining an electrochemical potential, which is the chemical potential μ_k plus the specific potential energy ψ_k, see Eq. (2.26) (de Groot and Mazur, 1962; Demirel, 2002). We shall not further elaborate on this electrochemical potential since it is not relevant for the purpose of this book. Instead, we simply neglect external forces in (2.59), apply the local-equilibrium hypothesis to the equation of state $\mu = \mu(p, T, c)$ and collect all terms proportional to the temperature gradient, so that Eq. (2.59) transforms into:

$$\Psi = -\left\{ \mathbf{Q} + \left[T\left(\frac{\partial \mu}{\partial T}\right)_{p,c} - \mu \right] \mathbf{J} \right\} \cdot \frac{\nabla T}{T} + \Pi : (\nabla \mathbf{v})^{(\mathrm{s})} - \mathbf{J} \cdot \nabla_T \mu, \qquad (2.60)$$

where $\nabla_T \mu$ indicates that the gradient has to be taken at constant T, or:

$$\nabla_T \mu = \left(\frac{\partial \mu}{\partial c}\right)_{T,p} \nabla c + \left(\frac{\partial \mu}{\partial p}\right)_{T,c} \nabla p, \qquad (2.61)$$

where, for practical purposes, the concentration derivative of μ is conveniently expressed in terms of the osmotic compressibility. In Eq. (2.61), the field $c(\mathbf{r}, t)$ indicates the single concentration needed to describe the composition of a binary mixture. Thus $c = c_1$, and the component 1 of the mixture is referred to as "solute". Notice that the election of which one of the components is the solute is completely arbitrary.

We identify in Eq. (2.60) three different types of dissipative fluxes contributing to Ψ. First a dissipative energy flux: $\mathbf{Q}'(\mathbf{r}, t) = \mathbf{Q} + [T(\partial \mu / \partial T) - \mu] \mathbf{J}$, which includes the heat flux plus the contributions to the energy flux from nonisothermal diffusion. The second dissipative flux is the deviatoric (symmetric) stress tensor Π, which was also present in a one-component fluid. The third dissipative flux is the diffusion $\mathbf{J}(\mathbf{r}, t)$, which is a new flux that does not exist in one-component fluids. To formulate the appropriate phenomenological laws, we note that $\mathbf{Q}'$ and the diffusion flux have the same tensorial character, so that we need to consider coupling among them, even in an isotropic fluid. However, the deviatoric stress tensor will not couple with the other fluxes, so that the result of the previous section regarding Newton's viscosity law, Eq. (2.50), continues to be valid for two-component fluids. Phenomenological laws for the dissipative energy flux $\mathbf{Q}'(\mathbf{r}, t)$ and the diffusion flux $\mathbf{J}(\mathbf{r}, t)$ need to be considered together, and the most general (linear) relations among the flux components have the form:

$$Q_i + \left[T \left(\frac{\partial \mu}{\partial T} \right)_{p,c} - \mu \right] J_i = -\sum_j \left\{ M_{ij}^{(11)} \frac{\partial_j T}{T} + M_{ij}^{(12)} [\partial_j \mu]_T \right\}, \qquad (2.62)$$

$$J_i = -\sum_j \left\{ M_{ij}^{(21)} \frac{\partial_j T}{T} + M_{ij}^{(22)} [\partial_j \mu]_T \right\}. \qquad (2.63)$$

In the phenomenological equations (2.62) and (2.63) we have decomposed the tensor of phenomenological Onsager coefficients in four $M^{(\alpha\beta)}$ square matrices. This, initially, amounts to a total of 36 phenomenological coefficients. Onsager's reciprocal relations make the two "diagonal" matrices symmetric: $M_{ij}^{(\alpha\alpha)} = M_{ji}^{(\alpha\alpha)}$, while the "non-diagonal" matrices satisfy: $M_{ij}^{(12)} = M_{ji}^{(21)}$. These relations reduce the number of independent coefficients to 24. If we further consider isotropy, the three independent phenomenological matrices become proportional to the identity matrix and we are finally left with three scalar phenomenological coefficients. Hence:

$$\mathbf{Q}' = \mathbf{Q} + \left[T \left(\frac{\partial \mu}{\partial T} \right)_{p,c} - \mu \right] \mathbf{J} = -L^{(11)} \frac{\boldsymbol{\nabla} T}{T} - L^{(12)} \boldsymbol{\nabla}_T \mu,$$

$$\mathbf{J} = -L^{(12)} \frac{\boldsymbol{\nabla} T}{T} - L^{(22)} \boldsymbol{\nabla}_T \mu. \qquad (2.64)$$

where Onsager's reciprocal relationship makes the two cross-coefficients equal. Next, we eliminate $\boldsymbol{\nabla}_T \mu$ from the equation for $\mathbf{Q}$ using the second equation, and then substitute (2.61) in the second equation, so that:

$$\mathbf{Q} = -\left\{ L^{(11)} - \frac{[L^{(12)}]^2}{L^{(22)}} \right\} \frac{\boldsymbol{\nabla} T}{T} + \left[\mu - T \left(\frac{\partial \mu}{\partial T} \right)_{p,c} + \frac{L^{(12)}}{L^{(22)}} \right] \mathbf{J},$$

$$\mathbf{J} = -L^{(12)} \frac{\boldsymbol{\nabla} T}{T} - L^{(22)} \left[\left(\frac{\partial \mu}{\partial c} \right)_{T,p} \boldsymbol{\nabla} c + \left(\frac{\partial \mu}{\partial p} \right)_{T,c} \boldsymbol{\nabla} p \right], \qquad (2.65)$$

which, for practical purposes, are usually written in the form (Landau and Lifshitz, 1959)

$$\mathbf{Q} = -\lambda \, \boldsymbol{\nabla} T + \left[\mu - T \left(\frac{\partial \mu}{\partial T} \right)_{p,c} + k_T \left(\frac{\partial \mu}{\partial c} \right)_{p,T} \right] \mathbf{J},$$

$$\mathbf{J} = -\rho D \left\{ \boldsymbol{\nabla} c + \frac{k_T}{T} \, \boldsymbol{\nabla} T + \frac{k_p}{p} \, \boldsymbol{\nabla} p \right\}, \qquad (2.66)$$

in terms of three transport coefficients that are used in practice: the thermal conductivity λ, the binary diffusion coefficient D, and the dimensionless thermal-diffusion ratio k_T. They are related to the Onsager coefficients by:

$$D = \frac{L^{(22)}}{\rho} \left(\frac{\partial \mu}{\partial c} \right)_{p,T}, \qquad \lambda = \frac{1}{T} \left\{ L^{(11)} - \frac{[L^{(12)}]^2}{L^{(22)}} \right\},$$

$$k_T = \left(\frac{\partial \mu}{\partial c} \right)_{p,T}^{-1} \frac{L^{(12)}}{L^{(22)}}, \qquad k_p = p \left(\frac{\partial \mu}{\partial c} \right)_{p,T}^{-1} \left(\frac{\partial \mu}{\partial p} \right)_{T,c}. \qquad (2.67)$$

In Eq. (2.66) we have also introduced a barodiffusion ratio k_p. As can be seen from Eq. (2.67), k_p is a thermodynamic property, not a transport property. It is often advantageous to replace, in the expression (2.67) for k_p, the derivative of the chemical potential with respect to pressure by the derivative of the density with respect to concentration (Landau and Lifshitz, 1959). In this book, we shall neglect the barodiffusion contribution in (2.66); it is only important in centrifuges, geophysics and in systems where pressure gradients are extremely large, which is not the case here. This assumption also means that we neglect any heat flux driven by pressure gradients, that would appear when the second of Eqs. (2.66) is substituted into the first. This latter phenomenon (for isothermal fluids) is often referred to as *mechanocaloric* effect. In addition, the same substitution shows that there will be some heat flux driven by concentration gradients, which is called the Dufour effect.

Mass diffusion resulting from a temperature gradient in fluids is known as thermal diffusion or thermodiffusion. In liquids the phenomenon is also referred to as Soret (1880) effect and instead of k_T one may define a Soret coefficient S_T such that:

$$k_T = T S_T \, c(1 - c). \qquad (2.68)$$

In addition to the thermal diffusion k_T and barodiffusion k_p ratios introduced in Eq. (2.66), the so-called thermal diffusion and barodiffusion coefficients, Dk_T and Dk_p respectively, are often employed in the literature. The latter is sometimes referred to as pressure diffusivity (Demirel, 2002). As a final comment on transport coefficients, notice that for an isothermal and isobaric process, we reproduce from the second of the phenomenological equations (2.66) the well-known Fick's law.

We are now in a position to obtain the hydrodynamic equations for a binary mixture by substituting the corresponding phenomenological laws (2.66) into the balance equations. As mentioned above, the phenomenological law for the deviatoric stress tensor is the same as for a one-component fluid, *i.e.*, Eq. (2.50), so that the Navier-Stokes equation (2.51) is also valid for mixtures.

In a binary mixture, mass balance will generate two independent equations for each component. It is customary to use the same continuity equation (2.5) for the total density ρ, plus the mass balance equation (2.13) of the component chosen to define the diffusion flux $\mathbf{J}$:

$$\rho\left(\frac{\partial c}{\partial t} + \mathbf{v} \cdot \boldsymbol{\nabla} c\right) = -\boldsymbol{\nabla} \cdot \mathbf{J}, \tag{2.69}$$

written in terms of the concentration $c = \rho_1/\rho$. Upon substituting the phenomenological equations (2.66) into Eq. (2.69) we obtain:

$$\frac{\partial c}{\partial t} + \mathbf{v} \cdot \boldsymbol{\nabla} c = D \left\{\nabla^2 c + \frac{k_T}{T} \nabla^2 T\right\}, \tag{2.70}$$

where we have again neglected any spatiotemporal dependence of the transport coefficients D and k_T . Equation (2.70) is the so-called *diffusion equation* with the term on the RHS proportional to $\nabla^2 T$ representing the Soret effect. Note that we have neglected barodiffusion, $k_p \simeq 0$.

As in Sect. 2.4.1 for a one-component fluid, the heat equation for a binary mixture can be deduced from the entropy balance, Eqs. (2.40) and (2.41). For a binary mixture it may be written as:

$$\rho T\left(\frac{\partial s}{\partial t} + \mathbf{v} \cdot \boldsymbol{\nabla} s\right) = -\boldsymbol{\nabla} \cdot \mathbf{Q}' + T\, \boldsymbol{\nabla} \cdot \left[\left(\frac{\partial \mu}{\partial T}\right)_{p,c} \mathbf{J}\right]$$
$$- \mathbf{J} \cdot \boldsymbol{\nabla}_T \mu + \Pi : (\boldsymbol{\nabla}\mathbf{v})^{(\mathrm{s})}, \tag{2.71}$$

in terms of the same dissipative fluxes $\mathbf{Q}'$ and $\mathbf{J}$ appearing in the LHS of Eq. (2.64). In principle, the heat equation for a mixture can be obtained by substituting the phenomenological relations (2.64) into the RHS of Eq. (2.71), which yields a lengthy expression. For this reason, following Landau and Lifshitz (1959), it is customary to introduce a number of simplifications. First, the viscous heating is neglected which, as for a one-component fluid, is second order in the fluxes. For the same reason, the term $\mathbf{J} \cdot \boldsymbol{\nabla}_T \mu$ will also be neglected. Inverting Eqs. (2.64), we can express $\boldsymbol{\nabla}_T \mu$ as a linear combination of the fluxes: $\boldsymbol{\nabla}_T \mu = [-L^{(1,2)}\mathbf{Q}' + L^{(1,1)}\mathbf{J}]/[L^{(1,1)}L^{(2,2)} + (L^{(1,2)})^2]$, so that the term $\mathbf{J} \cdot \boldsymbol{\nabla}_T \mu$ is indeed quadratic in the fluxes. Furthermore, in the second term in the RHS of Eq. (2.71) one assumes $\boldsymbol{\nabla}(\partial\mu/\partial T) \simeq 0$, since it will involve second derivatives of the chemical potential that may be neglected for systems not too far from equilibrium. With these simplifications, Eq. (2.71) becomes:

$$\rho T\left(\frac{\partial s}{\partial t} + \mathbf{v} \cdot \boldsymbol{\nabla} s\right) = -\boldsymbol{\nabla} \cdot \mathbf{Q}' + T\left(\frac{\partial \mu}{\partial T}\right)_{p,c} \boldsymbol{\nabla} \cdot \mathbf{J}. \tag{2.72}$$

The canonical form for the heat equation for a binary mixture is now obtained by substituting the phenomenological relations as given by Eq. (2.66) into (2.72), so that (Landau and Lifshitz, 1959; Cohen et al., 1971):

$$\rho T \frac{ds}{dt} = \lambda\, \nabla^2 T + \rho D \left[k_T \left(\frac{\partial \mu}{\partial c}\right)_{p,T} - T \left(\frac{\partial \mu}{\partial T}\right)_{p,c}\right]\left[\nabla^2 c + \frac{k_T}{T}\, \nabla^2 T\right], \tag{2.73}$$

where, to simplify notation, we have written the LHS in terms of a material time derivative. In deriving Eq. (2.73) we have again neglected any spatiotemporal dependence of the transport coefficients λ and k_T. As in Sect. 2.4.1 for a one-component fluid, one usually prefers the equation in terms of $T(\mathbf{r}, t)$ rather than $s(\mathbf{r}, t)$. This can be performed by applying the local-equilibrium principle to one of the thermodynamic Tds equations. If we use the equivalent of Eq. (2.55) for a binary system:

$$T ds = c_p \, dT - \frac{\alpha_p T}{\rho} \, dp - T \left(\frac{\partial \mu}{\partial T} \right)_{p,c} dc, \qquad (2.74)$$

we obtain from Eq. (2.73):

$$\frac{dT}{dt} = \frac{\alpha_p T}{\rho c_p} \frac{dp}{dt} + [a_T + D\epsilon_D] \, \nabla^2 T + \frac{DT}{k_T} \, \epsilon_D \, \nabla^2 c, \qquad (2.75)$$

where the diffusion equation (2.69) has been employed neglecting the advection terms, consistent with having neglected $\mathbf{v} \cdot \nabla s$ in Eq. (2.71). In Eq. (2.75) we have introduced a dimensionless Dufour-effect ratio ϵ_D defined by:

$$\epsilon_D = \frac{k_T^2}{T \, c_p} \left(\frac{\partial \mu}{\partial c} \right)_{T,p}. \qquad (2.76)$$

If the dimensionless Dufour effect ratio is negligibly small, the heat equation (2.75) for a binary mixture becomes identical to the heat equation (2.56) for a one-component fluid. Hence, taking $\epsilon_D \simeq 0$ is equivalent to neglecting the Dufour effect. For liquid mixtures, the Dufour effect can be indeed neglected in most practical circumstances, but not for gas mixtures.

2.5 Boundary conditions

For many applications the hydrodynamic equations need to be supplemented with appropriate conditions for the relevant physical fields at the boundaries of the fluid. We shall first consider conditions at an arbitrary interface between two continuous media and then apply the conditions to a fluid in the presence of a horizontal boundary.

2.5.1 Conditions at the interface between two continuous media

We consider an arbitrary interface, also to be referred to as boundary, between two continuous media. The interface or boundary between the two continua is a two-dimensional surface embedded in three-dimensional space.

Let $\mathbf{r}(x, y, z)$ designate a point on this surface and let us choose a local Cartesian system of reference with the z-axis perpendicular to the surface. The boundary surface in the neighborhood of $\mathbf{r}$ at time t can then be represented mathematically by a function of the form:

$$z = h(x, y, t), \tag{2.77}$$

so that a vector $\mathbf{n}$ normal to the surface, directed from continuum 1 (below the interface) to continuum 2 (above the interface), is given by:

$$\mathbf{n} = \begin{pmatrix} -\dfrac{\partial h}{\partial x} \\ -\dfrac{\partial h}{\partial y} \\ 1 \end{pmatrix}. \tag{2.78}$$

The thermodynamic fields in continuum 1 will be designated by $\mathbf{v}_1$, T_1, etc. and those in continuum 2 by $\mathbf{v}_2$, T_2, etc. Later we shall identify continuum 1 with the fluid under consideration and continuum 2 with a "wall". For the moment, however, we consider the more general case of two fluid media. At the interface, the velocities of both fluids are equal:

$$\mathbf{v}_1(\mathbf{r}, t) = \mathbf{v}_2(\mathbf{r}, t), \qquad \text{when } z = h(x, y, t), \tag{2.79}$$

so that no voids are formed by a differential movement of the two fluids. The motion of the interface is governed by a simple kinematic condition:

$$\frac{\partial h}{\partial t} = \mathbf{v}_1 \cdot \mathbf{n} = \mathbf{v}_2 \cdot \mathbf{n}, \qquad \text{when } z = h(x, y, t), \tag{2.80}$$

with $\mathbf{n}$ given by Eq. (2.78). Local thermodynamic equilibrium between the two continua may be visualized as local equilibrium between two phases. This means equality of temperatures:

$$T_1(\mathbf{r}, t) = T_2(\mathbf{r}, t), \qquad \text{when } z = h(x, y, t), \tag{2.81}$$

as well as equality of chemical potentials.

In addition to the conditions (2.79)-(2.81) for the fields, nonequilibrium thermodynamics also implies conditions for the dissipative fluxes. These are obtained by applying the balance laws to the flux of mass, momentum and energy across the boundary. In the case of mass balance, a flux of mass causes the movement of the boundary in accordance with Eq. (2.80). In a multi-component system diffusion fluxes occur. If mass cannot accumulate at the interface, then:

$$\mathbf{J}_{1,k} \cdot \mathbf{n} = \mathbf{J}_{2,k} \cdot \mathbf{n}, \qquad \text{when } z = h(x, y, t) \tag{2.82}$$

for any component k of the mixture. The more complicated case of adsortion/desorption at the interface has been considered by Nepomnyashchy et al. (2002).

While the movement of the interface, Eq. (2.80), is caused by mass flux, momentum balance requires equality of forces on both sides of the boundary. The balance of forces is complicated by the presence of surface tension. As is well known, the interface between two fluids behaves like an elastic membrane, and a surface force acts when the surface is deformed. The magnitude of the surface force is determined by the surface tension $\sigma(\mathbf{r}, t)$ for points $\{\mathbf{r}, t\}$ on the interface. We do not go here into details, but the interested reader may consult a recent monograph of Nepomnyashchy et al. (2002). It turns out that, taking into account surface tension, the following boundary condition is obtained for the deviatoric stress tensor (Landau and Lifshitz, 1959; Nepomnyashchy et al., 2002):

$$\left[(p_1 - \rho_1 gh) - (p_2 - \rho_2 gh) - \sigma \left(\frac{\partial^2 h}{\partial x^2} + \frac{\partial^2 h}{\partial y^2}\right)\right] \mathbf{n} =$$
$$\boldsymbol{\nabla}_{\|}\sigma + (\Pi_1 - \Pi_2) \cdot \mathbf{n}, \quad (2.83)$$

for points $\{\mathbf{r}, t\}$ on the interface, specified by $z = h(\mathbf{r}, t)$. The symbol $\boldsymbol{\nabla}_{\|}$ indicates that the gradient is taken along the surface. In Eq. (2.83) it is assumed that the only external force is gravity. The combination of double derivatives of $h(\mathbf{r}, t)$ in the LHS of Eq. (2.83) represents the effect of curvature of the interface. Hence, the LHS of Eq. (2.83) represents the classical Laplace surface overpressure (capillary effect). The term $\boldsymbol{\nabla}_{\|}\sigma$ in the RHS gives rise to a force tangent to the surface and is due to the variation of the surface tension with temperature and the (possible) presence of surfactants. This tangential force is normally referred to as a Marangoni force and it is known to play a role in the drying of paint, surfactant spreading, and, of course, in Marangoni convection. Again, more details may be found in Nepomnyashchy et al. (2002).

The presence of surface tension implies that some energy is stored in the interface. Hence surface tension will also affect the energy balance at the interface (Napolitano, 1978). However, in most practical situations the surface-tension contribution may be neglected (Nepomnyashchy et al., 2002). Then, energy balance at the interface is equivalent to equality of energy fluxes at both sides, so that:

$$\mathbf{Q}_1' \cdot \mathbf{n} = \mathbf{Q}_2' \cdot \mathbf{n}, \qquad \text{when } z = h(x, y, t), \qquad (2.84)$$

where $\mathbf{Q}'$ is the dissipative energy flow, given by the LHS of Eq. (2.62), which includes energy flow due to non-isothermal diffusion in a binary fluid. Equations (2.82)-(2.84) constitute boundary conditions (BC) for the dissipative fluxes. Usually, the phenomenological laws are substituted into these

BC, yielding equivalent BC expressed in terms of the derivatives of the thermodynamic fields. Sometimes one distinguishes between BC of the first kind, that specify thermodynamic fields at the boundaries, and BC of the second kind, that specify derivatives of these fields.

2.5.2   A flat horizontal fluid boundary

For the problems to be discussed in the present book we only need to consider boundary conditions for flat and stationary horizontal interfaces. Thus we consider a fluid (continuum 1) with a boundary located at $z = L$ with $\hat{z}$ as the unit vector perpendicular to the horizontal interface. For a one-component fluid $\mathbf{Q}' = \mathbf{Q}$, so that upon substituting Fourier's law into Eq. (2.84) we obtain the simpler condition:

$$\frac{1}{\lambda_2}\frac{\partial T_1}{\partial z} = \frac{1}{\lambda_1}\frac{\partial T_2}{\partial z}, \qquad \text{when } z = L. \qquad (2.85)$$

If continuum 2 is a perfectly heat-conducting wall, then $\lambda_2 \simeq \infty$, so that $\partial T_2/\partial z \simeq 0$. Thus at the interface with a perfectly conducting wall, Eq. (2.85) holds independently of the value of $\partial T_1/\partial z$. Hence, a perfectly conducting wall means that the temperature gradient of the fluid at the boundary may have any value. For perfectly conducting walls the only remaining boundary condition in the temperature field is Eq. (2.81), with T_2 being the uniform temperature of the wall.

In the case of a flat and stationary boundary Eq. (2.80) requires the velocities to be parallel to the interface, so that:

$$v_{1,z} = v_{2,z} = 0, \qquad \text{when } z = L. \qquad (2.86)$$

When the wall is solid (rigid) $\mathbf{v}_2 = 0$, so that, in accordance with Eq. (2.79), all components of the fluid velocity at the wall must vanish (not just the vertical one). Hence for a flat solid wall, dropping the index 1, we require for the fluid velocity $\mathbf{v}$:

$$\mathbf{v} = 0, \qquad \text{when } z = L. \qquad \text{(rigid)} \qquad (2.87)$$

As a consequence, the derivatives of the fluid velocity along the boundary must also vanish: $\partial_x \mathbf{v} = \partial_y \mathbf{v} = 0$. The derivatives with respect to the vertical z-direction have to be determined from the balance of forces, Eq. (2.83), without any contribution from a surface tension, since we are dealing now with solid walls. For the two horizontal velocity components one usually employs an argument similar to that elucidated above for the temperature at the interface with a perfectly conducting wall. That is, a solid wall is visualized as a continuum with infinite viscosity $\eta \simeq \infty$, so that the horizontal force balance at the interface holds, independently of the values of the z-derivatives of the fluid velocity: $\partial_z v_x$ and $\partial_z v_y$. The latter derivatives are undetermined and may have any value at the boundary. Regarding the z-component of Eq. (2.83) for a rigid wall, in many instances one assumes the fluid to be divergence free or incompressible. Then, combining Eq. (2.57) with Eq. (2.87) implies that:

$$\partial_z v_z(x, y, z = L, t) = 0. \qquad \text{(rigid and incompressible)} \quad (2.88)$$

In this way, the z-component of Eq. (2.83): $p - \rho g L = 0$ implies a redefinition of the pressure origin, which is always possible in the case of an incompressible fluid.

Another important case is that of free boundaries, meaning that the wall is a vapor or a gas, typically air. In general, the Marangoni and Laplace terms in Eq. (2.83) will cause the interface to curve, making this case quite complicated (Nepomnyashchy et al., 2002). However, if we neglect the Marangoni term, it is possible to imagine a stationary and flat free boundary. Then we first need to impose Eq. (2.86). Furthermore, if we neglect the stresses Π_2 caused by the gas and drop the index 1 in the deviatoric stress tensor Π of the fluid, we find from Eq. (2.83) that the horizontal components of Π (neglecting the Marangoni force) should satisfy:

$$\Pi_{xz} = \Pi_{yz} = 0, \qquad \text{when } z = L. \qquad \text{(free)} \quad (2.89)$$

Combining Eq. (2.89) with Eq. (2.86) and Newton's viscosity law (2.50), one readily concludes that at the boundary with a flat free surface, $\partial_z v_x = \partial_z v_y = 0$. Combining this fact with the divergence-free condition (2.57) for an incompressible fluid, it implies:

$$\partial_z^2 v_z(x, y, z = L, t) = 0. \qquad \text{(free and incompressible).} \quad (2.90)$$

Equation (2.90) is to be compared with Eq. (2.88) for a divergence free (incompressible) fluid with a rigid boundary. In this book we shall follow the convention of referring to Eq. (2.88) as "stick" boundary condition, and to Eq. (2.90) to as "free-slip" boundary condition. When any surface-tension effects are neglected, the vertical z-component of Eq. (2.83) for a free boundary becomes the same as that for a rigid boundary.

For the boundary condition (2.82) for the mass flux, the most realistic case to be considered is that of an impermeable wall. For a binary mixture, this means that the solute cannot diffuse in the wall, so that $\mathbf{J}_2 = 0$. Then, substituting the phenomenological equation (2.66) for a binary mixture into Eq. (2.82) , neglecting barodiffusion and dropping the index 1 for the fields in the fluids, we obtain:

$$\partial_z c + \frac{k_T}{T} \partial_z T = 0. \qquad \text{when } z = L. \qquad \text{(impermeable)} \quad (2.91)$$

When we combine Eq. (2.91) with Fourier's law for a binary mixture, we obtain from Eq. (2.84) the same condition, Eq. (2.85) for a binary fluid as for a one-component fluid. Hence, the discussion of perfectly conducting walls after Eq. (2.85) continues to be valid for binary mixtures, provided that the walls are impermeable.

Chapter 3

Fluctuations in fluids in thermodynamic equilibrium

In this chapter we present the theory of fluctuations in fluids that are in thermodynamic equilibrium. To understand fluctuations in fluids in non-equilibrium steady states to be developed in successive chapters, one needs first to be familiar with concepts for dealing with fluctuations in equilibrium. One important characteristic we shall discuss in this chapter is that thermal fluctuations in fluids in thermodynamic equilibrium are spatially short ranged, except when the fluid is near a critical point. This feature is in sharp contrast to what occurs in nonequilibrium steady states where the thermal fluctuations will always be long ranged.

Thermal fluctuations are characterized by space-time functions that account for the correlation between the value of a quantity at position $\mathbf{r}$ and time t with its value at $\mathbf{r}'$ and time t'. The intensity of the fluctuations is determined by equal-time correlation functions and the dynamical properties of the fluctuations by time-dependent correlation functions. The Fourier transforms of these correlations functions for the fluctuating density are commonly referred to as static and dynamic structure factors, respectively. These concepts, to be introduced in the present chapter, will be of great importance throughout the book.

There are two general methods to deduce the appropriate expressions for the time-dependent correlations of the fluctuating hydrodynamic variables in fluids in thermodynamic-equilibrium states. The first one proceeds by following the evolution of the fluctuating variables from an unspecified initial condition in terms of the (deterministic) linearized evolution equations, where in accordance with Onsager's regression hypothesis fluctuations evolve like deterministic deviations from equilibrium. The resulting expressions are then averaged over all possible initial conditions, by using statistical mechanics and the requirement that the equilibrium entropy be a maximum (Einstein hypothesis) (Mountain, 1966; Berne and Pecora, 1976; Boon and Yip, 1980; Forster, 1975). A second method computes the correlation functions for the hydrodynamic variables from fluctuating hydrodynamics in conjunction with the fluctuation-dissipation theorem (Landau and Lifshitz, 1958, 1959; Fox and Uhlenbeck, 1970a). We shall refer to this second method as stochastic forcing. As we shall discuss in Sect. 3.4, the fluctuation-dissipation theorem includes the Einstein hypothesis and both methods yield the same results for fluctuations in equilibrium fluids. A recent discussion of the relationship between the two procedures can be found in Vázquez and López de Haro (2001) [see also Landau and Lifshitz (1958)]. In this book we use the method of fluctuating hydrodynamics since we find it conceptually easier to extend this method to fluctuations in nonequilibrium states. While fluctuating hydrodynamics was originally proposed by Landau and Lifshitz (1958, 1959) on a phenomenological basis, various authors have shown subsequently how it can be derived from kinetic theory (Bixon and Zwanzig, 1969; Fox and Uhlenbeck, 1970b; Mashiyama and Mori, 1978).

We begin the discussion of stochastic forcing by introducing in Sect. 3.1 fluctuating dissipative fluxes and the fluctuation-dissipation theorem (FDT) for the correlation functions among such various random fluxes. The FDT will be specified for one-component fluids and binary fluid mixtures in Sect. 3.2. In Sect. 3.3 we derive the static and dynamic structure factor of one-component fluids and recover the classical expressions (Boon and Yip, 1980; Berne and Pecora, 1976) for these quantities. The relationship between the static structure factor and fluctuations of the entropy will be elucidated in Sect. 3.4, thus providing an a posteriori justification of the FDT. We conclude this chapter by outlining in Sect. 3.5 how the method of fluctuating hydrodynamics can be extended to fluids in nonequilibrium steady states by assuming a local version of the FDT.

3.1   Fluctuating hydrodynamics

The method of fluctuating hydrodynamics for dealing with fluctuations at hydrodynamic spatiotemporal scales in fluids in thermodynamic equilibrium is originally due to Landau and Lifshitz (1958, 1959). The central idea of fluctuating thermodynamics is to treat the dissipative fluxes as stochastic variables. As discussed in the previous chapter, the dissipative fluxes are the macroscopic manifestation of the existence of microscopic degrees of freedom in a thermodynamic system. For instance, in the case of the energy balance discussed in Sect. 2.2.4, we had to introduce a heat flux $\mathbf{Q}(\mathbf{r}, t)$ to reflect the fact that mechanical energy (potential or kinetic) can be transferred to internal degrees of freedom. In general, dissipation in a fluid is caused by the transfer of energy and momentum to random (through collisions) molecular motions. From this point of view it seems natural to assume that, at a mesoscopic level, dissipative fluxes will become random variables, reflecting the random nature of molecular motion.

The next important ingredient is the assumption that the phenomenological relations between dissipative fluxes and thermodynamic forces continue to be valid in the presence of fluctuations, but only on the average. This means that the actual value of the dissipative fluxes $J_\alpha(\mathbf{r}, t)$ will be equal to an average value specified by the phenomenological relations, plus some "fluctuation":

$$J_\alpha(\mathbf{r}, t) = \sum_\beta M_{\alpha\beta}(\mathbf{r}, t) X_\beta(\mathbf{r}, t) + \delta J_\alpha(\mathbf{r}, t), \tag{3.1}$$

where the random parts $\delta J_\alpha(\mathbf{r}, t)$ are called the *fluctuating dissipative fluxes*. Hence, instead of considering a value for a dissipative flux at a given spatiotemporal point, we shall treat $J_\alpha(\mathbf{r}, t)$ as consisting of an average part plus an (infinite) set of possible values for the random part with a certain functional probability $P[\delta J_\alpha]$, to be specified below for a global-equilibrium state. For the phenomenological relations to be valid on the average, the mean value of $\delta J_\alpha(\mathbf{r}, t)$ with probability distribution $P[\delta J_\alpha(\mathbf{r}, t)]$ has to be zero: $\langle \delta J_\alpha(\mathbf{r}, t) \rangle = 0$. In the language of van Kampen (1982), the random dissipative fluxes are a stochastic process.

Substitution of the hypothesis (3.1) into the corresponding balance laws yields a set of stochastic partial differential equations that describe the evolution of the thermodynamic fields and densities under the influence of fluctuating dissipative fluxes. In this way hydrodynamic equations transform into a Langevin-like set of partial differential equations, where the fluctuating dissipative fluxes play the role of independent random forces. The goal of fluctuating hydrodynamics is then to deduce the statistical properties of the thermodynamic fields and densities, such as their mean values or correlations. For this purpose one needs the statistical properties of the random dissipative fluxes. For a fluid in thermodynamic equilibrium, it is assumed that they have the general form:

$$\begin{aligned} \langle \delta J_\alpha(\mathbf{r}, t) \rangle &= 0 \\ \langle \delta J_\alpha(\mathbf{r}, t) \cdot \delta J_\beta(\mathbf{r}', t') \rangle &= C_{\alpha\beta}(\mathbf{r}, \mathbf{r}')\, \delta(t - t'). \end{aligned} \tag{3.2}$$

The first of Eqs. (3.2) states the obvious fact that, by definition, the average value of the random dissipative fluxes has to be zero. In the second of Eqs. (3.2) it is assumed that, as a consequence of the randomness of the molecular motion, the values of the fluctuating part of the dissipative fluxes at two different times are uncorrelated, independent of the spatial points where they are evaluated. In the language of stochastic-process theory (van Kampen, 1982; Gardiner, 1985), Eq. (3.2) implies that the random dissipative fluxes, δJ_α, are a set of Markov processes. For a system in equilibrium it can be demonstrated that the functions $C_{\alpha\beta}(\mathbf{r}, \mathbf{r}')$, specifying the spatial correlations between the random dissipative fluxes, have the form:

$$C_{\alpha\beta}(\mathbf{r}, \mathbf{r}') = k_{\mathrm{B}}T \left[M_{\alpha\beta} + M_{\beta\alpha}\right] \delta(\mathbf{r} - \mathbf{r}')$$

$$= 2k_{\mathrm{B}}T \; M_{\alpha\beta} \; \delta(\mathbf{r} - \mathbf{r}'), \tag{3.3}$$

where k_{B} is Boltzmann's constant, T the uniform temperature corresponding to a global-equilibrium state and $M_{\alpha\beta}$ the components of the same Onsager's dissipation matrix, introduced in Sect. 2.3. Equation (3.3) is the so called *fluctuation-dissipation theorem* (FDT). The FDT is traditionally derived by the following chain of arguments: first the spatial correlations $C_{\alpha\beta}(\mathbf{r}, \mathbf{r}')$ in Eq. (3.2) are left undetermined; then the fluctuations of the thermodynamic fields are calculated by substituting Eq. (3.1) into the balance laws; and finally each function $C_{\alpha\beta}(\mathbf{r}, \mathbf{r}')$ is determined by imposing that the entropy must be a maximum, so that the equal-time (static) correlation functions among the fluctuating fields have the entropy as probability generating functional[‡] (Landau and Lifshitz, 1958, 1959; Fox and Uhlenbeck, 1970a). In this book we adopt a less formal approach and in Eq. (3.3) we have already anticipated the result to be obtained. In Sect. 3.3 we evaluate the statistical properties of the fluctuating thermodynamic fields using Eq. (3.3). We then show in Sect. 3.4 how the FDT indeed implies that the static correlations among fluctuating fields have the entropy as probability generating functional. Thus, we shall not "derive" the FDT, but we shall "illustrate" how the FDT yields the correct results to be expected on the basis of thermodynamics and statistical mechanics.

Upon substitution of Eq. (3.3) into Eq. (3.2) we see that, in a global-equilibrium state, the values of the fluctuating fluxes have a multivariate spatiotemporal gaussian probability distribution. Equations (3.2) and (3.3) completely specify the functional probability $P[J_\alpha(\mathbf{r}, t)]$ of the random dissipative fluxes, because one can obtain all moments of this distribution from the second moments. Hence, from now on, when we mention in this book *averaging over fluctuations* we shall mean averaging with the probability distribution $P[J_\alpha(\mathbf{r}, t)]$, specified by Eqs. (3.2) and (3.3). It should be noted that the validity of the FDT (*i.e.*, normal gaussian distribution of thermal fluctuations) has only been established from statistical mechanics for fluctuations in systems that are in a global-equilibrium state. Extending the FDT to nonequilibrium steady states will be discussed in Sect. 3.5

To finalize, we comment that the present formulation of the FDT ceases to be valid for systems with slow relaxation processes like glassy systems,

---

[‡]Usually referred to as Einstein hypothesis, see Eq. (3.38)

polymer blends, etc. In those cases it seems that the FDT may be reformulated by using suitable kernels substituting the temporal delta function in Eq. (3.2) (Pérez-Madrid et al., 2003; Santamaría-Holek et al., 2004). Alternatively, Cugliandolo et al. (1997) have proposed the introduction of an "effective" temperature, which for glassy systems differs from the bath temperature. However, we believe that the need to define an effective temperature is an idea far from settled (Villar and Rubí, 2001). Furthermore, remarks about the formulation of the FDT when there are chemical reactions in the system will be presented later in Sect. 11.4.1.

3.2 Fluctuation-dissipation theorem for fluids and fluid mixtures

In the previous section we introduced the general form (3.2)-(3.3) of the fluctuation-dissipation theorem. In this section we specify the FDT for one-component fluids and for binary fluid mixtures that obey the hydrodynamic equations presented in Sect. 2.4.

3.2.1 Fluctuation-dissipation theorem for a one-component fluid

The first step in establishing the FDT for any thermodynamic system is to identify the relevant dissipative fluxes from the corresponding expression for the entropy production $\dot{S}$, or the dissipation function Ψ. For a one-component fluid the dissipation function was given by Eq. (2.46) with two dissipative fluxes: a vectorial heat flow $\mathbf{Q}(\mathbf{r}, t)$ and a second-rank (symmetric) deviatoric stress tensor $\Pi(\mathbf{r}, t)$. Therefore, the set of dissipative fluxes and their corresponding thermodynamic forces are in this case:

$$
\begin{aligned}
J_i^{(1)} &= Q_i, & X_i^{(1)} &= -\frac{1}{T}\frac{\partial T}{\partial x_i}, \\
J_{ij}^{(2)} &= \Pi_{ij}, & X_{ij}^{(2)} &= \frac{1}{2}\left(\frac{\partial v_i}{\partial x_j} + \frac{\partial v_i}{\partial x_j}\right).
\end{aligned}
\tag{3.4}
$$

The next step in fluctuating hydrodynamics is to identify the dissipation matrix from the phenomenological laws. For a one-component fluid the phenomenological laws are Fourier's law, Eq. (2.49), and Newton's viscosity law, Eq. (2.50), from which we readily identify the components of the dissipation matrix M as:

$$
M_{i,j}^{(11)} = \lambda T\,\delta_{ij}, \qquad M_{i,jk}^{(12)} = 0, \tag{3.5}
$$

$$
M_{ij,k}^{(21)} = 0, \qquad M_{ij,kl}^{(22)} = \eta\left(\delta_{ik}\delta_{jl} + \delta_{il}\delta_{jk}\right) + \left(\eta_\mathrm{v} - \frac{2}{3}\eta\right)\delta_{ij}\delta_{kl}.
$$

Indeed, multiplying the matrix M with the two thermodynamic forces, we have

$$J_i^{(1)} = \sum_j M_{ij}^{(11)} \, X_j^{(1)} + \sum_{kl} M_{i,kl}^{(12)} \, X_{kl}^{(2)},$$

$$J_{ij}^{(2)} = \sum_k M_{ij,k}^{(21)} \, X_k^{(1)} + \sum_{kl} M_{ijkl}^{(22)} \, X_{kl}^{(2)},$$

which, upon substitution of Eqs. (3.4) and (3.5), reproduce both Fourier's law and Newton's viscosity law.

Therefore, in accordance with the general considerations of Sect. 3.1, to develop the fluctuating hydrodynamics equations for a newtonian viscous fluid, we must consider two fluctuating dissipative fluxes: a random heat flow, $\delta \mathbf{Q}(\mathbf{r}, t)$, and a random deviatoric stress tensor $\delta \Pi(\mathbf{r}, t)$. These two random dissipative fluxes will be the sources of hydrodynamic fluctuations (thermal noise). According to the general version (3.2)-(3.3) of the FDT and the specification (3.5) of Onsager's dissipation matrix, the correlation functions between the random fluxes for a one-component fluid are given by (Landau and Lifshitz, 1959; Fox and Uhlenbeck, 1970a; Schmitz and Cohen, 1985a,b):

$$\langle \delta Q_i(\mathbf{r}, t) \cdot \delta Q_j(\mathbf{r}', t') \rangle = 2k_{\mathrm{B}} \lambda T^2 \delta_{ij} \, \delta(\mathbf{r} - \mathbf{r}') \, \delta(t - t') \tag{3.6a}$$

$$\langle \delta Q_i(\mathbf{r}, t) \cdot \delta \Pi_{jk}(\mathbf{r}', t') \rangle = \langle \delta \Pi_{ij}(\mathbf{r}, t) \cdot \delta Q_k(\mathbf{r}', t') \rangle = 0 \tag{3.6b}$$

$$\langle \delta \Pi_{ij}(\mathbf{r}, t) \cdot \delta \Pi_{kl}(\mathbf{r}', t') \rangle = 2k_{\mathrm{B}} T \left[\eta \left(\delta_{ik} \delta_{jl} + \delta_{il} \delta_{jk} \right) + \left(\eta_{\mathrm{v}} - \frac{2}{3} \eta \right) \delta_{ij} \delta_{kl} \right]$$
$$\times \, \delta(\mathbf{r} - \mathbf{r}') \, \delta(t - t'). \tag{3.6c}$$

Note that Eq. (3.6a) is invariant under permutation of the indices i and j, while Eq. (3.6c) is invariant if, simultaneously, i is permuted by k and j by l. This is to be expected from the space-time translational invariance of any equilibrium property, which shows the connection between space-time invariance and the Onsager's reciprocal relations (see Sect. 2.4). In addition, Eq. (3.6c) is also invariant under permutation of indices in any of the two tensors in the LHS. This is a consequence of the assumed symmetry of the deviatoric stress tensor. We may also mention that, by virtue of the Curie principle, the random heat flow and the random pressure tensor do not couple, so that the corresponding cross correlations are zero. The averages in Eqs. (3.6) are to be understood to be taken over the equilibrium probability functional distribution for the fluctuations, as was mentioned in Sect. 3.1. As was noted there, the fluctuations of the dissipative fluxes in an equilibrium fluid have a multivariate normal gaussian distribution. Hence, all the odd moments vanish, while the even moments can be expressed in terms of the second moment given by the FDT, Eqs. (3.6). Equations (3.6) will be used frequently in the remainder of this book.

An important particular case is an incompressible fluid, which is divergence-free as specified by Eq. (2.57). For an incompressible fluid Newton's viscosity law becomes simpler, since the term multiplying the divergence of velocity does not contribute. As a consequence, the equations for the FDT of an incompressible fluid simplify to:

$$\langle \delta Q_i(\mathbf{r}, t) \cdot \delta Q_j(\mathbf{r}', t') \rangle = 2k_B \lambda T^2 \delta_{ij} \, \delta(\mathbf{r} - \mathbf{r}') \, \delta(t - t') \tag{3.7a}$$

$$\langle \delta Q_i(\mathbf{r}, t) \cdot \delta \Pi_{jk}(\mathbf{r}', t') \rangle = 0 \tag{3.7b}$$

$$\langle \delta \Pi_{ij}(\mathbf{r}, t) \cdot \delta \Pi_{kl}(\mathbf{r}', t') \rangle = 2k_B T \eta \left(\delta_{ik}\delta_{jl} + \delta_{il}\delta_{jk} \right) \delta(\mathbf{r} - \mathbf{r}') \, \delta(t - t'). \tag{3.7c}$$

Note that the FDT for the random heat flow is unaffected by the incompressibility assumption.

To close this section we note that the statistical properties of the random fluxes, as given by Eqs. (3.6), are compatible with the time-correlation-function-expressions for the transport coefficients. For instance, integrating both sides of Eq. (3.6a) for indices i and j corresponding to the x-axis, we readily obtain:

$$\lambda = \frac{1}{k_B T^2 V} \int_V d\mathbf{r} \int_V d\mathbf{r}' \int_0^\infty d\tau \, \langle \delta Q_x(\mathbf{r}, t + \tau) \cdot \delta Q_x(\mathbf{r}', t) \rangle, \tag{3.8}$$

independent of t. Equation (3.8) is one example of the so-called Green-Kubo formulas (Boon and Yip, 1980). Formulas like (3.8) can be derived from the statistical mechanics of molecular fluids (Zwanzig, 1965; Kubo et al., 1991). We thus see that fluctuating hydrodynamics is consistent with the statistical mechanics of molecular fluids.

3.2.2 Fluctuation-dissipation theorem for a binary mixture

Just as in Sect. 3.2.1, we start from the corresponding expression (2.60) for the dissipation function Ψ that contains in this case three dissipative fluxes (see Sect 2.4.2). Consequently, to specify the equations of fluctuating hydrodynamics for a binary fluid mixture, we have to consider three random terms: a random energy flow $\delta \mathbf{Q}'$, a random deviatoric stress tensor $\delta \Pi$ and a random diffusion flow $\delta \mathbf{J}$. As was the case for a one-component fluid, $\delta \Pi$ does not couple with the other two dissipative fluxes, so the FDT for $\delta \Pi$ is the same as Eq. (3.6c) or (3.7c) for a one-component fluid, depending on whether or not the fluid is assumed to be incompressible. The other two dissipative fluxes do couple, and the relevant phenomenological laws are given by Eqs. (2.64). Hence, applying to a binary mixture the generic version of the FDT, Eqs. (3.2)-(3.3), we obtain:

$$\langle \delta Q'_i(\mathbf{r}, t) \cdot \delta Q'_j(\mathbf{r}', t') \rangle = 2k_B T \ L^{(11)} \ \delta_{ij} \ \delta(\mathbf{r} - \mathbf{r}') \ \delta(t - t')$$

$$\langle \delta Q'_i(\mathbf{r}, t) \cdot \delta J_j(\mathbf{r}', t') \rangle = 2k_B T \ L^{(12)} \ \delta_{ij} \ \delta(\mathbf{r} - \mathbf{r}') \ \delta(t - t') \qquad (3.9)$$

$$\langle \delta J_i(\mathbf{r}, t) \cdot \delta J_j(\mathbf{r}', t') \rangle = 2k_B T \ L^{(22)} \ \delta_{ij} \ \delta(\mathbf{r} - \mathbf{r}') \ \delta(t - t'),$$

where the coefficients $L^{(11)}$, etc. are the same as those in the phenomeno-logical laws, Eqs. (2.64). Of course, in practice, instead of using the phe-nomenological coefficients $L^{(11)}$, etc. it is again convenient to use the trans-port coefficients, defined by Eqs. (2.67). Solving Eqs. (2.67) for the Onsager phenomenological coefficients,we obtain

$$L^{(11)} = \lambda T + \rho D k_T^2 \left(\frac{\partial \mu}{\partial c} \right)_{T,p} = \lambda T \left[1 + \frac{D}{a_T} \epsilon_D \right]$$

$$L^{(22)} = \rho D \left(\frac{\partial \mu}{\partial c} \right)_{T,p}^{-1} \qquad\qquad (3.10)$$

$$L^{(12)} = \rho D k_T.$$

Substitution of Eq. (3.10) into Eq. (3.9), yields the equations for the FDT of a binary mixture (Cohen et al., 1971; Law and Nieuwoudt, 1989). Note that the barodiffusion ratio k_p does not appear in the FDT, since it depends only on equilibrium properties. Thus, barodiffusion is not a dissipative process and does not contribute to the FDT, in contrast to a suggestion by Foch (1971). However, the Dufour and Soret effects do contribute to the FDT leading to the appearance of ϵ_D and k_T in Eqs. (3.10). For liquids the Dufour effect can be neglected in practice, but not for gases. Sometimes, the cross correlation due to the Soret effect, given by the third line of Eq. (3.10), is also neglected (Hollinger et al., 1998).

3.3   Hydrodynamic fluctuations in a one-component fluid

To apply the method of fluctuating hydrodynamics we start from the bal-ance laws formulated in Chapter 2. For a one-component newtonian vis-cous fluid we have three balance laws derived in Sect. 2.2: conservation of mass (2.5), conservation of momentum (2.23), and conservation of en-ergy (2.34). For the present purpose we prefer to start from the entropy balance, Eq. (2.52), rather than from the energy balance (2.34). For the convenience of the reader, we display here again the balance laws for a one-component fluid in the absence of external volumetric forces (Landau and

Lifshitz, 1959; Batchelor, 1967):

$$\frac{\partial \rho}{\partial t} = -\boldsymbol{\nabla} \cdot (\rho \mathbf{v}) \tag{3.11a}$$

$$\frac{\partial}{\partial t}(\rho v_i) = -\frac{\partial}{\partial x_k}(p\delta_{ik} + \rho v_i v_k - \Pi_{ik}) \tag{3.11b}$$

$$\rho T \left(\frac{\partial s}{\partial t} + \mathbf{v} \cdot \boldsymbol{\nabla} s \right) = -\boldsymbol{\nabla} \cdot \mathbf{Q}, \tag{3.11c}$$

where a sum over repeated indices is understood. In Eq. (3.11c) we have already neglected viscous heating. Since we shall later linearize the hydrodynamic equations, viscous heating (being second order in the dissipative fluxes) will indeed not contribute to the theory. For the calculations of this section it is more convenient to use the mass density ρ and the temperature T as independent thermodynamic variables. Therefore, the entropy-density variable $s(\mathbf{r}, t)$ in Eq. (3.11c) will be replaced by the temperature variable $T(\mathbf{r}, t)$, as was done in Sect. 2.4.1 in deriving the heat equation. Moreover, the pressure will be expressed in terms of the equation of state.

In thermodynamic equilibrium the dissipative fluxes vanish: $\mathbf{Q} = \Pi = 0$, and from Eqs. (3.11) it can be shown that equilibrium implies a uniform temperature $T = T_0$, a uniform density $\rho = \rho_0$ and a zero velocity $\mathbf{v} = 0$. To obtain the equations of fluctuating hydrodynamics we proceed along the following steps:

1. We express the thermodynamic variables in the balance equations (3.11) as the sum of their equilibrium value and a fluctuating part: $T(\mathbf{r}, t) = T_0 + \delta T(\mathbf{r}, t)$, $\rho(\mathbf{r}, t) = \rho_0 + \delta\rho(\mathbf{r}, t)$ and $\mathbf{v}(\mathbf{r}, t) = \delta\mathbf{v}(\mathbf{r}, t)$.

2. In the phenomenological laws for the dissipative fluxes, to be substituted into the balance equations, we add a random contribution in accordance with Eq. (3.1). To apply the general theory of Sect. 3.1, we need to consider here a random heat flux $\delta\mathbf{Q}(\mathbf{r}, t)$ and a random deviatoric tensor $\delta\Pi(\mathbf{r}, t)$. For example, the flux $\mathbf{Q}$ in Eq. (3.11c) will be represented by a fluctuating Fourier law:

$$\mathbf{Q}(\mathbf{r}, t) = -\lambda \, \boldsymbol{\nabla} \delta T(\mathbf{r}, t) + \delta\mathbf{Q}(\mathbf{r}, t), \tag{3.12}$$

and the deviatoric stress tensor Π in Eq. (3.11b) by a similar fluctuating Newton's viscosity law.

Implementation of this procedure yields a set of stochastic differential equations for the fluctuating variables $\delta\rho$, $\delta\mathbf{v}$, and δT. Since the fluctuations around equilibrium are usually small, terms of second or higher order in the fluctuating variables can be neglected, so that we obtain:

$$\frac{\partial(\delta\rho)}{\partial t} = -\rho_0(\boldsymbol{\nabla} \cdot \delta\mathbf{v}), \tag{3.13a}$$

$$\rho_0 \frac{\partial(\delta \mathbf{v})}{\partial t} = -\nabla(\delta p) + \eta \nabla^2 \delta \mathbf{v} + \left(\eta_v + \frac{1}{3}\eta \right) \nabla(\nabla \cdot \delta \mathbf{v}) + \nabla \cdot \delta \Pi, \quad (3.13b)$$

$$\rho_0 c_V \frac{\partial(\delta T)}{\partial t} = \lambda \nabla^2 \delta T - \frac{(\gamma - 1)\rho_0 c_V}{\alpha_p}(\nabla \cdot \delta \mathbf{v}) - \nabla \cdot \delta \mathbf{Q}, \quad (3.13c)$$

where $\gamma = c_p/c_V$ is the adiabatic index. In deriving Eq. (3.13c) we have made use of the (equilibrium) thermodynamic relationship:

$$c_p - c_V = \frac{T_0 \alpha_p^2}{\rho_0 \varkappa_T}. \quad (3.14)$$

Notice that, as in Sect. 2.4.1, we are assuming all thermophysical properties as constants to be evaluated at the equilibrium temperature and density. This is a good approximation for fluids far from any critical point. We explicitly exclude critical fluctuations in this book. Having assumed all thermophysical properties as constants and having neglected second-order advection terms makes Eqs. (3.13) linear in the fluctuating fields. Nonlinear terms neglected in the derivation of (3.13) are responsible for the appearance of the so-called long-time tails in the time-correlation functions for transport coefficients (Dorfman et al., 1994; Kirkpatrick et al., 2002; Belitz et al., 2005). We mention that Eqs. (3.13), in the particular case of an isothermal and divergence-free (incompressible) fluid (i.e., $\delta \rho = \delta p = \delta T = 0$) and when advection is included, reduce to a single nonlinear stochastic differential equation for the fluid velocity usually referred to as the Burgers (1974) equation. We note that, since the random fluxes appear under a divergence in Eqs. (3.13), they do not add a source term to the corresponding balance laws, so that even in the presence of fluctuations balance of mass, energy and momentum continues to hold.

Upon averaging over fluctuations the random-force terms in the set of Eqs. (3.13) disappear, since, by virtue of Eq. (3.2), $\langle \delta \Pi \rangle = \langle \delta \mathbf{Q} \rangle = 0$. Thus in linear approximation fluctuations, on the average, evolve following the same deterministic hydrodynamic equations as for small perturbations from an equilibrium state. This result is known as *Onsager's regression hypothesis*. Here, Onsager's hypothesis appears as a consequence of the procedure followed in formulating fluctuating hydrodynamics; that is, items #1 and #2 above represent an alternative formulation of Onsager's regression hypothesis.

To solve the system (3.13) of linear stochastic differential equations, we first take the divergence of Eq. (3.13b) to obtain:

$$\rho_0 \frac{\partial(\delta \psi)}{\partial t} = -\frac{1}{\varkappa_T} \nabla^2 \left(\alpha_p \delta T + \frac{\delta \rho}{\rho_0} \right) + \left(\eta_v + \frac{4}{3}\eta \right) \nabla^2(\delta \psi) + \nabla \cdot (\nabla \cdot \delta \Pi),$$

$$(3.15)$$

where $\delta\psi(\mathbf{r}, t) = \boldsymbol{\nabla} \cdot \delta\mathbf{v}(\mathbf{r}, t)$ represents the divergence of the velocity fluctuations. In deriving Eq. (3.15), as anticipated, we have eliminated the pressure by using the (differential) equation of state $\delta p = (\alpha_p \delta T + \delta\rho/\rho)/\varkappa_T$. Equations (3.13a), (3.13c) and (3.15) constitute a set of three stochastic differential equations for the three fluctuating fields $\delta\rho(\mathbf{r}, t)$, $\delta T(\mathbf{r}, t)$ and $\delta\psi(\mathbf{r}, t)$.

Next, it is convenient to convert the set of differential equations into a set of algebraic equations by applying a Fourier transformation in space and time. The resulting equations for the Fourier transforms of the fluctuating variables as a function of the wave vector $\mathbf{q}$ and the frequency ω are conveniently expressed as:

$$\mathsf{G}^{-1}(\omega, q) \cdot \begin{pmatrix} \delta\rho(\omega, \mathbf{q}) \\ \delta\psi(\omega, \mathbf{q}) \\ \delta T(\omega, \mathbf{q}) \end{pmatrix} = \mathbf{F}(\omega, \mathbf{q}), \tag{3.16}$$

where the matrix $\mathsf{G}^{-1}(\omega, q)$ is the inverse of the *linear response function*, sometimes also referred to as inverse susceptibility. For our problem, the inverse response function appearing in the LHS of Eq. (3.16) is:

$$\mathsf{G}^{-1}(\omega, q) = \begin{pmatrix} \mathrm{i}\,\omega & \rho_0 & 0 \\ \dfrac{-c_{\mathrm{s}}^2 q^2}{\gamma\rho_0} & (\mathrm{i}\,\omega + D_V q^2) & \dfrac{-\alpha_p c_{\mathrm{s}}^2 q^2}{\gamma} \\ 0 & \dfrac{(\gamma - 1)}{\alpha_p} & (\mathrm{i}\,\omega + \gamma a_T q^2) \end{pmatrix}, \tag{3.17}$$

with $D_V = (\eta_v + (4/3)\eta)/\rho$, and $c_{\mathrm{s}}^2 = \gamma/(\rho\varkappa_T)$ being the longitudinal kinematic viscosity and the square of the adiabatic speed of sound of the fluid, respectively. The term $\mathbf{F}(\omega, \mathbf{q})$ in Eq. (3.16) represents a vector of Langevin-like random forces:

$$\mathbf{F}(\omega, \mathbf{q}) = \frac{-1}{\rho_0} \begin{pmatrix} 0 \\ q_i q_j \delta\Pi_{ij}(\omega, \mathbf{q}) \\ \mathrm{i}\,(c_V)^{-1} q_i \delta Q_i(\omega, \mathbf{q}) \end{pmatrix} \tag{3.18}$$

expressed in terms of the Fourier transforms of the derivatives of the fluctuating fluxes. In Eq. (3.18) for the random forces, the indices i and j run over $\{x, y, z\}$ and summation over repeated indices is understood. The random-force vector $\mathbf{F}(\omega, \mathbf{q})$ is often simply called the noise; the set of hydrodynamic fluctuating fields which are the unknowns in Eq. (3.16) are usually referred as *modes* or hydrodynamic modes.

Mathematically, the set of random forces $\mathbf{F}(\omega, \mathbf{q})$ act as additive noise. In the literature some attention has also been devoted to the possible role of multiplicative noise in the fluctuating-hydrodynamics equations (García Ojalvo and Sancho, 1999). We note that "natural" or thermal noise is always additive, while sources of multiplicative noise are usually associated with experimental uncertainties. In this book, all noise sources to be considered will be additive noise.

Equation (3.16) can be readily solved for the fluctuating variables by computing the actual linear response function through inversion of Eq. (3.17). This procedure yields:

$$
\begin{pmatrix} \delta\rho(\omega,\mathbf{q}) \\ \delta\psi(\omega,\mathbf{q}) \\ \delta T(\omega,\mathbf{q}) \end{pmatrix} = \begin{bmatrix} -\rho_0(i\omega + \gamma a_T q^2) \\ i\omega(i\omega + \gamma a_T q^2) \\ -i\omega\dfrac{(\gamma - 1)}{\alpha_p} \end{bmatrix} \dfrac{F_1(\omega,\mathbf{q})}{\det\left[\mathsf{G}^{-1}(\omega,\mathbf{q})\right]}
\tag{3.19}
$$

$$
+ \begin{bmatrix} -\alpha_p\rho_0\dfrac{c_s^2 q^2}{\gamma} \\ i\omega\rho_0\dfrac{c_s^2 q^2}{\gamma} \\ \dfrac{c_s^2 q^2}{\gamma} - \omega^2 + i\omega D_V q^2 \end{bmatrix} \dfrac{F_2(\omega,\mathbf{q})}{\det\left[\mathsf{G}^{-1}(\omega,\mathbf{q})\right]},
$$

where $\det\left[\mathsf{G}^{-1}(\omega,\mathbf{q})\right]$ indicates the determinant of the inverse linear response function (3.17), while $F_1(\omega,\mathbf{q})$ and $F_2(\omega,\mathbf{q})$ are the second and third components of the noise (3.18). To evaluate the determinant, one customarily adopts the approximations:

$$
c_s q \gg a_T q^2 \qquad \text{and} \qquad c_s q \gg D_V q^2.
\tag{3.20}
$$

To justify these approximations, we need to keep in mind that the solution of Eq. (3.19) for the fluctuating variables is specifically needed for the interpretation of light-scattering experiments. As will be discussed in Chapter 10, in light-scattering experiments, $\mathbf{q}$ is to be identified with the scattering vector, whose magnitude q is of the order of $10^2 - 10^3$ cm^{-1}. For normal liquids $c_s \approx 10^5$ cm s^{-1}, while both a_T and D_V are of the order of 10^{-3} cm^2 s^{-1}. Hence, the conditions specified in Eq. (3.20) are well justified. With these approximations, the roots in ω of $\det\mathsf{G}^{-1}(\omega,q)$ are expanded up to first order in the small parameters $a_T q^2$ and $D_V q^2$, so that (Berne and Pecora, 1976):

$$
\det\left[\mathsf{G}^{-1}(\omega,q)\right] = (i\omega + a_T q^2)\,[i(\omega - c_s q) + \hat{\Gamma}_s q^2]\,[i(\omega + c_s q) + \hat{\Gamma}_s q^2], \tag{3.21}
$$

where $\hat{\Gamma}_s = \frac{1}{2}[D_V + (\gamma - 1)a_T]$ is the sound attenuation or absorption coefficient of the fluid.

As noted before, the average of the fluctuations of the thermodynamic fields is indeed zero, *i.e.*, $\langle\delta\rho(\omega,\mathbf{q})\rangle = \langle\delta T(\omega,\mathbf{q})\rangle = \langle\delta\mathbf{v}(\omega,\mathbf{q})\rangle = 0$, as becomes obvious upon averaging Eq. (3.19) over fluctuations, and using the first moments of the random dissipative fluxes as specified by the first of Eqs. (3.2). To obtain the second moments of the fluctuating variables, such as the Fourier transform $\langle \delta \rho^*(\omega, \mathbf{q}) \cdot \delta \rho(\omega', \mathbf{q}') \rangle$ of the autocorrelation function for the density fluctuations, we need the correlation functions between the components of the random force $\mathbf{F}(\omega, \mathbf{q})$. They can be computed from the relationship between the random forces and the fluctuating dissipative fluxes, Eq. (3.18), and from the correlation functions of the different components of the fluctuating fluxes, as given by the fluctuation-dissipation theorem (3.6) for a newtonian viscous fluid. Double Fourier transforming Eqs. (3.6) one sees that such functions are conveniently expressed in terms of a correlation matrix $\mathsf{C}(q)$, namely:

$$\langle F_\alpha^*(\omega, \mathbf{q}) \cdot F_\beta(\omega', \mathbf{q}') \rangle = C_{\alpha\beta}(q) \, (2\pi)^4 \, \delta(\omega - \omega') \, \delta(\mathbf{q} - \mathbf{q}') \tag{3.22a}$$

with:

$$\mathsf{C}(q) = \frac{2k_B T_0}{\rho_0} \begin{pmatrix} 0 & 0 & 0 \\ 0 & D_V \, q^4 & 0 \\ 0 & 0 & \dfrac{T_0 \lambda}{\rho_0 c_V^2} \, q^2 \end{pmatrix}, \tag{3.22b}$$

where ρ_0 and T_0 again represent the equilibrium mass density and temperature, respectively.

3.3.1 Structure factor

From Eqs. (3.19) and (3.22), the correlation functions among the various fluctuating fields can now be readily calculated. From both an experimental and a theoretical point of view, the density autocorrelation function $\langle \delta \rho^*(\omega, \mathbf{q}) \cdot \delta \rho(\omega', \mathbf{q}') \rangle$ is the most important one, since it is related to the so-called dynamic structure factor $S(\omega, \mathbf{q})$ of the fluid by:

$$\langle \delta \rho^*(\omega, \mathbf{q}) \cdot \delta \rho(\omega', \mathbf{q}') \rangle = \rho_0 \, m_0 \, S(\omega, \mathbf{q}) \, (2\pi)^4 \, \delta(\omega - \omega') \, \delta(\mathbf{q} - \mathbf{q}'), \tag{3.23}$$

where ρ_0 is the average density and m_0 is the mass of one molecule. The mass of the molecules appears in Eq. (3.23) because in the theory of liquids (March and Tosi, 1976; Boon and Yip, 1980; Lifshitz and Pitaevskii, 1986; Hansen and McDonald, 1986) the dynamic structure factor is defined in terms of the number density of particles, while ρ here designates the mass density. Notice that the dynamic structure factor has units of time: for typical liquids it is usually reported in picoseconds (Boon and Yip, 1980). The appearance of the two delta functions in Eq. (3.23) reflects the fact that in an homogeneous liquid the real-space correlation functions between fluctuating quantities depends only on the differences $t - t'$ and $\mathbf{r} - \mathbf{r}'$.

The importance of the structure factor is twofold. From a theoretical point of view, many studies in liquid theory are devoted to obtain $S(\omega, \mathbf{q})$ from microscopic models (March and Tosi, 1976; Boon and Yip, 1980; Lifshitz and Pitaevskii, 1986; Hansen and McDonald, 1986). From an experimental point of view, $S(\omega, \mathbf{q})$ is a quantity that can be actually measured

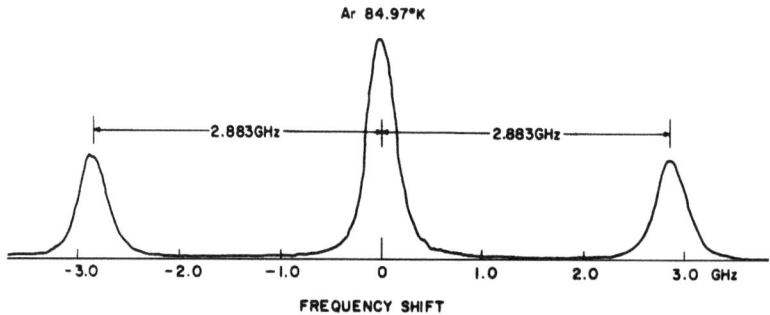

Figure 3.1: Rayleigh-Brillouin spectrum of liquid Argon at 85 K, for scattering wave number $q \approx 213\,000$ cm^{-1}. From Fleury and Boon (1969).

by light scattering (see Chapter 10), by X-ray diffraction (scattering), and by neutron diffraction (scattering). Specifically, as is discussed in detail in Sect. 10.1.1, the spectrum of light scattered by a fluid results proportional to $S(\omega, \mathbf{q})$ where $\mathbf{q}$ is to be identified with the so-called scattering wave vector, which is related to the geometry of the scattering experiment and the wavelength of the incident light, see Eq. (10.4).

The dimensionless static structure factor $S(\mathbf{q})$ is defined upon integration over the frequency of the dynamic structure factor:

$$S(\mathbf{q}) = \frac{1}{2\pi} \int_{-\infty}^{\infty} S(\omega, \mathbf{q}) \, d\omega. \tag{3.24}$$

Applying a double inverse Fourier transform to Eq. (3.23), the static structure factor can be related to the equal-time mass density autocorrelation function, namely:

$$\langle \delta\rho^*(t, \mathbf{q}) \cdot \delta\rho(t, \mathbf{q}') \rangle = \rho_0 m_0 \, S(\mathbf{q}) \, (2\pi)^3 \, \delta(\mathbf{q} - \mathbf{q}'). \tag{3.25}$$

The static structure factor is also an experimental accessible quantity, being directly proportional to the total intensity scattered by a fluid. Different experimental techniques probe different ranges in the scattering vector $\mathbf{q}$. For the purpose of this book, we shall be most interested in light-scattering, which probes $\mathbf{q}$ values in the hydrodynamic range: $q \simeq 10 - 10^5$ cm^{-1}. Typical wave numbers probed with X-ray or neutron scattering are of the order of nm^{-1} and one would need to extrapolate those data to $q \to 0$ to make contact with the hydrodynamic regime.

By combining Eqs. (3.19) and Eq. (3.22), the density-density autocorrelation function $\langle \delta\rho^*(\omega, \mathbf{q}) \cdot \delta\rho(\omega', \mathbf{q}') \rangle$ can be readily calculated. As expected, it can be cast in the form of Eq. (3.23), i.e., it is proportional to the product of two delta functions; specifically, the dynamic structure factor $S(\omega, \mathbf{q})$ is:

$$m_0 \ S(\omega, q) = \frac{2k_B T q^2 \left[(\gamma - 1)c_s^2 a_T q^4 + (\omega^2 + \gamma^2 a_T^2 q^4) D_V q^2\right]}{(\omega^2 + a_T^2 q^4) \left[(\omega - c_s q)^2 + \hat{\Gamma}_s^2 q^4\right] \left[(\omega + c_s q)^2 + \hat{\Gamma}_s^2 q^4\right]}, \qquad (3.26)$$

In deriving Eq. (3.26) for the dynamic structure factor $S(\omega, q)$ of a fluid in equilibrium, use has been made again of the relationship (3.14). Furthermore, since there is no possible confusion, we replaced the averages mass density ρ_0 and temperature T_0 by simply ρ and T, respectively. Note that for the isotropic fluid considered in this section, the dynamic structure factor only depends on the magnitude q of the wave vector $\mathbf{q}$. For an easier interpretation of the expression (3.26) for the dynamic structure factor, it is again convenient to use the approximations (3.20). Equation (3.26) can then be rewritten as (Berne and Pecora, 1976; Boon and Yip, 1980):

$$S(\omega, \mathbf{q}) = S_E \left\{ \frac{\gamma - 1}{\gamma} \frac{2a_T q^2}{\omega^2 + a_T^2 q^4} \right.$$

$$\left. + \frac{1}{\gamma} \left[\frac{\hat{\Gamma}_s q^2}{(\omega + c_s q)^2 + \hat{\Gamma}_s^2 q^4} + \frac{\hat{\Gamma}_s q^2}{(\omega - c_s q)^2 + \hat{\Gamma}_s^2 q^4} \right] \right\}, \qquad (3.27)$$

where the dimensionless quantity S_E is given by

$$m_0 \ S_E = \frac{k_B T \gamma}{c_s^2} = \rho \varkappa_T k_B T = \frac{k_B \alpha_p^2 T^2}{c_p - c_V}. \qquad (3.28)$$

Equation (3.27) shows that to a very good approximation the spectrum of scattered light contains three separately identifiable lorentzians. As an example, we show in Fig. 3.1 the spectrum of light scattered by liquid argon at $T = 85$ K, as measured by Fleury and Boon (1969). The central component in Fig 3.1, commonly referred to as Rayleigh line, arises from entropy or, equivalently, temperature fluctuations at constant pressure and the width of this Rayleigh line is proportional to the decay rate $a_T q^2$ of these fluctuations. The two spectral lines that are shifted by $-c_s q$ and $+c_s q$ with respect to the frequency of the incident light source, are the Stokes and anti-Stokes components of the Brillouin doublet [or Brillouin-Mandelstam in the Russian literature (Fabelinskii, 1965)]. The Brillouin lines arise from thermally excited propagating sound waves (modes) associated with adiabatic pressure fluctuations. The width of the Brillouin lines is proportional to $\hat{\Gamma}_s q^2$. Because of the different physical origins and because of the validity of the approximations (3.20) (or equivalently $c_s q \gg \hat{\Gamma}_s q^2$), the Rayleigh line and the Brillouin lines are usually investigated separately, both in theory and in experiments.

For the interpretation of photon-counting experiments, as further discussed in Chapter 10, the relevant quantity is the time-dependent autocorrelation function $\langle \delta\rho^*(\mathbf{q}, t) \cdot \delta\rho(\mathbf{q}', t') \rangle$ of the density fluctuations. It can be readily evaluated by applying a double inverse Fourier transform in the frequencies ω and ω' to Eq. (3.23):

$$\langle \delta\rho^*(\mathbf{q}, t) \cdot \delta\rho(\mathbf{q}', t') \rangle = \rho m_0 \, S(q, |t - t'|) \, (2\pi)^3 \, \delta(\mathbf{q} - \mathbf{q}'), \qquad (3.29)$$

where again the average density is identified with the equilibrium density. Applying an inverse Fourier transform on the frequency ω to the dynamic structure factor $S(\omega, q)$, as given by Eq. (3.27), one obtains the function $S(q, \tau)$:

$$S(q, \tau) = S_{\mathrm{E}} \left\{ \frac{\gamma - 1}{\gamma} \, \exp\left(-a_T q^2 \tau\right) + \frac{1}{\gamma} \cos(c_s q \tau) \, \exp(-\hat{\Gamma}_s q^2 \tau) \right\}, \qquad (3.30)$$

which depends on the (absolute) time difference $\tau = |t - t'|$. Comparing Eq. (3.29) with Eq. (3.25), we notice that the static structure factor equals the function $S(q, \tau)$ at $\tau = 0$. We conclude that the static structure factor of a fluid in equilibrium is equal to the quantity S_{E} defined by Eq. (3.28)

The fact, Eq. (3.29), that $S(q, \tau)$ depends only on the difference $|t - t'|$ is a consequence of the time-translation symmetry of equilibrium states. However, the observation that the autocorrelation (3.30) is exponentially decaying as a function of τ, with decay rates $\Gamma_T = a_T q^2$ and $\Gamma_s = \hat{\Gamma}_s q^2$, is a consequence of having adopted a linear approximation to fluctuating hydrodynamics. Although we have considered only density fluctuations here, the fact that, in a linear approximation time correlation functions decay exponentially, is completely general. It also applies to autocorrelation functions of the velocity or the concentration fluctuations, to be discussed later.

The structure factor of a binary fluid mixture in equilibrium can be evaluated by a similar procedure to the one used in the present section for a one-component fluid (Cohen et al., 1971). As already explained in Sect. 3.2.2, in the case of a binary fluid mixture one needs to consider an additional fluctuating diffusion flux, $\delta\mathbf{J}$, and an additional fluctuating concentration field, δc. Just as for a one-component fluid, the structure factor of a binary mixture will contain Rayleigh and Brillouin components (Berne and Pecora, 1976; Boon and Yip, 1980). The Brillouin components of a binary mixture are related to the sound velocity and attenuation, similar to Eq. (3.27). However, due to the presence of concentration fluctuations, the Rayleigh component of a binary mixture differs from that of a one-component fluid (Berne and Pecora, 1976; Boon and Yip, 1980). A detailed discussion of the Rayleigh component of the structure factor of a binary mixture in equilibrium is incorporated in Sect. 5.3, see Eqs. (5.33) and (5.37) in particular.

3.3.2   Equal-time correlations

The so-called static or equal-time correlation functions are of special theoretical and experimental interest. These static correlation functions can be readily obtained from the dynamic correlation functions discussed in the previous section. For instance, the equal-time correlation function of the density fluctuations, $\langle \delta\rho^*(\mathbf{q}, t) \cdot \delta\rho(\mathbf{q}', t) \rangle$, can be readily obtained by substituting Eq. (3.30) into (3.29) and setting $\tau = 0$:

$$\langle \delta\rho^*(\mathbf{q}, t) \cdot \delta\rho(\mathbf{q}', t) \rangle = \rho m_0 \, S_E \left\{ \frac{\gamma - 1}{\gamma} + \frac{1}{\gamma} \right\} (2\pi)^3 \, \delta(\mathbf{q} - \mathbf{q}'). \qquad (3.31)$$

We note that Eq. (3.31) can be also obtained by integrating Eq. (3.27) over the frequency ω. From this second procedure, we see that the first term inside the curly brackets $I_R = S_E(\gamma - 1)/\gamma$ represents the intensity of the Rayleigh line, while the second term $2I_B = S_E/\gamma$ represents the intensity of the Brillouin doublet. The ratio of the integrated intensities of the Rayleigh peak to those of the Brillouin peaks,

$$\frac{I_R}{2I_B} = \gamma - 1,$$

is usually known as the Landau-Placzek ratio. Applying a double inverse Fourier transform to Eq. (3.31), we obtain for the equal-time autocorrelation function of the density fluctuations in real space:

$$\langle \delta\rho(\mathbf{r}, t) \cdot \delta\rho(\mathbf{r}', t) \rangle = \rho m_0 \, S_E \, \delta(\mathbf{r} - \mathbf{r}'). \qquad (3.32)$$

The presence of a delta function in Eq. (3.31) causes the real-space formula (3.32) to depend only on the difference $\mathbf{r} - \mathbf{r}'$, reflecting the spatial-translational symmetry of equilibrium states. However, the fact that the static structure factor does not depend on q causes the presence of the delta function in Eq. (3.32), confirming that the equal-time autocorrelation function for a system in thermodynamic equilibrium is spatially short-ranged relative to any (longer-ranged) hydrodynamic scale. Although the present expression for S_E has been obtained with the approximations (3.20), the spatial delta-function character of $\langle \delta\rho(\mathbf{r}, t) \cdot \delta\rho(\mathbf{r}', t) \rangle$ continues to hold even without these approximations, except that the calculation would have been longer and more involved. The spatially short-ranged nature of the static correlation functions at hydrodynamic scales is a universal feature of equilibrium systems. An exception is a fluid near the critical point where the equal-time density fluctuations become spatially long-ranged because of the divergent behavior of the compressibility (Fisher, 1964). As explained above, critical behavior of the fluctuations is not included here because we are using *linear* fluctuating hydrodynamics.

If we define an "average" mass density fluctuation in a given volume element ΔV by:

$$\overline{\delta\rho}(t) = \frac{1}{\Delta V} \int_{\Delta V} \delta\rho(\mathbf{r}, t) \, dr, \qquad (3.33)$$

then, from Eqs. (3.32) and (3.28), we readily obtain that $\langle \overline{\delta\rho}^2(t) \rangle$ is independent of t and equals:

$$\langle \overline{\delta\rho}^2 \rangle = \rho m_0 \frac{S_{\mathrm{E}}}{\Delta V} = k_{\mathrm{B}} T \frac{\rho^2 \varkappa_T}{\Delta V}. \tag{3.34}$$

The average density in a given volume fluctuates because the average number of particles contained in ΔV fluctuates. If $N = \rho \Delta V / m_0$ indicates the average number of particles in ΔV, then from Eq. (3.34) we obtain for the fluctuations δN in the number of particles:

$$\frac{\langle \delta N^2 \rangle}{N^2} = k_{\mathrm{B}} T \frac{\varkappa_T}{\Delta V}, \tag{3.35}$$

which is simply the fluctuations in the number of particles as calculated from the macrocanonical ensemble. Equations (3.34) or (3.35) reflect the well-known fact that fluctuations are more important in small volumes.

In the traditional derivation of the Rayleigh-Brillouin spectrum following the Mountain (1966) method (Berne and Pecora, 1976; Boon and Yip, 1980), the same Eq. (3.27) is obtained, except that the dimensionless amplitude S_{E} is left undetermined. This amplitude is obtained in a later stage by using the general result (3.35) from statistical physics. In our derivation of the Rayleigh-Brillouin spectrum we have closely followed the method of stochastic forcing (Landau and Lifshitz, 1958, 1959; Fox and Uhlenbeck, 1970a). For systems in equilibrium both methods give the same results. However, as earlier mentioned in Sect. 3.5, the advantage of the Landau and Lifshitz (1958, 1959) method is that it can be readily extended to nonequilibrium states, while the extension of the Mountain (1966) method to nonequilibrium fluid states is less obvious.

3.4   Equilibrium correlation functions and entropy probability functional

As discussed by many investigators, starting with Einstein (1910), the static correlation functions of a system in equilibrium may be computed from a field theory where entropy plays the role of probability functional (Fox and Uhlenbeck, 1970a). A given fluctuation $\delta\mathbf{\Phi}(\mathbf{r}, t)$ of the set of thermodynamic fields at time t, will cause the actual entropy of the system to deviate from the (maximum) equilibrium value. This deviation ΔS will, in general, be a functional of the fluctuating fields: $\Delta S[\delta\mathbf{\Phi}; t] = \int_V \rho s[\delta\mathbf{\Phi}(\mathbf{r}, t)] \, d\mathbf{r}$, where the integral extends over the volume V occupied by the system. The particular case of a one-component newtonian viscous fluid has been analyzed in detail by Rubí and Mazur (2000). From purely thermodynamic considerations, it follows that the entropy deviation associated with a given thermodynamic fluctuation $\delta\boldsymbol{\Phi}(\mathbf{r},t) = \{\delta\rho(\mathbf{r},t), \delta T(\mathbf{r},t), \delta\mathbf{v}(\mathbf{r},t)\}$ around an equilibrium state $\boldsymbol{\Phi}_0 = \{\rho_0, T_0, 0\}$, is given by (Rubí and Mazur, 2000):

$$\Delta S[\delta\boldsymbol{\Phi};t] = -\frac{1}{2}\int_V d\mathbf{r}\left\{\frac{[\delta\rho(\mathbf{r},t)]^2}{\rho_0^2 T_0 \varkappa_T} + \frac{\rho_0 c_V}{T_0^2}[\delta T(\mathbf{r},t)]^2 + \frac{\rho_0}{T_0}[\delta\mathbf{v}(\mathbf{r},t)]^2\right\}.$$

(3.36)

Note that this entropy deviation functional is quadratic in the fluctuating fields and does not contain cross-terms. This is to be expected, since the entropy is maximum in equilibrium so that the derivatives of the entropy with respect to the temperature or the volume cancel when evaluated for an equilibrium state (Rubí and Mazur, 2000). The entropy deviations close to equilibrium can generally be approximated by a quadratic form, that by virtue of the second law, has to be definite positive. Cross-terms can be eliminated by using the appropriate conservation laws (Fox and Uhlenbeck, 1970a; Rubí and Mazur, 2000). We should mention that the presence of $\delta\mathbf{v}$ in Eq. (3.36) has led to some debate in the thermodynamics community. Some authors (Glansdorff and Prigogine, 1971) claim that the velocity is not a proper thermodynamic variable, and have proposed a new nonequilibrium thermodynamic potential in which $\delta\mathbf{v}$ does not appear explicitly. However, it seems that this modified thermodynamic potential is equivalent to just a change of variables (Oono, 1976). Hence, we continue to adopt Eq. (3.36).

Apart from these subtleties, entropy fluctuations around a equilibrium state are expected to be generically expressed as a volume integral of a bilinear form in the fluctuating fields, as in Eq. (3.36) for a one-component viscous fluid. Thus, we expect that $\Delta S[\delta\boldsymbol{\Phi};t]$ is generically given by an expression of the form (Fox and Uhlenbeck, 1970a; Rubí, 1984):

$$\Delta S[\delta\boldsymbol{\Phi};t] = -\frac{1}{2}\int_V d\mathbf{r}\,\delta\boldsymbol{\Phi}^\mathsf{T}(\mathbf{r},t)\cdot\mathsf{S}\cdot\delta\boldsymbol{\Phi}(\mathbf{r},t),$$

(3.37)

where S is a symmetric positive definite matrix, *i.e.*, a matrix with real and positive eigenvalues. The equal-time (static) correlation functions between the fluctuating thermodynamic fields may be then calculated from a probability functional given by (Fox and Uhlenbeck, 1970a; Rubí, 1984):

$$P[\delta\boldsymbol{\Phi};t] = \frac{1}{\mathcal{Z}}\exp\left\{\frac{1}{k_\mathrm{B}}\,\Delta S[\delta\boldsymbol{\Phi};t]\right\},$$

(3.38)

where $\mathcal{Z}$ is a normalization constant. Equation (3.38) is usually referred to as the Einstein (1910) hypothesis, or the Boltzmann-Einstein hypothesis. Note from Eq. (3.37) that entropy deviations, ΔS, from an equilibrium state are negative (entropy is a maximum in equilibrium). Equation (3.38) means that fluctuations $\delta\boldsymbol{\Phi}$ with large associated ΔS will have a small probability, while fluctuations with small ΔS have a higher probability. The equal-time correlation functions between a pair of fluctuating variables

$\langle \delta \Phi_\alpha^*(\mathbf{r}, t) \cdot \delta \Phi_\beta(\mathbf{r}', t) \rangle$ can be inferred from the probability functional, defined by Eq.(3.38), by simply evaluating the path integral:

$$\langle \delta \Phi_\alpha^*(\mathbf{r}, t) \cdot \delta \Phi_\beta(\mathbf{r}', t) \rangle =$$
$$\frac{1}{\mathcal{Z}} \int \mathcal{D}[\delta \Phi] \; \delta \Phi_\alpha^*(\mathbf{r}, t) \; \delta \Phi_\beta(\mathbf{r}', t) \; e^{\frac{1}{k_{\mathrm{B}}} \Delta S[\delta \Phi; t]}. \quad (3.39)$$

Substitution of Eq. (3.37) into Eq. (3.39) yields gaussian path integrals, that can be evaluated with standard methods. Precisely because the resulting path integrals are gaussian, we anticipate that the corresponding static autocorrelation functions will be spatially short ranged, *i.e.*, proportional to delta functions. Furthermore, since S in Eq. (3.37) can be diagonalized with positive eigenvalues, there exist in general a certain (linear) combination of statistically independent fluctuating fields, *i.e.*, fluctuating fields with vanishing cross-correlations. For the particular case of a one-component newtonian fluid, we conclude from Eq. (3.36) that density, temperature and velocity fluctuate independently.

As an example of an explicit calculation of the path integral (3.39), we consider here a simple one-dimensional thermodynamic system, confined to a segment $[0, L]$. Furthermore, we assume for simplicity, that the density fluctuations vanish at both boundaries $x = 0$ and $x = L$ (so that we do not need to deal with complex numbers). Then a generic density fluctuation at a given time, $\delta \rho(x)$, may be represented by a Fourier series:

$$\delta \rho(x) = \sum_{N=1}^\infty \rho_N \sin \left(\frac{N\pi}{L} x \right), \quad (3.40)$$

where ρ_N are an stochastic real numbers, varying from $-\infty$ to ∞. Substituting Eq. (3.40) into Eq. (3.36), we obtain for the entropy deviation:

$$\Delta S(t) = \frac{-1}{2T_0} \int_0^L dx \; \frac{1}{\rho_0^2 \varkappa_T} \left[\sum_{N=1}^\infty \rho_N \sin \left(\frac{N\pi}{L} x \right) \right]^2 + \Delta S'[\delta T, \delta \mathbf{v}]$$

$$= \frac{-L}{4T_0 \rho_0^2 \varkappa_T} \sum_{N=1}^\infty \rho_N^2 + \Delta S'[\delta T, \delta \mathbf{v}], \quad (3.41)$$

where $\Delta S'[\delta T, \delta \mathbf{v}]$ denotes the contribution to the entropy from simultaneous independent temperature and velocity fluctuations, that do not need to be considered here. In Eq. (3.41) the multivariate gaussian character of the path integral is now evident. The normalization constant $\mathcal{Z}$ is to be determined from:

$$\mathcal{Z} = \mathcal{Z}'[\delta T, \delta \mathbf{v}] \prod_{N=1}^\infty \sqrt{\frac{L}{4\pi k_{\mathrm{B}} T_0 \rho_0^2 \varkappa_T}}, \quad (3.42)$$

where $\mathcal{Z}'[\delta T, \delta \mathbf{v}]$ represents the independent normalization of temperature and velocity fluctuations. Observe that the normalization constant $\mathcal{Z}$ in Eq. (3.42) is formally divergent. This problem can be overcome (as usual with path integrals) by restricting the summation (3.40) to a finite large value of N. This is equivalent to introducing a cutoff in the wave number of the fluctuations. Using such a cutoff, we can evaluate the path integral (3.39) for the density-density autocorrelation, so as to obtain:

$$\langle \delta \rho^*(x) \cdot \delta \rho(x') \rangle = \sum_{N=1}^{\infty} \frac{2k_B T_0 \rho_0^2 \varkappa_T}{L} \sin\left(\frac{N\pi}{L}x\right) \sin\left(\frac{N\pi}{L}x'\right) \qquad (3.43)$$
$$= k_B T_0 \rho_0^2 \varkappa_T \; \delta(x - x'),$$

which is the one-dimensional counterpart of Eq. (3.32) obtained from fluctuating hydrodynamics. As anticipated, the result does not depend on the temperature or velocity fluctuations, whose contribution to the path integral cancels with $\mathcal{Z}'[\delta T, \delta \mathbf{v}]$. The explicit calculations of this section refer to the most general case when all possible fluctuations in a newtonian fluid are present. Later in this book we shall make a series of approximations (incompressibility, for instance) that restrict the fluctuating degrees of freedom. In those cases, one has to incorporate the approximations in the expression equivalent to (3.36) for the fluctuations in the entropy. For a more detailed discussion of the entropy as a generating functional of the equilibrium equal-time correlation functions, with more examples and applications than the simple one considered here, the reader is referred to the recent book of Barrat and Hansen (2003).

The fact that equal-time correlation functions may be calculated from the probability functional (3.38) is a direct consequence of the FDT (3.3), the derivation of which is actually based on this assumption. As explained in detail by Landau and Lifshitz (1958, 1959) the function $C_{\alpha\beta}(\mathbf{r}, \mathbf{r}')$ in Eq. (3.2) is determined in a such a way that the static correlations among the fluctuating thermodynamic fields have the entropy as a probability generating functional, in accordance with Eq. (3.38). In this book we have not presented a systematic derivation of the FDT, but we have shown how, for a one-component viscous fluid, the equal-time density fluctuations calculated from the FDT (3.6) reproduce the expected result based on the Einstein hypothesis (3.38). For a more formal presentation and rigorous derivation of the FDT, we refer to books by Landau and Lifshitz (1958, 1959), by Kubo et al. (1991), or to Fox and Uhlenbeck (1970a). Furthermore, as we have seen, the formulation (3.6) of the FDT reproduces the well-known Eqs. (3.34)-(3.35) from statistical physics; a fact that shows the consistency of fluctuating hydrodynamics.

Without any doubt, one of the most fundamental properties of entropy is precisely its ability to be the generating functional for the equilibrium equal-time (static) fluctuations. We can even consider the Einstein hypothesis (3.38) as the "definition" of equilibrium entropy. Actually, what we have discussed here is the well-known principle of thermodynamic stability of a system in equilibrium, but in terms of space-dependent fluctuations (which leads to the use of path integrals). An important line of research in nonequilibrium fluctuations is to look for "lagrangian" functionals that will generate the nonequilibrium static fluctuations (Derrida et al., 2001, 2002). We suspect that this is the correct path to arrive at a definition of an entropy outside of thermodynamic equilibrium, or to demonstrate that a nonequilibrium entropy does not exist.

We note that the generic short-ranged spatial character of equal-time correlations depends crucially on the entropy deviation being a quadratic form of the fluctuating fields, like in Eq. (3.36). If higher-order terms, neglected so far, become important, the correlations can become spatially long-ranged, even in thermodynamic equilibrium. This happens, for instance, in a fluid near a critical point (Fisher, 1964). At the critical point the isothermal compressibility $\varkappa_T$ diverges, so that the coefficient multiplying the square of density fluctuations in (3.36) for ΔS vanishes, and higher-order terms in the fluctuating fields cannot be neglected. Higher-order deviations will also appear for non-newtonian fluids. In this book we restrict ourselves to dense newtonian fluids and liquids that are not close to a critical point.

3.5   Extension of fluctuating hydrodynamics to nonequilibrium steady states

As was elucidated in Chapter 2, both nonequilibrium thermodynamics and fluid dynamics are based on the assumption of local equilibrium. That is, the behavior of a fluid that is not in thermodynamic equilibrium, can be formulated at any given time t in terms of locally defined thermodynamic properties, like $\rho(\mathbf{r}, t)$, $T(\mathbf{r}, t)$, $p(\mathbf{r}, t)$ that continue to be interrelated by the laws of equilibrium thermodynamics, see Sect. 2.1.2. As mentioned in Chapter 1, an attempt to provide a theoretical basis for the principle of local equilibrium was presented by Bogoliubov (1946, 1962). This theory would imply that in a nonequilibrium fluid, at least far away from any hydrodynamic instability, the expressions for the hydrodynamic correlation functions for a system in equilibrium derived in this chapter would continue to hold but with the local values of the thermodynamic properties. For instance, this would imply that the factor S_E appearing in Eq. (3.32) for the equal-time density autocorrelation function would continue to be given by Eq. (3.26) in terms of the corresponding local thermodynamic properties that now depend on $\mathbf{r}$ and t. However, the picture of Bogoliubov (1946, 1962) has turned out to be too simple. As discussed in Chapter 1 it turns out that when a system is in a nonequilibrium state, the presence of nonzero (on average) dissipative fluxes induces a coupling between hydrodynamic modes leading to long-ranged correlations among the fluctuating fields (Kirkpatrick et al., 1982b; Ronis and Procaccia, 1982). In this book we shall show how these long-ranged nonequilibrium fluctuations can be described by fluctuating hydrodynamics.

The theory of fluctuating hydrodynamics was originally formulated by Landau and Lifshitz (1958, 1959) to deal with fluctuations in fluids in thermodynamic equilibrium. The extension of fluctuating hydrodynamics to fluids in nonequilibrium states is based on the assumption that the correlations functions (3.2)-(3.3) among the random dissipative fluxes are now given by a local-equilibrium version of the FDT. This extension is based on the assumption that the correlation functions of the nonconserved quantities, in contrast to those of the conserved quantities, remain short ranged in nonequilibrium.

In this extension of fluctuating hydrodynamics one can identify two sources that cause the fluctuations in nonequilibrium states to be long ranged:

1. A first source is the spatial dependence of the noise correlations resulting from the local application of the FDT. For instance, from Eq. (3.3) we see that the noise correlation functions are generically proportional to the temperature T, that in nonequilibrium depends on the position $T(\mathbf{r})$. As a consequence, the thermodynamic fluctuations at a given location will depend on the temperature at neighboring locations (Proccacia et al., 1979; Ronis et al., 1980; Tremblay et al., 1981; Tremblay, 1984). We shall refer to this source of nonequilibrium effects as *inhomogeneously correlated thermal noise*.

2. A second source is the presence of mode-coupling phenomena in nonequilibrium fluids (Kirkpatrick et al., 1982b; Kirkpatrick and Cohen, 1983). Nonequilibrium states are characterized by the presence of average thermodynamic fluxes different from zero. These non-zero fluxes cause couplings between hydrodynamic modes that can be described by the extension of fluctuating hydrodynamics to nonequilibrium states (Ronis and Procaccia, 1982; Schmitz and Cohen, 1985a,b; Law and Sengers, 1989). We shall refer to this source of nonequilibrium effects as *mode coupling*.

In this book we shall be mostly concerned with the use of fluctuating hydrodynamics to derive expressions for the fluctuations in newtonian fluids subjected to a temperature gradient, usually referred to as the Rayleigh-Bénard (RB) problem. With the RB problem as an example, we shall elucidate how the two mechanisms mentioned above cause long-ranged fluctuations in fluids in stationary nonequilibrium states. The two sources generating long-ranged nonequilibrium fluctuations have been usually considered in the literature somewhat independently. In the subsequent chapter, we shall evaluate the effects of both sources on the nonequilibrium fluctuations for the RB problem. As elaborated in Sect. 4.3, the spatial dependence of the FDT has a similar effect to other spatial inhomogeneities arising from temperature dependence of the thermophysical properties. In both cases, they lead to nonlinear contributions to the hydrodynamics. A major conclusion is that such inhomogeneities have a negligible effect on the temperature fluctuations that determine the Rayleigh component of the structure factor, while they have a major effect on the pressure fluctuations that determine the Brillouin components. On the other hand, advective coupling will turn out to have a major effect on the temperature fluctuations and a secondary effect on the pressure fluctuations.

Mode-coupling also exist in fluctuations around equilibrium states, related to nonlinear terms in the hydrodynamic equations (Fixman, 1967; Kawasaki, 1970, 1976). However, for equilibrium states, such nonlinear effects are only relevant in some particular circumstances. For instance, mode-coupling is responsible for the enhancement of the thermal conductivity and the viscosity of a fluid near the critical point (Hohenberg and Halperin, 1977; Sengers, 1985). Mode coupling also causes an algebraic time decay of the correlation-functions expressions for the transport coefficients, referred to as "long-time tails" (Alder and Wainwright, 1970; Ernst et al., 1971; Dorfman and Cohen, 1970, 1975; Dorfman, 1975; Ernst et al., 1976a,b; Ernst, 2005). Nonlinear effects also cause the presence of a long-time tail in the autocorrelation functions of the velocity of a Brownian particle (Zwanzig and Bixon, 1970; Widom, 1971; Fox, 1984). They also lead to renormalized transport coefficients in the expressions for the hydrodynamic correlation functions that depend on frequency (Zwanzig et al., 1972; Bedeaux and Mazur, 1974). In molecular fluids the long-time tails affect fluctuations at scales that are larger than molecular scales but smaller than hydrodynamic scales. They have been observed in computer simulations and in some neutron-scattering experiments. For a detailed discussion of these issues the reader is referred to some recent reviews (Dorfman et al., 1994; Kirkpatrick et al., 2002; Belitz et al., 2005).

In molecular fluids the spatial and time scales associated with these long-time tails are short compared to any hydrodynamic scales. Thus the major difference between fluctuations in equilibrium and nonequilibrium is that hydrodynamic fluctuations in equilibrium states are generally short ranged except in fluids near critical points, while hydrodynamic fluctuations in nonequilibrium states are *always* spatially long ranged, even when calculated with a linear theory.

Chapter 4

Thermal nonequilibrium fluctuations in one-component fluids

After reviewing hydrodynamic fluctuations in an equilibrium fluid, we now proceed with a study of fluctuations when a fluid is subjected to a stationary temperature gradient. This situation is realized in practice by confining the fluid between two horizontal plates with two different temperatures, a situation commonly referred to as the Rayleigh-Bénard (RB) problem. In this chapter we focus on fluctuations with sufficiently large wave numbers q that the corresponding wave lengths are much smaller than the distance L between the plates, but that still correspond to hydrodynamic length scales, *i.e.*, length scales much larger than the intermolecular distances. For such wave numbers the nonequilibrium fluctuations will only depend on the thermodynamic and transport properties of the fluid (and the gravity g through its appearance in the Navier-Stokes equation) and we shall refer to these fluctuations as "bulk" nonequilibrium fluctuations. This chapter also introduces tools and methods that will be needed in a subsequent treatment of nonequilibrium fluctuations with wave numbers for which finite-size (or *confinement*) effects are important.

The material in this chapter is organized as follows: In Sects. 4.1 and 4.2, we present the theory of "bulk" nonequilibrium fluctuations for incompressible one-component fluids on the basis of the so-called fluctuating Oberbeck-Boussinesq equations, in the sequel to be called fluctuating Boussinesq equations. Use of the Boussinesq equations implies that we are neglecting pressure fluctuations, so that these sections will only deal with nonequilibrium effects on the Rayleigh component of the structure factor. In this first case nonequilibrium fluctuations arise in leading order from a coupling between

temperature and viscous fluctuations through the presence of a temperature
gradient (second source of nonequilibrium effects mentioned in Sect. 3.5).
In Sect. 4.3, we consider fluctuations in a fluid subjected to a temperature
gradient that has a vanishingly small thermal expansion coefficient, which
means that the density will only depend on pressure. Hence, this section
will only deal with nonequilibrium effects on the Brillouin components of
the structure factor. In this second case nonequilibrium fluctuations arise
in leading order from the spatial dependence of the prefactor appearing in
the fluctuation-dissipation theorem (first source of nonequilibrium effects
mentioned in Sect. 3.5). In using a local-equilibrium version of the FDT
we are introducing an assumption which, ultimately, requires justification
from microscopic nonequilibrium statistical mechanics (Keizer, 1978; Garrido et al., 1990). An analysis of the nonequilibrium fluctuations in these
two independent steps is justified by their very different nature of these fluctuations as will be further discussed at the end of Sect. 4.3. We conclude this
chapter with an assessment of the relative importance of the two sources for
long-ranged fluctuations on the Rayleigh component of the nonequilibrium
structure factor in Sect. 4.4.

4.1 Boussinesq approximation

We consider a one-component fluid layer between two horizontal plates separated by a distance L. The fluid layer is subjected to a temperature gradient
in the vertical direction by maintaining the plates at two different temperatures, T_1 at $z = -L/2$ and T_2 at $z = L/2$. The size of the system in the
two horizontal x- and y-directions is much larger than the size L in the vertical z-direction. Then the imposed external temperature gradient may be
parallel or antiparallel to gravity, depending on the sign of $\Delta T = T_2 - T_1$.

The relevant hydrodynamic equations for this problem were derived in
Chapter 2. For the discussion of the Boussinesq approximation it is more
convenient to use temperature and pressure as independent thermodynamic
variables with mass density as a dependent variable. Then, the relevant
set of hydrodynamic equations consists of the mass balance Eq. (2.5), the
momentum balance Eq. (2.23), and the version (2.56) of the heat equation.
For the convenience of the reader, they are reproduced below:

$$\frac{\partial \rho}{\partial t} = -\boldsymbol{\nabla} \cdot (\rho \mathbf{v}) \tag{4.1a}$$

$$\rho \left[\frac{\partial \mathbf{v}}{\partial t} + (\mathbf{v} \cdot \boldsymbol{\nabla}) \, \mathbf{v} \right] = -\boldsymbol{\nabla} p + \boldsymbol{\nabla} \Pi - \rho g \hat{\mathbf{z}} \tag{4.1b}$$

$$\rho c_p \left[\frac{\partial T}{\partial t} + \mathbf{v} \cdot \boldsymbol{\nabla} T \right] = -\boldsymbol{\nabla} \cdot \mathbf{Q} - \frac{T}{\rho} \left(\frac{\partial \rho}{\partial T} \right)_p \left[\frac{\partial p}{\partial t} + \mathbf{v} \cdot \boldsymbol{\nabla} p \right]. \tag{4.1c}$$

Viscous heating is neglected in Eq. (4.1c). The only volumetric force considered in Eqs. (4.1) is the gravity directed in the negative direction of the vertical z-axis. The corresponding term $-\rho g \hat{\mathbf{z}}$ in the Navier-Stokes Eq. (4.1b) will become very important and is referred to as the buoyancy term.

The boundary conditions are perfectly conducting walls at $z = -L/2$ and $z = L/2$, meaning that:

$$
\begin{aligned}
T(\mathbf{r}, t) &= T_1, \quad \text{when} \quad z = -L/2, \\
T(\mathbf{r}, t) &= T_2, \quad \text{when} \quad z = L/2,
\end{aligned}
\tag{4.2}
$$

while the temperature gradients at the walls remain undetermined, as discussed after Eq. (2.85). For the boundary conditions on the velocity we shall consider both rigid and free walls in accordance with Eqs. (2.87) and (2.89), respectively.

In this book we adopt the so-called Boussinesq approximation as is commonly done in the literature for dealing with Rayleigh-Bénard convection (Chandrasekhar, 1961; Tritton, 1988; Swift and Hohenberg, 1977; Cross and Hohenberg, 1993). We shall closely follow here the rigorous presentation of the Boussinesq approximation due to Mihaljan (1962). First, in the Boussinesq approximation the dependence of the density of the fluid on pressure is neglected and it is assumed that the density depends only on the temperature:

$$
\rho(\mathbf{r}, t) = \bar{\rho}_0 \{ 1 - \alpha_p [T(\mathbf{r}, t) - \bar{T}_0] \},
\tag{4.3}
$$

where $\bar{T}_0 = \frac{1}{2}(T_2 + T_1)$ is the average temperature at the center of the fluid layer and $\bar{\rho}_0$ the density corresponding to this average temperature. In the Boussinesq approximation, all other thermophysical properties, including the thermal expansion coefficient α_p, are treated as constants (independent of the local temperature). Since Eq. (4.3) implies that we only consider density fluctuations resulting from temperature fluctuations (at constant pressure) and not from pressure fluctuations, we use the heat equation (4.1c) with temperature and pressure as the relevant variables. After substitution of the equation of state (4.3) into Eqs. (4.1), they become:

$$
\alpha_p \frac{dT}{dt} = \left[1 - \alpha_p (T - \bar{T}_0) \right] \boldsymbol{\nabla} \cdot \mathbf{v}
\tag{4.4a}
$$

$$
\bar{\rho}_0 \left[1 - \alpha_p (T - \bar{T}_0) \right] \frac{d\mathbf{v}}{dt} = -\boldsymbol{\nabla} p + \boldsymbol{\nabla} \Pi - \bar{\rho}_0 \left[1 - \alpha_p (T - \bar{T}_0) \right] g \hat{\mathbf{z}}
\tag{4.4b}
$$

$$
\bar{\rho}_0 \left[1 - \alpha_p (T - \bar{T}_0) \right] c_p \frac{dT}{dt} = \frac{\alpha_p T}{1 - \alpha_p (T - \bar{T}_0)} \frac{dp}{dt} - \boldsymbol{\nabla} \cdot \mathbf{Q}.
\tag{4.4c}
$$

The partial derivatives in Eqs. (4.1) have been converted into material derivatives to simplify the notation.

To compare the magnitude of the different terms in Eqs. (4.4) so as to justify the Boussinesq approximation, it is advantageous to consider a set of dimensionless working equations. Following Mihaljan (1962), we scale the various fields by:

$$\tilde{\mathbf{r}} = \frac{\mathbf{r}}{L}, \qquad \tilde{t} = \frac{t a_T}{L^2}, \qquad \tilde{T} = \frac{T}{\Delta T}, \quad \tilde{p} = \frac{p L^2}{\bar{\rho}_0 a_T^2}, \quad \tilde{\mathbf{v}} = \frac{\mathbf{v} L}{a_T},$$

$$\tilde{\Pi} = \frac{\Pi L^2}{\bar{\rho}_0 a_T^2}, \quad \tilde{\mathbf{Q}} = \frac{L \mathbf{Q}}{\bar{\rho}_0 c_p a_T \ \Delta T}, \tag{4.5}$$

where a_T is the (average) thermometric diffusivity (thermal diffusivity) of the fluid in the layer. In terms of dimensionless variables, the hydrodynamic equations become:

$$\epsilon_1 \frac{d\theta}{d\tilde{t}} = [1 - \epsilon_1 \theta] \, \tilde{\boldsymbol{\nabla}} \tilde{\mathbf{v}} \tag{4.6a}$$

$$[1 - \epsilon_1 \theta] \frac{d\tilde{\mathbf{v}}}{d\tilde{t}} = -\tilde{\boldsymbol{\nabla}}(\tilde{p} + G\tilde{z}) - Pr \ Ra \ \theta \ \hat{\mathbf{z}} + \tilde{\boldsymbol{\nabla}} \tilde{\Pi} \tag{4.6b}$$

$$[1 - \epsilon_1 \theta] \frac{d\theta}{d\tilde{t}} = \frac{\epsilon_2 \tilde{T}}{1 - \epsilon_1 \theta} \frac{d\tilde{p}}{d\tilde{t}} - \tilde{\boldsymbol{\nabla}} \tilde{\mathbf{Q}}, \tag{4.6c}$$

In Eqs. (4.6) we have introduced a dimensionless variable $\theta = \tilde{T} - \tilde{\tilde{T}}_0$, a symbol $\tilde{\boldsymbol{\nabla}}$ indicating a derivative with respect to dimensionless (spatial) variables, various dimensionless numbers: $\epsilon_1 = \alpha_p \Delta T$, $\epsilon_2 = \alpha_p a_T^2 / c_p L^2$, $G = L^3 g / a_T^2$, and Pr indicating the Prandtl number:

$$Pr = \frac{\nu}{a_T}, \tag{4.7}$$

with $\nu = \eta / \bar{\rho}_0$ being the kinematic viscosity of the fluid. The final dimensionless quantity introduced in Eqs. (4.6) is the Rayleigh number Ra, defined by:

$$Ra = -\frac{\alpha_p L^3 g \Delta T}{\nu a_T} = -\frac{\alpha_p L^4 g \nabla T_0}{\nu a_T}, \tag{4.8}$$

with $\nabla T_0 = \Delta T / L$. Notice that the Rayleigh number is negative when gravity and temperature gradient are antiparallel (heating from above) and positive otherwise (heating from below). As discussed in detail by Pérez Cordón and Velarde (1975), for usual liquid layers $\epsilon_1 \simeq 10^{-4}$, while $\epsilon_2 \simeq 10^{-11}$. The Boussinesq approximation is thus obtained by taking the limits $\epsilon_1 \to 0$ and $\epsilon_2 \to 0$ in the balance equations (4.6) for given values of G, Pr and Ra. If we then revert to dimensional variables, and introduce the linear phenomenological laws for the deviatoric stress tensor (Newton viscosity law) and for the heat flux (Fourier law), we obtain the Boussinesq equations in their conventional form:

$$0 = \nabla \mathbf{v} \tag{4.9a}$$

$$\bar{\rho}_0 \left[\frac{\partial \mathbf{v}}{\partial t} + (\mathbf{v} \cdot \nabla) \, \mathbf{v} \right] = -\nabla p + \eta \, \nabla^2 \mathbf{v} - \bar{\rho}_0 [1 - \alpha_p (T - \bar{T}_0)] g \hat{\mathbf{z}} \tag{4.9b}$$

$$\bar{\rho}_0 c_p \left[\frac{\partial T}{\partial t} + \mathbf{v} \cdot \nabla T \right] = \lambda \, \nabla^2 T. \tag{4.9c}$$

Equation (4.9a) indicates that, within the Boussinesq approximation, the fluid can be considered to be divergence-free, a fact that has been used to simplify the Navier-Stokes equation (4.9b). As mentioned earlier, any dependence of c_p and of the transport coefficients on temperature has been neglected in Eqs. (4.9).

Equations (4.9) have a simple stationary solution that satisfies the boundary conditions:

$$
\begin{aligned}
\mathbf{v}(\mathbf{r}, t) &= \mathbf{v}_0 = 0 \\
T(\mathbf{r}, t) &= T_0(z) = \bar{T}_0 + \nabla T_0 \, z \\
\nabla p_0(z) &= -\bar{\rho}_0 g [1 - \alpha_p \nabla T_0 z].
\end{aligned} \tag{4.10}
$$

The stationary solution given by Eqs. (4.10) is called the *conductive solution* and represents the case of a quiescent fluid subjected to a uniform temperature gradient in the vertical direction. It represents a nonequilibrium stationary or steady state, in which the local values of the thermodynamic fields do not change with time. The main difference with equilibrium is that in nonequilibrium steady states at least one of the dissipative fluxes, in the case (4.10) the heat flow, is different from zero. For this reason they are often referred to as *dissipative states*, and the dissipation function (integrated over the thickness) has a finite nonzero value, which has to be positive by virtue of the second law of thermodynamics. Furthermore, it can be demonstrated (Glansdorff and Prigogine, 1971) that the entropy production associated with the steady state (4.10) has the minimum (nonzero and positive) value compatible with the BC (4.2). This is the so-called *minimum entropy production* theorem, which we shall not further discuss in this book.

Our goal is to study fluctuations around this conductive solution. As earlier elucidated in Sect. 3.5, to formulate the appropriate fluctuating-hydrodynamics equations we proceed in two steps. First, we write each thermodynamic field in Eqs. (4.1) as the sum of its average stationary value and a fluctuating part, *i.e.*, $T(\mathbf{r}, t) = T_0(z) + \delta T(\mathbf{r}, t)$, $\mathbf{v}(\mathbf{r}, t) = \delta \mathbf{v}(\mathbf{r}, t)$, etc. Secondly, we add fluctuating dissipative fluxes to the phenomenological laws, namely a random stress tensor $\delta \Pi(\mathbf{r}, t)$ and a random heat flux $\delta \mathbf{Q}(\mathbf{r}, t)$. Finally, we linearize in the fluctuations (Law and Sengers, 1989;

Segrè et al., 1993b). In the Boussinesq approximation, the fluctuating version of the continuity equation then reduces to divergence-free velocity fluctuations: $\nabla \cdot \delta \mathbf{v} = 0$. This allows us to further simplify the fluctuating version of the Navier-Stokes equation, Eq. (4.9b), by taking a double curl, thus eliminating the pressure fluctuations. The z-component v_z of the fluctuating velocity is coupled to the temperature fluctuations in linear order through the advection term in the heat equation and the buoyancy term in the Navier-Stokes equation. The coupled equations for the fluctuations of the temperature and the vertical velocity component are:

$$\frac{\partial}{\partial t}\left(\nabla^2 \delta v_z\right) = \nu \, \nabla^2 \left(\nabla^2 \delta v_z\right) + \alpha_p g \left(\partial_x^2 + \partial_y^2\right) \delta T - F_0, \tag{4.11a}$$

$$\frac{\partial}{\partial t}\, \delta T = a_T \, \nabla^2 \delta T - \delta v_z \, \nabla T_0 + F_1, \tag{4.11b}$$

which we call the *linearized fluctuating Boussinesq equations*. The terms F_0 and F_1 in Eqs. (4.11) represent the two corresponding components of the thermal noise $\mathbf{F}(\mathbf{r}, t)$, related to the random stress tensor $\delta \Pi$ and random heat flux $\delta \mathbf{Q}$ by (Hohenberg and Swift, 1992):

$$\mathbf{F}(\mathbf{r}, t) = \frac{1}{\bar{\rho}_0}\left(\begin{array}{c} \{\nabla \times [\nabla \times (\nabla \cdot \delta \Pi(\mathbf{r}, t))]\}_z \\ -\dfrac{1}{c_p}\,\nabla \cdot \delta \mathbf{Q}(\mathbf{r}, t) \end{array}\right). \tag{4.12}$$

The subscript z in Eq. (4.12) indicates that the first component of the thermal noise $F_0(\mathbf{r}, t)$ has to be identified with the z-component of the vector between the curly brackets. The fluctuating Boussinesq equations were originally adopted by Zaitsev and Shliomis (1971) and by Swift and Hohenberg (1977) to study thermal fluctuations close to the convective Rayleigh-Bénard instability. Since $\langle \delta T \rangle = 0$ and $\langle \delta v_z \rangle = 0$ everywhere in the fluid layer, we note that the solution of the fluctuating Boussinesq equations will be independent of the sign adopted for δT or δv_z (Chandrasekhar, 1961; Swift and Hohenberg, 1977; Hohenberg and Swift, 1992; Cross and Hohenberg, 1993). Since for an incompressible fluid $c_p \simeq c_V$, the second component of the thermal noise in Eq. (4.11b) may be identified with the same random force F_2 introduced in Sect. 3.3 for a compressible fluid in equilibrium.

Both $\delta T(\mathbf{r}, t)$ and $\delta v_z(\mathbf{r}, t)$ are obtained by solving the pair of inhomogeneous coupled differential equations (4.11). The x and y components of the fluctuating velocity satisfy the equivalent of Eq. (4.11a), but with the buoyancy term replaced by the horizontal components of the double curl of $\delta T \hat{\mathbf{z}}$. Then, the solution $\delta T(\mathbf{r}, t)$ of (4.11) acts as an additional source of noise for the x and y components of the fluctuating velocity. Since in this book we are mainly interested in the density autocorrelation (proportional to the structure factor), we shall not discuss fluctuations in the horizontal velocity. They do not contribute to the structure factor, see Eq. (4.19) below. To our knowledge, no systematic study of nonequilibrium effects on the horizontal components of the velocity has been so far been reported in the literature.

There exists an alternative derivation of the Boussinesq equations for compressible fluids in which the temperature gradient in Eq. (4.11b) is to be interpreted as the sum of the externally imposed temperature gradient ∇T_0 and the adiabatic temperature gradient $(\alpha \bar{T}_0/c_P)g$ (Spiegel and Veronis, 1960; Chasnov and Lee, 2001). The effect of this adiabatic temperature gradient on the Rayleigh component of the dynamic structure factor has been investigated by Segrè et al. (1993b). Here we adopt the Boussinesq equations in their more classical form, which in addition to neglecting any pressure dependence of the density also neglects the adiabatic temperature gradient compared to the magnitude of the imposed ∇T_0. In practice, this is a very good approximation (Gray and Giorgini, 1976; Tritton, 1988).

In Eqs. (4.11) we have neglected the nonlinear advection terms, that are usually included in the standard presentation of the Boussinesq equations (Chandrasekhar, 1961). This approximation is only justified when the fluctuations are "small". Whether the nonequilibrium fluctuations are small depends on the linear stability of the "conductive" solution, as discussed in more detail in Chapter 8. As is well known (Chandrasekhar, 1961; Manneville, 1990), linear stability depends on the value of the dimensionless Rayleigh number, Ra, defined in Eq. (4.8). The conductive solution is stable for both negative and positive values of the Rayleigh number Ra as long as Ra is smaller than a positive critical value Ra_c. When a one-component fluid is heated from above, the "conductive" state is always stable (provided $\alpha_p > 0$). To assure that fluctuations are indeed small we restrict ourselves in this chapter to that situation, *i.e.*, to $Ra < Ra_c$.

Studies can be found in the literature that deal with possible non-Boussinesq effects (Gray and Giorgini, 1976; Normand et al., 1977; Tritton, 1988), and that are concerned with the effects resulting from spatial variation of some thermophysical properties, neglected in the standard Boussinesq approximation, Eqs. (4.11). Deviations of the Boussinesq approximation do affect the characteristic nature of pattern formation upon the appearance of convection (Busse, 1967; Ahlers, 1980; Bodenschatz et al., 1991; Cross and Hohenberg, 1993). However, the intensity of the nonequilibrium fluctuations in convection-free states in liquids below the convective instability appears to be much less sensitive to deviations from the Boussinesq approximation. Indeed, as discussed in Chapter 10, Segrè et al. (1992) have analyzed and measured the temperature fluctuations in liquid toluene subjected to temperature gradients up to 220 K cm^{-1}. At the higher value of the temperature gradient some of the thermophysical properties varied significantly over the height of the fluid layer. Nevertheless, the intensity of the observed nonequilibrium temperature fluctuations was equal within one percent to the intensity predicted when all thermophysical-property values are taken at their average value in the fluid layer. Similar results were obtained by Ahlers (1980), measuring the critical Rayleigh number for the onset of Rayleigh-Bénard convection and the heat transfer slightly above the threshold. However, non-Boussinesq effects are important in dealing with fluctuations in a fluid near a critical point (Oh et al., 2004).

4.2   Bulk structure factor in the presence of a stationary temperature gradient

Starting from the random Boussinesq equations, Eqs. (4.11), we derive here an expression for the structure factor of a fluid subjected to a stationary temperature gradient in the presence of gravity. When the contribution of gravity is neglected we shall recover the expressions obtained by Kirkpatrick et al. (1982b) who used a combination of kinetic theory and mode-coupling theory. Subsequently, the same results have been reproduced, on the basis of Landau's fluctuating hydrodynamics, by Ronis and Procaccia (1982) and others (Schmitz and Cohen, 1985a,b; Law and Sengers, 1989). Our derivation is similar to the original derivation of Ronis and Procaccia (1982), but instead of the full fluctuating-hydrodynamics equation we use here the fluctuating Boussinesq equations that simplifies the treatment of the evolution equations for the fluctuating variables considerably. It will also enable us to make later contact with work of other investigators dealing with fluctuations close to the convective instability.

To calculate the fluctuations around nonequilibrium steady states, in addition to the fluctuating Boussinesq Eqs. (4.11), we need the fluctuation-dissipation theorem (Landau and Lifshitz, 1959). As was explained in Sect. 3.5, in nonequilibrium fluctuating hydrodynamics it is assumed that, due to local equilibrium, the FDT still holds *locally* (Ronis et al., 1980; Schmitz and Cohen, 1985a). Thus, as in equilibrium, both $\delta\Pi$ as $\delta\mathbf{Q}$ will act as *white noise*, their autocorrelation functions being short ranged in time and in space, proportional to delta functions. However, the fact that FDT is now *local* means that, for instance, in the presence of an external temperature gradient in the z-direction, the FDT for the random heat flow will be obtained by substituting the local temperature into Eq. (3.6a), so that:

$$\langle \delta Q_i(\mathbf{r},t) \cdot \delta Q_j(\mathbf{r}',t') \rangle = 2k_{\mathrm{B}}\lambda \left[T_0(z)\right]^2 \delta_{ij}\; \delta(\mathbf{r}-\mathbf{r}')\;\delta(t-t') \tag{4.13}$$

$$= 2k_{\mathrm{B}}\lambda\bar{T}_0^2 \left[1 + \frac{\nabla T_0\, z}{\bar{T}_0}\right]^2 \delta_{ij}\;\delta(\mathbf{r}-\mathbf{r}')\;\delta(t-t'),$$

where Eq. (4.10) has been used to represent the local value of the temperature $T_0(z)$. In accordance with the Boussinesq approximation, we have neglected any local dependence (through the temperature) of the thermal conductivity λ. As will be discussed in Sect. 4.4, the presence of ∇T_0

in the expression (4.13) for the local equilibrium FDT does cause some long-ranged contribution to nonequilibrium fluctuations. However, as also demonstrated in Sect. 4.4, nonequilibrium contributions arising from inhomogeneously correlated noise are in practice negligibly small compared with the nonequilibrium contributions arising from the coupling of the fields. Hence, in this section, we shall apply Eq. (4.13) for the FDT, neglecting the term with ∇T_0. Of course, in cases where there is no coupling among the fields, nonequilibrium effects arising from inhomogeneously correlated noise cannot be neglected, as will shall see in Sect. 4.3.

To determine the bulk structure factor, *i.e.*, the structure factor in the absence of any boundary conditions, we apply a temporal and spatial Fourier transformation to the fluctuating Boussinesq Eqs. (4.11). We thus obtain a set of equations for the fluctuations in the vertical component of the velocity $\delta v_z(\omega, \mathbf{q})$ and for the fluctuations in the temperature $\delta T(\omega, \mathbf{q})$ as a function of the frequency ω and the wave vector $\mathbf{q}$:

$$\begin{pmatrix} q^2(\mathrm{i}\,\omega + \nu\,q^2) & -\alpha_p\,g\,q_\parallel^2 \\ \nabla T_0 & \mathrm{i}\,\omega + a_T\,q^2 \end{pmatrix} \begin{pmatrix} \delta v_z(\omega, \mathbf{q}) \\ \delta T(\omega, \mathbf{q}) \end{pmatrix} = \mathbf{F}(\omega, \mathbf{q}), \qquad (4.14)$$

where $q_\parallel{}^2 = q_x^2 + q_y^2$ represents the magnitude of the vector $\mathbf{q}_\parallel$ indicating the component of the fluctuations wave vector in the xy-plane, *i.e.*, the component of $\mathbf{q}$ perpendicular to the temperature gradient. We note that the parallel and perpendicular directions are here defined with respect to the horizontal plane, as done by most researchers in the field (Kirkpatrick and Cohen, 1983; Schmitz and Cohen, 1985a,b; van Beijeren and Cohen, 1988b), but unlike the notation used by Segrè et al. (1993b) or Segrè and Sengers (1993). The matrix appearing in the LHS of (4.14) is identified with the inverse response function for our problem. The Fourier transformation $\mathbf{F}(\omega, \mathbf{q})$ of the noise (4.12) is related to the Fourier transforms of the random stress tensor and the random heat flow as:

$$\mathbf{F}(\omega, \mathbf{q}) = \frac{-1}{\rho_0} \begin{pmatrix} \mathrm{i}\left[q_z\, q_i q_j \delta \Pi_{ij}(\omega, \mathbf{q}) - q^2\, q_i \delta \Pi_{iz}(\omega, \mathbf{q})\right] \\ \mathrm{i}\,(c_p)^{-1}\, q_i \delta Q_i(\omega, \mathbf{q}) \end{pmatrix}. \qquad (4.15)$$

The solution of (4.14) for $\delta v_z(\omega, \mathbf{q})$ and $\delta T(\omega, \mathbf{q})$ can be readily obtained as:

$$\begin{pmatrix} \delta v_z(\omega, \mathbf{q}) \\ \delta T(\omega, \mathbf{q}) \end{pmatrix} = \mathsf{G}(\omega, \mathbf{q})\, \mathbf{F}(\omega, \mathbf{q}), \qquad (4.16a)$$

where, inverting the matrix in Eq. (4.14), we express the linear response function $\mathsf{G}(\omega, \mathbf{q})$ for the current problem as:

$$\mathsf{G}(\omega, \mathbf{q}) = \frac{\begin{pmatrix} \mathrm{i}\,\omega + a_T\,q^2 & \alpha_p g q_\parallel^2 \\ -\nabla T_0 & q^2\left(\mathrm{i}\,\omega + \nu\,q^2\right) \end{pmatrix}}{q^2\,[\mathrm{i}\omega + \Gamma_+(\mathbf{q})]\,[\mathrm{i}\omega + \Gamma_-(\mathbf{q})]}, \qquad (4.16b)$$

with two decay rates given by:

$$\Gamma_{\pm}(\mathbf{q}) = \frac{1}{2} \, q^2 \left[(\nu + a_T) \pm \sqrt{(\nu - a_T)^2 - 4g\alpha_p \nabla T_0 \frac{q_{\parallel}^2}{q^6}} \right]. \qquad (4.17)$$

As already mentioned in the previous section, in the expression (4.17) for the decay rates $\Gamma_{\pm}(\mathbf{q})$ the temperature gradient ∇T_0 should actually be identified with the effective temperature gradient $\nabla T_0 + (\alpha \bar{T}_0/c_P)g$, as was done by Segrè et al. (1993b). However, the contribution $(\alpha \bar{T}_0/c_P)g$ from the adiabatic temperature gradient is neglected in the classical Boussinesq approximation that we have adopted here. In the limit $g \to 0$, we have $\Gamma_+ = \nu q^2$ and $\Gamma_- = a_T q^2$, which are the decay rates of the transverse-velocity fluctuations δv_z and the temperature fluctuations δT, respectively, when the system is in global thermodynamic equilibrium (see Chapter 3). We see that gravity causes a (small) mixing of these decay rates. Notice also from Eq. (4.16) that ∇T_0 enters into the expression for $\delta v_z(\omega, \mathbf{q})$ only through the decay rates $\Gamma_{\pm}(\mathbf{q})$.

For large and positive ∇T_0 (*i.e.*, when heating from above), depending on the wave vector $\mathbf{q}$, the decay rates $\Gamma_{\pm}(\mathbf{q})$ given by Eq. (4.17) may have a nonzero imaginary part which implies the presence of *propagating modes*. In terms of dimensionless variables, propagating modes appear when:

$$-4 \, Ra \, \frac{\tilde{q}_{\parallel}^2}{\tilde{q}^2} > \frac{(Pr - 1)^2}{Pr}, \qquad (4.18)$$

where the height L of the layer is used to make the wave vector $\mathbf{q}$ dimensionless: $\tilde{\mathbf{q}} = \mathbf{q}L$. Condition (4.18) may be satisfied for large and negative Ra. The presence of propagating modes in a fluid layer driven out of thermal equilibrium is a consequence of the coupling between decay rates (4.17) due to the presence of both gravity and a temperature gradient; the phenomenon was first predicted by Boon et al. (1979), and it has also been discussed by Segrè et al. (1993b).

Combining the general definition (3.23) of $S(\omega, \mathbf{q})$ with the equation of state (4.3) in the Boussinesq approximation, we conclude that the dynamic structure factor $S(\omega, \mathbf{q})$ is related to the autocorrelation function of the temperature fluctuations by:

$$\langle \delta T^*(\omega, \mathbf{q}) \, \delta T(\omega', \mathbf{q}') \rangle = \frac{m_0}{\alpha_p^2 \rho} \, S(\omega, \mathbf{q}) \, (2\pi)^4 \, \delta(\omega - \omega') \, \delta(\mathbf{q} - \mathbf{q}'), \qquad (4.19)$$

where we have assumed that the average mass density appearing in the general definition (3.23) of the dynamic structure factor can be identified with $\rho = \bar{\rho}_0$, the density in the middle of the layer. Equation (4.19) shows that the Boussinesq approximation only yields the Rayleigh component of the structure factor. To determine the effects of a temperature gradient on the

Brillouin components one needs a different hydrodynamic approximation to be considered in Sect. 4.3.

To deduce the autocorrelation function $\langle \delta T^*(\omega, \mathbf{q})\, \delta T(\omega', \mathbf{q}')\rangle$ of the temperature fluctuations from Eqs. (4.16), we need the correlation functions among the various components of the Langevin noise term $\mathbf{F}(\omega, \mathbf{q})$, defined by Eq. (4.15). Fourier transforming the FDT for a one-component newtonian fluid, given by Eq. (3.6), and using the definition (4.15), we see that these correlation functions can be conveniently expressed in terms of a correlation matrix $C(\mathbf{q})$, defined by:

$$\langle F_\alpha^*(\omega, \mathbf{q})\, F_\beta(\omega', \mathbf{q}')\rangle = (2\pi)^4\, C_{\alpha\beta}(\mathbf{q})\, \delta(\omega - \omega')\, \delta(\mathbf{q} - \mathbf{q}'). \tag{4.20}$$

Equation (3.7) for the FDT of a one-component newtonian fluid allows us to evaluate the correlation matrix for this problem as:

$$C(\mathbf{q}) = 2k_B \bar{T}_0 \begin{bmatrix} \dfrac{\nu}{\rho}\, q_\parallel^2\, q^4 & 0 \\[2ex] 0 & \dfrac{\bar{T}_0 a_T}{\rho c_P}\, q^2 \end{bmatrix}. \tag{4.21}$$

Notice that in obtaining (4.21), we have replaced the temperature by its average value $\bar{T}_0$. For consistency, we have neglected any spatial variation of the various thermophysical properties, so that the prefactors appearing in Eq. (3.6) correspond to their average values in the layer. This means that we are using here an "average" version of the local fluctuation-dissipation theorem. Use of the average FDT turns out to be an excellent approximation. This issue is further discussed in Sect. 4.4.

Combining Eqs. (4.16), (4.19) and (4.21) we can obtain the dynamic structure factor $S(\omega, \mathbf{q})$ of a nonequilibrium fluid in the Boussinesq approximation. When there are no propagating modes in the fluid $\Gamma_\pm(q)$ are real numbers; in that case $S(\omega, \mathbf{q})$ can be conveniently expressed as the sum of two lorentzians:

$$S(\omega, \mathbf{q}) = S_E \frac{\gamma - 1}{\gamma}\left\{ A_+(\mathbf{q}) \frac{2\,\Gamma_+(\mathbf{q})}{[\omega^2 + \Gamma_+^2(\mathbf{q})]} + A_-(\mathbf{q}) \frac{2\,\Gamma_-(\mathbf{q})}{[\omega^2 + \Gamma_-^2(\mathbf{q})]} \right\}, \tag{4.22}$$

where the dimensionless amplitudes $A_\pm(\mathbf{q})$ are given by:

$$A_\pm(\mathbf{q}) = \pm \frac{a_T q^2 [\Gamma_\pm^2(\mathbf{q}) - \nu^2 q^4] - \nu(c_P/T)(\nabla T_0)^2 q_\parallel^2}{\Gamma_\pm(\mathbf{q})\, [\Gamma_+^2(\mathbf{q}) - \Gamma_-^2(\mathbf{q})]} \tag{4.23}$$

with

$$m_0\, S_E \frac{\gamma - 1}{\gamma} = \frac{k_B \alpha_p^2 \bar{T}_0^2}{c_p}. \tag{4.24}$$

In Eq. (4.22), $S_E(\gamma - 1)/\gamma$ represents the dimensionless amplitude of the Rayleigh line for a fluid in thermodynamic equilibrium at the average temperature $\bar{T}_0$, in accordance with Eq. (3.28). In deriving Eq. (4.22) we have employed the thermodynamic relation Eq. (3.14) for the average values in the fluid layer to facilitate a comparison with the expression for the equilibrium fluctuations in Chapter 3.

For Ra and $\mathbf{q}$ values for which there are propagating modes in the fluid, Eq. (4.22) continues to be valid. However, since in this case the decay rates $\Gamma_\pm(\mathbf{q})$ form a pair of complex conjugate numbers, it is more convenient to express the structure factor as:

$$S(\omega, \mathbf{q}) = S_E \frac{\gamma - 1}{\gamma} \left\{ A_1(\mathbf{q}) \left[\frac{2\,\mathrm{Re}(\Gamma_+)}{[\omega + \mathrm{Im}(\Gamma_+)]^2 + [\mathrm{Re}(\Gamma_+)]^2} \right. \right.$$
$$\left. + \frac{2\,\mathrm{Re}(\Gamma_+)}{[\omega - \mathrm{Im}(\Gamma_+)]^2 + [\mathrm{Re}(\Gamma_+)]^2} \right] \qquad (4.25)$$
$$\left. + A_2(\mathbf{q}) \left[\frac{2\,[\omega + \mathrm{Im}(\Gamma_+)]}{[\omega + \mathrm{Im}(\Gamma_+)]^2 + [\mathrm{Re}(\Gamma_+)]^2} - \frac{2\,[\omega - \mathrm{Im}(\Gamma_+)]}{[\omega - \mathrm{Im}(\Gamma_+)]^2 + [\mathrm{Re}(\Gamma_+)]^2} \right] \right\},$$

where $A_1(\mathbf{q})$ and $A_2(\mathbf{q})$ are real coefficients with lengthy expressions not given here. Equation (4.25) shows that, when propagating modes are present in the system, the spectrum of the fluctuations consists of two lorentzians symmetrically displaced from $\omega = 0$. The asymmetric component of Eq. (4.25) can in most circumstances be neglected ($A_2(\mathbf{q}) \simeq 0$). Moreover, it should be noticed that, when integrating over ω, the asymmetric part does not contribute to the static structure factor.

Equation (4.22), with Eqs. (4.17) and (4.23), determine the Rayleigh line for a newtonian viscous fluid in the presence of both gravity and a temperature gradient ∇T_0. This set of equations was first correctly derived by Segrè et al. (1993b). In the limit $g \to 0$, we recover the celebrated expression for the Rayleigh component of the structure factor first obtained by Kirkpatrick et al. (1982b) and subsequently reproduced by Ronis and Procaccia (1982) and others (Schmitz and Cohen, 1985a; Law and Sengers, 1989). In modern light-scattering experiments one does not measure a structure factor as a function of frequency, but a time-correlation function $S(\mathbf{q}, \tau)$ depending on the difference ($\tau = |t - t'|$) between the two times for which the correlation among the fluctuating variables is determined (see Chapter 10). Applying an inverse Fourier transform in ω to Eq. (4.22), we obtain in the the limit $g \to 0$ (no propagating modes):

$$S(\mathbf{q}, \tau) = S_E \frac{\gamma - 1}{\gamma} \left\{ [1 + A_T(\mathbf{q})] \, \exp(-a_T q^2 |\tau|) \right.$$
$$\left. - A_\nu(\mathbf{q}) \, \exp(-\nu q^2 |\tau|) \right\}, \qquad (4.26)$$

where $A_T(\mathbf{q})$ and $A_\nu(\mathbf{q})$ represent the dimensionless nonequilibrium enhancements in the amplitude of the thermal and the viscous fluctuations, respectively. They are given by:

$$A_\nu(\mathbf{q}) = \frac{1}{Pr} A_T(\mathbf{q}) = \frac{\nu(c_P/T)(\nabla T_0)^2}{a_T(\nu^2 - a_T^2)} \frac{q_\parallel^2}{q^6}. \tag{4.27}$$

In equilibrium $A_T(\mathbf{q}) = A_\nu(\mathbf{q}) = 0$, and Eq. (4.26) reduces to the expression for the Rayleigh component of the dynamic structure factor of a fluid in equilibrium as obtained in Chapter 3, i.e., the first term in the RHS of Eq. (3.30). As already anticipated, the decay rates of the nonequilibrium thermal and viscous fluctuations are the same as in equilibrium when gravity effects are neglected. However, the presence of a temperature gradient produces an enhancement of the amplitude of the fluctuations. Since for usual liquids $\nu \gg a_T$, the amplitude of the nonequilibrium viscous fluctuations is negative. The nonequilibrium enhancements of the temperature and viscous fluctuations are anisotropic and depend on the magnitude of the horizontal component $q_\parallel$ of the scattering vector. They reach a maximum for $\mathbf{q} \perp \nabla T_0$ and they vanish for $\mathbf{q} \parallel \nabla T_0$. The amplitudes increase with the square of the temperature gradient and decrease as q^{-4} with increasing wave number of the fluctuations.

Next we consider the static structure factor of the nonequilibrium fluid:

$$S(\mathbf{q}) = S(\mathbf{q}, \tau = 0) = (2\pi)^{-1} \int_{-\infty}^{\infty} d\omega\, S(\omega, \mathbf{q}), \tag{4.28}$$

which determines the total intensity of Rayleigh scattering (Berne and Pecora, 1976). For comparison with subsequent chapters, we find it convenient to introduce again dimensionless wave numbers $\tilde{q} = qL$ and $\tilde{q}_\parallel = q_\parallel L$, where L is the finite height of the fluid layer. Integrating the expression (4.22) for $S(\omega, \mathbf{q})$ and using Eq. (4.23) for $A_\pm(\mathbf{q})$, we obtain:

$$S(\mathbf{q}) = S_\mathrm{E} \frac{\gamma - 1}{\gamma} \left\{ 1 + \tilde{S}_\mathrm{NE}^0 \frac{\tilde{q}_\parallel^2}{(\tilde{q}^6 - Ra\, \tilde{q}_\parallel^2)} \right\}. \tag{4.29}$$

In Eq. (4.29) $\tilde{S}_\mathrm{NE}^0$ represents the strength of the nonequilibrium enhancement of the structure factor, which is given by:

$$\tilde{S}_\mathrm{NE}^0 = \frac{Pr\, Ra}{Pr + 1} + \frac{(Pr - 1)(c_P/T)L^4}{\nu^2 - a_T^2} (\nabla T_0)^2, \tag{4.30}$$

where we have retained the effect of gravity. Since we have not yet considered any confinement effects, Eqs. (4.29) and (4.30) do not depend explicitly on the height L. Indeed, Eqs. (4.29) and (4.30) are identical to Eq. (2.35) in Segrè et al. (1993b). The term in Eq. (4.30) proportional to Ra is related to the adiabatic temperature gradient and is in practice negligibly small. Hence, the intensity of the nonequilibrium fluctuations continues to be proportional to $(\nabla T_0)^2$ in the presence of gravity. After Eq. (3.32) we discussed how, the fact that the equilibrium static structure factor does not depend on $\mathbf{q}$, causes the equal-time correlations to be spatially short-ranged. In a similar way, we shall discuss in Sect. 7.5 how, the fact that the nonequilibrium static structure factor (4.29) depends explicitly on $\mathbf{q}$, causes the equal-time nonequilibrium correlations to be spatially long ranged.

Note that Eq. (4.29) can only be valid for negative Ra, *i.e.*, when the fluid layer is heated from above so that $\nabla T_0 > 0$. For any $Ra > 0$, the nonequilibrium contribution to the structure factor will always diverge at some finite value of the horizontal component $q_\parallel$ of the scattering vector. As will be demonstrated in Chapter 6, when boundary conditions are taken into account, we shall obtain an expression for $S(\mathbf{q})$ that is valid not only for negative Rayleigh numbers, but also for a finite interval of positive Rayleigh numbers up to a Rayleigh number Ra_c corresponding to the onset of Rayleigh-Bénard convection.

We conclude this section by remarking that newtonian viscous fluid behavior is included in the Boussinesq approximation. Nonequilibrium fluctuations in non-newtonian viscoelastic fluids subjected to a stationary temperature gradient have been considered by López de Haro et al. (2002).

4.3   Nonequilibrium effects on the Brillouin doublet

As we mentioned in Sect. 3.3, there are in principle two sources that cause (long-ranged) nonequilibrium fluctuations. The first source is a coupling between fluctuating fields through the temperature gradient considered in the previous section. The second cause is a spatial inhomogeniety of the noise correlation functions resulting from the local application of the FDT. As an example of the second source we consider the effect of a temperature gradient on the Brillouin components of the dynamic structure factor, since for the Brillouin doublet spatial inhomogeneities induced by the temperature gradient play a major role (Schmitz and Cohen, 1985a, 1987).

To determine the effects of a temperature gradient on the Rayleigh component of the dynamic structure factor we neglected any dependence of the density on the pressure and treated the density as a function of temperature only, in accordance with Eq. (4.3). By this procedure we reduced the fluctuations spectrum to only the Rayleigh line. To determine the effects of the temperature gradient on the Brillouin components we treat the density as a function of pressure, while neglecting the dependence of the density on temperature. That is, we consider a fluid with a finite isothermal compressibility $\varkappa_T$ but with a vanishingly small thermal expansion coefficient ($\alpha_p = 0$). By this procedure we reduce the fluctuations spectrum to only the Brillouin doublet.

If $\alpha_p = 0$, temperature fluctuations do not couple with velocity fluctuations. Then the spatiotemporal evolution of the fluctuations continues to be given by the same set of equations as in equilibrium, namely Eqs. (3.13a) and (3.15). Specifically (Tremblay et al., 1981; García-Colín and Velasco, 1982; Schmitz and Cohen, 1987; López de Haro et al., 2002):

$$\frac{\partial(\delta\rho)}{\partial t} = -\rho_0(\delta\psi) \tag{4.31a}$$

$$\frac{\partial(\delta\psi)}{\partial t} = -\frac{1}{\rho_0\varkappa_T}\nabla^2\left(\frac{\delta\rho}{\rho_0}\right) + \frac{1}{\rho_0}\left(\eta_v + \frac{4}{3}\eta\right)\nabla^2(\delta\psi) + F_1(\mathbf{r}, t), \tag{4.31b}$$

with $F_1(\mathbf{r}, t) = (\rho_0)^{-1}\nabla \cdot (\nabla \cdot \delta\Pi)$, as in Eq. (3.18). We can readily apply a Fourier transformation in space and time on Eqs. (4.31), and solve for $\delta\rho(\omega, \mathbf{q})$, so as to obtain:

$$\begin{aligned}\delta\rho(\omega, \mathbf{q}) &= \frac{\rho_0\, F(\omega, \mathbf{q})}{(\omega^2 - c_s^2 q^2) - i\omega D_V q^2}, \\ &\simeq \frac{\rho_0\, F(\omega, \mathbf{q})}{(\omega - c_s q + i\Gamma_s q^2)(\omega + c_s q + i\Gamma_s q^2)}, \end{aligned} \tag{4.32}$$

where in the second line, as in Eq. (3.21) for a fluid in equilibrium, we have used the approximation (3.20) for determining the ω-roots of the denominator. Since $\alpha_p \simeq 0$, we simply have $\Gamma_s = \frac{1}{2}D_V$. Note that the temperature gradient ∇T_0 does not appear explicitly in Eq. (4.32), but it will appear when we apply the local FDT to evaluate the correlation function $\langle F_1^*(\omega, \mathbf{q}) \cdot F_1(\omega', \mathbf{q}')\rangle$.

The Fourier transform $F_1(\omega, \mathbf{q})$ is given as the double divergence of the random stress tensor $\delta\Pi_{ij}$, *cfr.* Eq. (3.18):

$$F(\omega, \mathbf{q}) = -\frac{1}{\rho_0}q_i q_j \delta\Pi_{ij}(\omega, \mathbf{q}). \tag{4.33}$$

The correlation function $\langle\delta\Pi_{ij}^*(\mathbf{r}, t) \cdot \delta\Pi_{kl}(\mathbf{r}', t')\rangle$ is given by Eq. (3.6b) in terms of the local temperature $T(\mathbf{r})$, namely:

$$\begin{aligned}\langle\delta\Pi_{ij}^*(\mathbf{r}, t) \cdot \delta\Pi_{kl}(\mathbf{r}', t')\rangle = 2k_B T(\mathbf{r})\, \delta(\mathbf{r} - \mathbf{r}')\, \delta(t - t') \\ \times \left[\eta\left(\delta_{ik}\delta_{jl} + \delta_{il}\delta_{jk}\right) + \left(\eta_v - \frac{2}{3}\eta\right)\delta_{ij}\delta_{kl}\right], \end{aligned} \tag{4.34}$$

with

$$T(\mathbf{r}) = \bar{T}_0 + \mathbf{r} \cdot \nabla T_0. \tag{4.35}$$

Notice that a linear profile of temperature continues to be the stationary solution of the heat equation, even in the $\alpha_p \simeq 0$ approximation.

To obtain $\langle F_1^*(\omega, \mathbf{q}) \cdot F_1(\omega', \mathbf{q}') \rangle$ we need to apply a double spatial and temporal Fourier transformation to Eq. (4.34), what is complicated due to the explicit dependence on $\mathbf{r}$ of the prefactor multiplying the delta functions. To overcome this difficulty, we follow the method of Tremblay et al. (1981) in which the temperature field is represented by:

$$T(\mathbf{r}) = \bar{T}_0 + \Delta T \, \sin(\mathbf{k} \cdot \mathbf{r}), \qquad (4.36)$$

where $\mathbf{k}$ is an auxiliary vector and ΔT an auxiliary (scalar) temperature difference ΔT, such that: $\boldsymbol{\nabla} T_0 = \Delta T \mathbf{k}$. To first order in $\mathbf{k}$, Eq. (4.36) reduces to Eq. (4.35). It is easier to perform the calculation starting from Eq. (4.36) with the final result to be expanded to first order in $\mathbf{k}$. Then, substituting Eq. (4.36) into Eq. (4.34) and taking into account the definition (4.33) of $F_1(\omega, \mathbf{q})$, one obtains for the correlation function of interest (Tremblay et al., 1981):

$$\langle F^*(\omega, \mathbf{q}) \cdot F(\omega', \mathbf{q}') \rangle = \frac{2\pi}{\rho_0^2} \, \delta(\omega - \omega') \left[2\eta(\mathbf{q} \cdot \mathbf{q}')^2 + (\eta_v - \tfrac{2}{3}\eta)(q')^2 q^2 \right] (2\pi)^3$$

$$\times \left\{ 2 k_\mathrm{B} \bar{T}_0 \, \delta(\mathbf{q} - \mathbf{q}') + i k_\mathrm{B} \Delta T \left[\delta(\mathbf{q} - \mathbf{q}' - \mathbf{k}) - \delta(\mathbf{q} - \mathbf{q}' + \mathbf{k}) \right] \right\}. \quad (4.37)$$

Now, from Eqs. (4.32) and (4.37), we readily deduce the autocorrelation function for the density fluctuations. It can be cast in the form:

$$\langle \delta\rho^*(\omega, \mathbf{q}) \cdot \delta\rho(\omega', \mathbf{q}') \rangle = (2\pi)^4 \, \delta(\omega - \omega') \left\{ S_\mathrm{E}(\omega, \mathbf{q}) \, \delta(\mathbf{q} - \mathbf{q}') \right.$$

$$\left. + i \, \Delta T \, N(\omega, \mathbf{q}, \mathbf{q}') \left[\delta(\mathbf{q} - \mathbf{q}' - \mathbf{k}) - \delta(\mathbf{q} - \mathbf{q}' + \mathbf{k}) \right] \right\}, \quad (4.38)$$

where

$$S_\mathrm{E}(\omega, \mathbf{q}) = \frac{2 k_\mathrm{B} \bar{T}_0 \rho_0 D_V q^4}{(\omega^2 - c_\mathrm{s}^2 q^2)^2 + \omega^2 D_V^2 q^4}, \qquad (4.39)$$

represents the equilibrium structure factor for the Brillouin doublet at the temperature $\bar{T}_0$. Indeed, with the approximations $c_\mathrm{s} q \gg a_T q^2$ and $c_\mathrm{s} q \gg D_V q^2$, adopted in accordance with Eq. (3.20), Eq. (4.39) can be identified with the second term on the RHS of Eq. (3.27).

The nonequilibrium contribution in Eq. (4.38) is represented by the function $N(\omega, \mathbf{q}, \mathbf{q}')$, given by:

$$N(\omega, \mathbf{q}, \mathbf{q}') = \frac{k_\mathrm{B} \left[2\eta(\mathbf{q} \cdot \mathbf{q}')^2 + (\eta_v - \tfrac{2}{3}\eta)(q')^2 q^2 \right]}{\left[\omega^2 - c_\mathrm{s}^2 q^2 + i\omega D_V q^2 \right] \left[\omega^2 - c_\mathrm{s}^2 (q')^2 - i\omega D_V (q')^2 \right]}. \quad (4.40)$$

We note from Eq. (4.38) that the correlation function $\langle \delta\rho^*(\omega, \mathbf{q}) \cdot \delta\rho(\omega', \mathbf{q}') \rangle$ is no longer proportional to a single delta function $\delta(\mathbf{q} - \mathbf{q}')$, as was found in Eqs. (3.23) or (4.19) for the equilibrium or nonequilibrium Rayleigh line, respectively. This fact has consequences for the calculation of the dynamic structure factor, $S(\omega, \mathbf{q})$, as actually measured in light-scattering experiments (see Chapter 10). This point has been also addressed by Tremblay et al. (1981), who showed that, when the density autocorrelation function is written in the form (4.38), the dynamic structure factor is given by:

$$S(\omega, \mathbf{q}) = S_E(\omega, \mathbf{q}) + i\, \Delta T[N(\omega, \mathbf{q}+\tfrac{1}{2}\mathbf{k}, \mathbf{q}-\tfrac{1}{2}\mathbf{k}) - N(\omega, \mathbf{q}-\tfrac{1}{2}\mathbf{k}, \mathbf{q}+\tfrac{1}{2}\mathbf{k})], \quad (4.41)$$

The demonstration of Eq. (4.41) is considered in Sect. 10.1.1.

Combining Eqs. (4.32) and (4.37)-(4.41) and expanding the result up to terms linear in $\mathbf{k}$, we finally obtain (Tremblay et al., 1981; Schmitz and Cohen, 1987):

$$S(\omega, \mathbf{q}) = S_E(\omega, \mathbf{q}) \left\{ 1 - \frac{2\omega^3 D_V}{(\omega^2 - c_s^2 q^2)^2 + \omega^2 D_V^2 q^4} \frac{\mathbf{q} \cdot \boldsymbol{\nabla} T_0}{T_0} \right\}, \quad (4.42)$$

with $S_E(\omega, \mathbf{q})$ given by Eq. (4.39). Equation (4.42) is our final result for the Brillouin spectrum of a liquid subjected to a temperature gradient; it no longer depends on the auxiliary quantities ΔT or $\mathbf{k}$. Instead of using the representation (4.36) for the temperature field, Eq. (4.42) can also be obtained by other mathematical techniques. For instance, Malek Mansour et al. (1988) solved the fluctuating-hydrodynamics equations for the Brillouin lines as a power series of trigonometric functions, similar to a calculation to be performed in Sect. 4.4. Both procedures give the same final result, what give us confidence in the current development. Equation (4.42) incorporates the leading correction resulting from an uniform temperature gradient; it is possible to obtain also higher-order terms, but for that purpose one needs a different calculation method as, for instance, the one discussed by Ronis et al. (1979).

For a simple interpretation of Eq. (4.42) we use again that for regular fluids $c_s q \gg D_V q^2$, see Eq. (3.20). By expressing $S_E(\omega, \mathbf{q})$ as in the second term of the RHS of Eq. (3.27), Eq. (4.42) reduces to:

$$S(\omega, \mathbf{q}) = \frac{S_E}{\gamma} \left\{ \frac{\hat{\Gamma}_s q^2 [1 - \epsilon_B(\omega, \mathbf{q})]}{(\omega - c_s q)^2 + \hat{\Gamma}_s^2 q^4} + \frac{\hat{\Gamma}_s q^2 [1 + \epsilon_B(\omega, \mathbf{q})]}{(\omega + c_s q)^2 + \hat{\Gamma}_s^2 q^4} \right\}, \quad (4.43)$$

where the nonequilibrium correction $\epsilon_B(\omega, \mathbf{q})$ is given by:

$$\epsilon_B(\omega, \mathbf{q}) = \frac{2 c_s\, \omega^2 D_V q^2}{(\omega^2 - c_s^2 q^2)^2 + \omega^2 D_V^2 q^4} \frac{\hat{\mathbf{q}} \cdot \boldsymbol{\nabla} T_0}{T_0}. \quad (4.44)$$

Here S_E represents again the intensity of the fluctuations in thermodynamic equilibrium (3.28), evaluated at the average temperature $\bar{T}_0$ as in Eq. (4.22). Notice that, when $\nabla T_0 = 0$, Eq. (4.43) reduces to the same expression for the Brillouin doublet in thermodynamic equilibrium presented in Sect. 3.3. In Eq. (4.43) the Stokes and anti-Stokes components of the

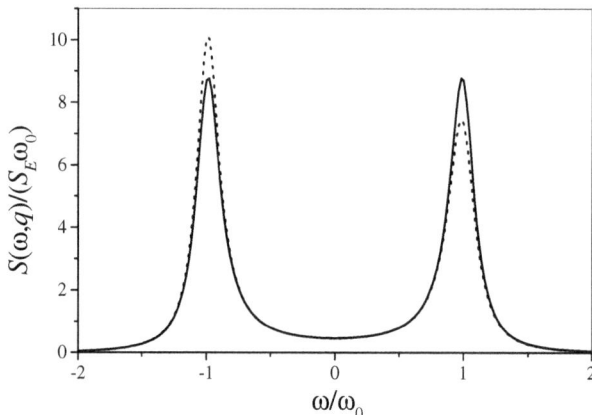

Figure 4.1: Brillouin spectrum in equilibrium, $\nabla T_0 = 0$, (solid line) and under the presence of a stationary temperature gradient (dashed line), from Eq. (4.42). Dimensionless parameters $D_V q^2/c_s q = 0.23$ and $(\mathbf{q} \cdot \nabla T_0)/(q^2 \bar{T}_0) = 0.018$.

Brillouin spectrum are easily identified. Then, we observe that the effect of the temperature gradient is to enhance (depending on the sign of ϵ_B) the Stokes peak and to suppress the anti-Stokes peak by the same amount ϵ_B. This phenomenon is shown graphically in Fig. 4.3 where, for some specific values of the dimensionless parameters $D_V q^2/c_s q$ and $(\mathbf{q}\cdot\nabla T_0)/(q^2 \bar{T}_0)$, Eq. (4.42) is plotted as a function of ω/ω_0 ($\omega_0 = c_s q$) for $\nabla T_0 = 0$ (solid curve) and for $\nabla T_0 \neq 0$ (dashed curve). As will be discussed in more detail in Chapter 10, asymmetry of the Brillouin spectrum in a fluid subjected to a temperature gradient has been observed experimentally (Beysens et al., 1980; Kiefte et al., 1984; Suave and de Castro, 1996), although in practice and for a complete interpretation of the experiments, one needs also to consider effects from sound-absorbing walls and bending of sound in an inhomogeneous medium (Schmitz and Cohen, 1987; Schmitz, 1988).

In Eq. (4.43), the first and second term inside the curly brackets can be independently integrated over ω, so as to obtain the intensity of the Stokes and anti-Stokes components of the Brillouin spectrum: I_S and I_{AS}, respectively. Retaining only the leading term in the small parameter $D_V q^2/c_s q$, one finds:

$$\frac{I_S - I_{AS}}{I_{B,0}} = \overline{\epsilon_B}(\mathbf{q}) = \frac{c_s}{D_V} \frac{\hat{\mathbf{q}} \cdot \nabla T_0}{q^2 \bar{T}_0}, \tag{4.45}$$

where $I_{B,0} = S_E/\gamma$ is the Brillouin-scattering intensity in the absence of a temperature gradient (equilibrium). In practice, the frequency dependence of $\epsilon_B(\omega, \mathbf{q})$ is unimportant, and is often replaced by its average value (Ronis et al., 1979): $\epsilon_B(\omega, \mathbf{q}) \simeq \overline{\epsilon_B}(\mathbf{q})$. The "average" is in the sense that Eq. (4.45) still holds when $\epsilon_B(\omega, \mathbf{q})$ in (4.43) is replaced by $\overline{\epsilon_B}(\mathbf{q})$.

The Brillouin lines are caused by scattering from propagating sound modes (phonons). The physical nature of the asymmetry in the Brillouin lines is readily understood since phonons with wave vector $-\mathbf{q}$ and phonons with wave vector $+\mathbf{q}$ arrive from regions with different temperatures. As can be seen from Eq. (4.45), the asymmetry of the Brillouin lines has a maximum when $\mathbf{q} \parallel \nabla T_0$ and vanishes when $\mathbf{q} \perp \nabla T_0$. This is just opposite to the effects of a temperature gradient on the Rayleigh line caused by mode coupling.

It is also interesting to note that $I_S + I_{AS} = I_{B,0}$, meaning that the presence of a temperature gradient does not change the total intensity of light scattered by phonons. This result is more general and it applies also to the full Eq. (4.42), without the approximation $c_s q \gg D_V q^2$; it applies even if higher-order terms are included in Eq. (4.42) (Schmitz and Cohen, 1987). As a consequence, the real-space equal-time density autocorrelation continues to be short ranged, *i.e*, proportional to a delta function $\delta(\mathbf{r} - \mathbf{r}')$ (van der Zwan et al., 1981). Of course, due to the propagation of phonons the density fluctuation here at a given time will be correlated with the density fluctuation at a distance d a time $\simeq d/c_s$ later, but such is a two-times correlation function, and the spatially short-ranged nature of correlation refers to equal-time fluctuations.

In this section we have accounted for the spatial dependence of the temperature only in the thermal noise term (4.34). More generally, one could also include the spatial dependence of the density ρ. This complication has been analyzed by van der Zwan et al. (1981), who have also included inhomogeneities in the viscosities. Their results up to first order in the gradient are the same as Eqs. (4.43) and (4.45) here, since these extra inhomogeneities only contribute in higher order.

The asymmetry in the Brillouin peaks due to the presence of a temperature gradient has a long history and was the first nonequilibrium effect on the structure factor of a fluid ever to be predicted, as reviewed by Fabelinskii (1994). Almost 60 years ago, Mandelstam (1934) already remarked that the intensity of light scattered at a particular point in a fluid should depend on the temperature at other points in the fluid. His idea was developed quantitatively by Leontovich (1935) and Vladimirskii (1943). Actually, Eq. (4.45) was first proposed by Leontovich (1935). However, some early experimental attempts to detect the asymmetry in the Brillouin peaks were carried out without success (Landsberg and Shubin, 1939; Chistyi, 1977). In the western literature, the interest in this topic arose around 1980 (Procaccia et al., 1979; Proccacia et al., 1979; Kirkpatrick et al., 1979, 1980, 1982b; van der

Zwan et al., 1981), independently of the previous Russian studies (Mandelstam, 1934; Leontovich, 1935; Vladimirskii, 1943). Subsequently, papers were published by authors familiar with the early Russian theoretical and experimental investigations (Tremblay et al., 1981; Fabelinskii, 1994), which helped to clarify the situation.

4.4   Nonequilibrium fluctuations in heat conduction

In the previous sections we derived the leading effects of a temperature gradient on the Rayleigh component of the dynamic structure factor (due to density fluctuations resulting from temperature fluctuations) and on the Brillouin components of the dynamic structure factor (due to density fluctuations resulting from pressure fluctuations). In the first case the major effect results from a coupling of the density fluctuations with the viscous fluctuations through the temperature gradient; in the second case a temperature gradient causes an asymmetry in the Brillouinn lines resulting from spatial inhomogeneity of the noise correlation functions. In this section we investigate the possible effects of a spatial inhomogeneity of the noise correlations on the temperature fluctuations associated to the Rayleigh line of the scattering spectrum.

   We continue to consider a thermodynamic system bounded by two horizontal planes maintained at different temperatures. As in Sect. 4.2, we assume the thermal conductivity to be independent of temperature, so that the temperature gradient is uniform. We no longer consider any coupling between temperature and velocity fluctuations so as to concentrate ourselves here only on nonequilibrium effects arising from inhomogeneously correlated noise. Without any coupling between temperature and velocity fluctuations, the spatiotemporal evolution of the temperature fluctuations will be given by the second fluctuating Boussinesq equation, Eq. (4.11b), which reduces to a simple heat-diffusion equation:

$$\frac{\partial}{\partial t}\, \delta T(\mathbf{r}, t) = a_T \nabla^2 \delta T(\mathbf{r}, t) - \frac{1}{\rho c_P} \boldsymbol{\nabla} \cdot \delta \mathbf{Q}(\mathbf{r}, t). \tag{4.46}$$

Equation (4.46) can also be considered as applying to heat conduction in a solid.

   The calculations in this section become simpler if we take the origin of the z-coordinate at the lower plate, so that the bounding plates are located at $z = 0$ and $z = L$. Then, for the correlations among the components of $\delta \mathbf{Q}(\mathbf{r}, t)$, we have instead of Eq. (4.13):

$$\langle \delta Q_i^*(\mathbf{r}, t) \cdot \delta Q_j(\mathbf{r}', t') \rangle = 2k_{\mathrm{B}} \lambda T_0^2(\mathbf{r})\, \delta(\mathbf{r} - \mathbf{r}')\, \delta(t - t')\, \delta_{ij}$$

$$= 2k_{\mathrm{B}}\lambda \overline{T_0}^{-2} \left[1 + \frac{\nabla T_0 L}{\overline{T_0}} \left(\frac{z}{L} - \frac{1}{2} \right) \right]^2 \delta(\mathbf{r} - \mathbf{r}')\, \delta(t - t')\, \delta_{ij}, \quad (4.48)$$

Equation (4.48) follows from the FDT (3.6a) for fluctuations around equilibrium when the equilibrium temperature is replaced with its local value $T_0(\mathbf{r})$ as given by Eq. (4.10), with $z = 0$ located at the lower plate. We interpret Eq. (4.46) as a linear Langevin equation where the inverse response function operator is $\mathcal{G}^{-1} = \partial_t - a_T \nabla^2$, and where the term $F = -1/(\rho c_p) \nabla \cdot \delta \mathbf{Q}(\mathbf{r}, t)$ plays the role of an inhomogeneously correlated random force.

We solve Eq. (4.46) assuming perfectly conducting plates at $z = 0$ and $z = L$. From Eq. (2.81), this means that temperature fluctuations vanish ($\delta T(\mathbf{r}, t) = 0$) at $z = 0$ and $z = L$, while the temperature gradient at the plates is undetermined. To incorporate this boundary condition, we do not apply a Fourier transformation to the Langevin Eq. (4.46) in the z-coordinate, but instead expand the solution in a Fourier sine series, also referred to as an expansion into (hydrodynamic) modes:

$$\delta T(\omega, \mathbf{q}_\parallel, z) = \sum_{N=1}^{\infty} T_N(\omega, \mathbf{q}_\parallel)\, \sin\left(\frac{N\pi}{L} z \right), \quad (4.49)$$

where ω is the frequency of the fluctuations and $\mathbf{q}_\parallel = \{q_x, q_y\}$ is the wave vector of the fluctuations in the horizontal xy-plane. Equation (4.49) shows why in the present case it is simpler to take $z = 0$ at the lower plate. If, instead, we had kept $z = 0$ at the center of the layer, Eq. (4.49) would had required sines for even N and cosines for odd N (see Chapter 6). Next, substituting Eq. (4.49) into Eq. (4.46), assuming again that the temperature dependence of the various thermophysical properties can be neglected, we readily solve for the coefficients $T_N(\omega, \mathbf{q}_\parallel)$ and obtain:

$$T_N(\omega, \mathbf{q}_\parallel) = \frac{F_N(\omega, \mathbf{q}_\parallel)}{\mathrm{i}\omega + \Gamma_N(q_\parallel)}, \quad (4.50)$$

where $q_\parallel^2 = q_x^2 + q_y^2$ is the magnitude of the horizontal wave vector and $\Gamma_N(q_\parallel)$ is the decay rate of the fluctuations, given by:

$$\Gamma_N(q_\parallel) = a_T \left(\frac{N^2 \pi^2}{L^2} + q_\parallel^2 \right). \quad (4.51)$$

Notice that the decay rates are always real numbers, independent of the index N or $q_\parallel$; this means that in our simple problem there are no propagating modes, see Sect. 6.2. In Eq. (4.50), $F_N(\omega, \mathbf{q}_\parallel)$ represents the projections of the random noise in the LHS of Eq. (4.46) onto the set of sine functions, namely:

$$F_N(\omega, \mathbf{q}_\parallel) = \frac{-1}{\rho c_p} \frac{2}{L} \int_0^L dz\, \sin\left(\tfrac{N\pi}{L} z \right)\, [\nabla \cdot \delta \mathbf{Q}](\omega, \mathbf{q}_\parallel, z), \quad (4.52)$$

with

$$[\boldsymbol{\nabla} \cdot \delta\mathbf{Q}]\,(\omega, \mathbf{q}_{\parallel}, z) = \mathrm{i}q_x \delta Q_x(\omega, \mathbf{q}_{\parallel}, z) + \mathrm{i}q_y \delta Q_y(\omega, \mathbf{q}_{\parallel}, z)$$
$$+ \partial_z \delta Q_z(\omega, \mathbf{q}_{\parallel}, z) \quad (4.53)$$

being the Fourier transform of $\boldsymbol{\nabla} \cdot \delta\mathbf{Q}(\mathbf{r}, t)$ in time and in the horizontal x and y coordinates.

To calculate the correlation function $\langle \delta T^*(\omega, \mathbf{q}_{\parallel}, z) \cdot \delta T(\omega', \mathbf{q}_{\parallel}', z') \rangle$ of the temperature fluctuations we need the correlation functions $\langle F_N^*(\omega, \mathbf{q}_{\parallel}) \cdot F_M(\omega', \mathbf{q}_{\parallel}') \rangle$ of the Fourier components of the random force. These correlation functions can be computed first by applying to the local-equilibrium FDT (4.48) a double Fourier transformation in t and t', and in the horizontal variables; then combining that result with Eq. (4.52). We find that these correlation functions are conveniently expressed in terms of a correlation matrix $\mathsf{C}(q_{\parallel})$ as:

$$\langle F_N^*(\omega, \mathbf{q}_{\parallel}) \cdot F_M(\omega', \mathbf{q}_{\parallel}') \rangle = \frac{2k_B \overline{T_0}^2}{\rho c_p} \frac{\nu}{L^2} \frac{2}{L} \, \tilde{C}_{NM}(q_{\parallel})$$
$$\times \; (2\pi)^3 \, \delta(\omega - \omega') \, \delta(\mathbf{q}_{\parallel} - \mathbf{q}_{\parallel}'), \quad (4.54)$$

where the components of the correlation matrix, also referred to as dimensionless noise correlation coefficients, are given by:

$$\tilde{C}_{NM}(q_{\parallel}) = \frac{2L}{Pr} \int_0^L \int_0^L dz \, dz' \, \sin\left(\tfrac{N\pi}{L}z\right) \, \sin\left(\tfrac{M\pi}{L}z'\right)$$
$$\times \left(q_{\parallel}^2 + \partial_z \partial_{z'}\right) \cdot \left\{ \left[1 + \tfrac{\nabla T_0 L}{T_0}\left(\tfrac{z}{L} - \tfrac{1}{2}\right)\right]^2 \delta(z - z') \right\}. \quad (4.55)$$

In Eqs. (4.54) and (4.55) we employed the definition of the thermal diffusivity and we introduced the Prandtl number, to facilitate comparison with later expressions. To evaluate the double integral of the derivatives of the delta function in Eq. (4.55), we perform a couple of integrations by parts to move the differential operators from the delta function to the preceding sine functions. The resulting integrals can be readily evaluated and the final expression for $\tilde{C}_{NM}(q_{\parallel})$ may be found in Ortiz de Zárate and Sengers (2004).

We now have all the information needed for the calculation of the autocorrelation function $\langle \delta T^*(\omega, \mathbf{q}_{\parallel}, z) \cdot \delta T(\omega', \mathbf{q}_{\parallel}', z') \rangle$ of the temperature fluctuations. Similarly to Eq. (4.19), we can express the correlation function in terms of a dynamic structure factor. However, in view of Eq. (4.54) we must consider here a partially Fourier transformed structure factor $S(\omega, \mathbf{q}_{\parallel}, z, z')$

such that[‡]:

$$\langle \delta T^*(\omega, \mathbf{q}_\parallel, z) \, \delta T(\omega', \mathbf{q}'_\parallel, z') \rangle = \rho m_0 \, \frac{(2\pi)^3}{\alpha_p^2 \rho^2}$$

$$\times \; S(\omega, q_\parallel, z, z') \, \delta(\omega - \omega') \, \delta(\mathbf{q}_\parallel - \mathbf{q}'_\parallel). \quad (4.56)$$

The presence of two delta functions is expected, since translational symmetries in time and in the horizontal plane imply that the correlation function $\langle \delta T^*(\mathbf{r}_\parallel, z, t) \cdot \delta T(\mathbf{r}'_\parallel, z', t') \rangle$ depends only on the differences $t - t'$ and $\mathbf{r}_\parallel - \mathbf{r}'_\parallel$. However, due to the presence of a temperature gradient and two boundaries, translational symmetry is broken in the vertical z-direction. In Eq. (4.56) the function $S(\omega, q_\parallel, z, z')$ can be represented by a double Fourier series in the variables z and z':

$$S(\omega, q_\parallel, z, z') = \frac{4k_{\rm B}\alpha_p^2 \overline{T}_0^2 \nu}{L^3 c_p m_0} \sum_{N,M=1}^{\infty} \frac{\tilde{C}_{NM}(q_\parallel) \, \sin\left(\frac{N\pi}{L}z\right) \, \sin\left(\frac{M\pi}{L}z'\right)}{[-i\omega + \Gamma_N(q_\parallel)] \, [i\omega + \Gamma_M(q_\parallel)]}. \quad (4.57)$$

Our goal here, as in previous sections, is to calculate the *equal-time* (or static) autocorrelation function, $\langle \delta T^*(\mathbf{q}_\parallel, z, t) \cdot \delta T(\mathbf{q}'_\parallel, z', t) \rangle$ of the temperature fluctuations. Applying a double inverse Fourier transformation in ω and ω' to Eq. (4.56), see Eq. (6.30), we obtain:

$$\langle \delta T^*(\mathbf{q}_\parallel, z, t) \cdot \delta T(\mathbf{q}'_\parallel, z', t) \rangle = \frac{m_0}{\rho \alpha_p^2} \, S(q_\parallel, z, z') \, (2\pi)^2 \, \delta(\mathbf{q}_\parallel - \mathbf{q}'_\parallel), \quad (4.58)$$

where $S(q_\parallel, z, z')$ is the static structure factor obtained by integration over the frequency of the dynamic structure factor, see Eq. (6.29). The ω integration of $S(\omega, q_\parallel, z, z')$ yields (Ortiz de Zárate and Sengers, 2004):

$$S(q_\parallel, z, z') = \frac{k_{\rm B}\alpha_p^2 \overline{T}_0^2}{c_p m_0} \left\{ \frac{T_0^2(z)}{\overline{T}_0^2} \, \delta(z - z') + \frac{\nabla T_0^2 L^2}{\overline{T}_0^2} \, S_{\rm NE}(q_\parallel, z, z') \right\}, \quad (4.59)$$

with

$$S_{\rm NE}(q_\parallel, z, z') = \frac{1}{L} \sum_{N=1}^{\infty} \frac{2}{N^2\pi^2 + \tilde{q}_\parallel^2} \, \sin\left(\frac{N\pi}{L}z\right) \, \sin\left(\frac{N\pi}{L}z'\right), \quad (4.60)$$

and where $\tilde{q}_\parallel = q_\parallel L$ is the same dimensionless wave number earlier introduced in Sect. 4.2. We observe in Eq. (4.59) that the equal-time autocorrelation function for the temperature fluctuations is expressed as a sum of a short-ranged local-equilibrium contribution and a long-ranged nonequilibrium contribution, the latter given by Eq. (4.60). The local-equilibrium

---

[‡]By adopting Eq. (4.46), we are implicitly assuming an equation of state of the from (4.3), *i.e.*, so that we do not consider any contribution from pressure fluctuations. See also Eq. (6.25)

contribution is the same as the one obtained from the theory of equilibrium fluctuations for the Rayleigh line, Sect. 3.3, but with the uniform equilibrium temperature replaced with the local nonequilibrium value, $T_0(\mathbf{r})$. The nonequilibrium contribution (4.60) to the temperature fluctuations arises from the fact that we have used a local version of the FDT, so that the right-hand side of Eq. (4.48) has become dependent on the coordinate z.

Since we have neglected here the presence of velocity fluctuations, the problem of fluctuations in a fluid subjected to a stationary ∇T_0 reduces to that of fluctuations in a system with simple heat conduction only, as has been considered by a number of authors. Actually, if we take the long wavelength $q_\| \to 0$ limit in Eq. (4.60), we reproduce previous results of Garcia et al. (1987) and of Malek Mansour et al. (1987), for the horizontal average of the temperature fluctuations. Alternatively, taking $q_\| = 0$ in Eq. (4.60) can be interpreted as reducing the original problem to one spatial dimension, in which case we reproduce the result of Breuer and Petruccione (1994), obtained by solving the one-dimensional Fokker-Planck equation corresponding to the Langevin equation (4.46) considered here.

The sum of the series in Eq. (4.60) may be performed exactly (Ortiz de Zárate and Sengers, 2004):

$$S_{\mathrm{NE}}(q_\|, z, z') = \frac{\cosh\left[\tilde{q}_\|\left(1 - \frac{|z-z'|}{L}\right)\right]}{2L\,\tilde{q}_\|\sinh(\tilde{q}_\|)} - \frac{\cosh\left[\tilde{q}_\|\left(1 - \frac{z+z'}{L}\right)\right]}{2L\,\tilde{q}_\|\sinh(\tilde{q}_\|)}. \tag{4.61}$$

We present here an exact expression for the horizontal wave number dependent correlation function, while in the literature some investigators have only considered the average of the correlation function over the horizontal plane (Garcia et al., 1987; Malek Mansour et al., 1987). The nonequilibrium contribution to the equal-time correlation function in real space is obtained by applying a double spatial inverse Fourier transformation to $\langle \delta T^*(\mathbf{q}_\|, z, t) \cdot \delta T^*(\mathbf{q}'_\|, z', t)\rangle$. As is to be expected from the translational symmetry of the problem, it will depend on the difference $\mathbf{r}_\| - \mathbf{r}'_\|$. Specifically, separating $\langle \delta T^*(\mathbf{r}, t) \cdot \delta T(\mathbf{r}', t)\rangle$ into an equilibrium part and a nonequilibrium part, we obtain for the nonequilibrium contribution to the intensity of the temperature fluctuations:

$$\langle \delta T^*(\mathbf{r}, t) \cdot \delta T(\mathbf{r}', t)\rangle_{\mathrm{NE}} = \tag{4.62}$$

$$\frac{2k_{\mathrm{B}}\nabla T_0^2 L}{\rho c_P}\int_0^\infty dq_\|\sum_{N=1}^\infty \frac{2\pi q_\| J_0(q_\| r_\|)}{N^2\pi^2 + \tilde{q}_\|^2}\,\sin\left(\tfrac{N\pi}{L}z\right)\,\sin\left(\tfrac{N\pi}{L}z'\right),$$

where $r_\|$ is the distance in the horizontal xy-plane between the two points $\mathbf{r}$ and $\mathbf{r}'$ at which the correlation function is evaluated. Equation (4.62) diverges when $\mathbf{r} = \mathbf{r}'$, indicating that, in three-dimensional heat conduction, some short-ranged nonequilibrium contribution is present in addition to the long-ranged contribution.

The divergence of $\langle \delta T^*(\mathbf{r}, t) \cdot \delta T(\mathbf{r}', t) \rangle_{\text{NE}}$ in real space at $\mathbf{r} = \mathbf{r}'$, is eliminated by an averaging procedure (Garcia et al., 1987; Malek Mansour et al., 1987). Horizontal averaging makes the three-dimensional problem equivalent to that of one-dimensional heat conduction in a segment $[0, L]$, studied by Breuer and Petruccione (1994), where such a divergence does not appear. The validity of Eq. (4.62) to describe the nonequilibrium temperature fluctuations, averaged over a horizontal plane ($q_\parallel = 0$), has been verified by numerically solving the stochastic heat-conduction equation (Garcia et al., 1987). For a further discussion of the literature on the subject the reader is referred to Ortiz de Zárate and Sengers (2004).

Our initial motivation for this section was to compare the contribution from the inhomogeneity of the thermal noise with the contribution from the nonequilibrium coupling between temperature and velocity fluctuations. We have deduced that, in the absence of coupling with the velocity, due exclusively to the fact that thermal noise is inhomogeneously correlated, the static structure factor $S(q_\parallel, z, z')$ of the nonequilibrium layer is given by Eq. (4.59), which can be rewritten as:

$$S(q_\parallel, z, z') = S_E \frac{\gamma - 1}{\gamma} \left\{ S_E(z, z') + \left(\frac{\nabla T_0 L}{\bar{T}_0} \right)^2 S_{\text{NE}}(q_\parallel, z, z') \right\}, \qquad (4.63)$$

where S_E is the same quantity defined by Eq. (3.28), but now referred to the average $\bar{T}_0$ temperature in the layer; $S_E(z, z')$ is a short-range local-equilibrium contribution, and $S_{\text{NE}}(q_\parallel, z, z')$ a dimensionless nonequilibrium enhancement. The enhancement of the intensity of the fluctuations due to mode-coupling was considered in Sect. 4.2 for the "bulk" fluid, and is proportional to the amplitude $\tilde{S}_{\text{NE}}^0$, defined by Eq. (4.30). In Chapter 7 we shall study the nonequilibrium enhancement of fluctuations due to mode-coupling and including effects due to the finite-size of the fluid layer. A technique similar to the one developed here shall be employed to obtain the same structure factor $S(q_\parallel, z, z')$ discussed in this section, namely, without Fourier transformation in the vertical variables. Comparing Eq. (7.4) with our current Eq. (4.63), we conclude that inhomogeneously correlated thermal noise causes an enhancement of nonequilibrium fluctuations proportional to $(\nabla T_0 L/\bar{T}_0)^2$, while the enhancement due to mode coupling is proportional to $\tilde{S}_{\text{NE}}^0$ like for the "bulk" fluid. Hence, the contribution from a nonequilibrium coupling between temperature and velocity fluctuations, as given by Eq. (4.30), relative to the contribution from the inhomogeneously correlated noise is determined by the dimensionless ratio:

$$\mathcal{R}_s = \tilde{S}_{\text{NE}}^0 \left(\frac{\bar{T}_0}{\nabla T_0 L} \right)^2 \simeq \frac{L^2 c_p \bar{T}_0}{a_T(\nu + a_T)}, \qquad (4.64)$$

where we have neglected the term proportional to the Rayleigh number in $\tilde{S}_{\text{NE}}^0$. If we consider a liquid layer with height $L = 2$ mm and if we adopt values of the physical quantities for toluene at 25°C (Li et al., 1994a), we find $\mathcal{R}_s \simeq 10^{13}$. We see that, while long-ranged correlations due to inhomogeneously correlated random forces exist in principle, they are totally negligible in practice as compared to the long-ranged correlations resulting from a coupling between the temperature and velocity fluctuations. This conclusion is supported by experimental measurements of the nonequilibrium temperature fluctuations in fluids (Law et al., 1990; Segrè et al., 1992), as reviewed in Sect. 10.2.

Chapter 5

Thermal nonequilibrium fluctuations in fluid mixtures

In the previous chapter we considered nonequilibrium fluctuations in a one-component fluid subjected to a stationary temperature gradient. In this chapter we shall consider the case of binary liquids. A stationary temperature gradient ∇T_0 in a liquid mixture induces a stationary concentration gradient ∇c_0 through the Soret effect (Tyrrell, 1961; de Groot and Mazur, 1962; Fitts, 1962; Haase, 1969; Lin et al., 1991). Just as a temperature gradient causes a coupling between temperature fluctuations and velocity fluctuations in the direction of the temperature gradient, so will a concentration gradient cause a coupling between concentration fluctuations and velocity fluctuations in the direction of the concentration gradient. Hence, in a liquid mixture, subjected to a temperature gradient, not only nonequilibrium temperature and viscous fluctuations will be present, but also nonequilibrium concentration fluctuations as first pointed out by Law and Nieuwoudt (Law and Nieuwoudt, 1989; Nieuwoudt and Law, 1990), and subsequently also analyzed by Segrè and Sengers (1993). Just as the nonequilibrium temperature fluctuations studied in Sect. 4.2 are proportional to the square of the temperature gradient ∇T_0 and inversely proportional to the fourth power of the wave number q, the new hydrodynamic nonequilibrium coupling in mixtures will cause the intensity of the nonequilibrium concentration fluctuations to be proportional to the square of the concentration gradient ∇c_0 and inversely proportional to q^4. At very small q, the nonequilibrium concentration fluctuations will be modified by gravity effects and by finite-size effects.

For the case of a binary mixture, the first source of nonequilibrium effects mentioned in Sect. 3.5, namely the spatial dependence of the noise correlation resulting from the local application of the FDT, has been considered by Velasco and García-Colín (1991). A nonequilibrium asymmetry in the Brillouin lines appears, similar to the one described in Sect. 4.3 for a one-component fluid. Again, this nonequilibrium effect depends on the scalar product $\mathbf{q} \cdot \nabla T_0$. Thus, the effect is maximum when $\mathbf{q} \parallel \nabla T_0$ and is zero when $\mathbf{q} \perp \nabla T_0$.

In this chapter we focus our attention on nonequilibrium temperature and concentration fluctuations that contribute to the Rayleigh components of the dynamic structure factor of mixtures. As in the derivation of the noise correlation matrix (4.21) for the Rayleigh line of a one-component fluid, we neglect the spatial variation of the thermophysical properties through their dependence on temperature, using again an "average" version of the FDT. For educational reasons, instead of considering the full (complicated) structure factor (Law and Nieuwoudt, 1989; Nieuwoudt and Law, 1990; Segrè and Sengers, 1993) and then adopt some simplifying approximations to retain only concentration and temperature fluctuations, we shall show how the intensity of the nonequilibrium Rayleigh line can be readily derived from the same linearized random Boussinesq equations commonly used for studying the convective instability in binary fluids (Schechter et al., 1972, 1974; Platten and Chavepeyer, 1975; Platten and Legros, 1984; Cross and Hohenberg, 1993; Hollinger and Lücke, 1998), including the buoyancy term and, thus, the effects of gravity. These Boussinesq equations for a binary mixture shall be presented in Sect. 5.1. Furthermore, this scheme helps making contact with the abundant literature dealing with the binary Boussinesq problem and its stability analysis, briefly reviewed in Sect. 5.1.1.

Moreover, if one is only interested in nonequilibrium concentration fluctuations, the Boussinesq equations can be further simplified by adopting a large-Lewis-number approximation, originally proposed by Velarde and Schechter (1972) while discussing the linear stability problem. Therefore, instead of solving the complete fluctuating hydrodynamics equations and then simplify the solution by retaining terms that are relevant experimentally, we shall make the simplification *a priori* by adopting a set of approximate fluctuating Boussinesq equations, as discussed in Sect. 5.2. The advantage of this approach is that elucidating the physical origin of the nonequilibrium concentration fluctuations and calculating their contribution to the structure becomes simple and straightforward. This approach also facilitates the inclusion of gravity effects to be discussed in this chapter, as well as the incorporation of finite-size effects on the nonequilibrium concentration fluctuations to be considered in Chapter 9. For completeness, we finalize this chapter in Sect. 5.3 by a comprehensive discussion of the Rayleigh component of the nonequilibrium structure factor of a binary mixture, including the contributions from viscous and thermal fluctuations. Such a comprehensive discussion will be based on the full Boussinesq equations for a binary mixture, without adopting the approximation of Velarde and Schechter (1972).

In the nonequilibrium thermodynamics of mixtures there does not exist a universally agreed upon nomenclature and sign convention for some of the thermophysical properties appearing in the theory. As mentioned in the Introduction, in this volume we are following the recommendations made in the book edited by Köhler and Wiegand (2002).

5.1 Linearized fluctuating Boussinesq equations for a binary liquid

As in Sect. 4.1 for a one-component fluid, we consider now a binary liquid between two parallel plates maintained at different temperatures. For a binary mixture we not only need an equation for the velocity fluctuations associated with the Navier-Stokes equation (2.51) and an equation for the temperature fluctuations associated with the heat equation (2.75), but also an equation for the concentration fluctuations associated with the diffusion equation (2.70). Hence, in addition to boundary conditions for the velocity and the temperature, we need a boundary condition for the concentration. The more realistic case is that of impermeable walls, Eq. (2.91), at $z = \pm\frac{1}{2}L$.

In the case of a binary mixture a Boussinesq approximation, similar to the one discussed in Sect. 4.1 for a one-component fluid, is usually introduced (Schechter et al., 1972, 1974; Platten and Chavepeyer, 1975; Cross and Hohenberg, 1993). The Boussinesq approximation for a mixture implies that one does not incorporate any dependence of the density on pressure, but only its dependence on the temperature and on the concentration. Thus instead of Eq. (4.3) for a one-component fluid, we now assume that the equation of state of the mixture is represented by:

$$\rho(\mathbf{r}, t) = \bar{\rho}_0\{1 - \alpha_p[T(\mathbf{r}, t) - \bar{T}_0] + \beta[c(\mathbf{r}, t) - \bar{c}_0]\}, \tag{5.1}$$

where β is the solutal expansion coefficient, defined by:

$$\beta = \frac{1}{\rho}\left(\frac{\partial\rho}{\partial c}\right)_{T,p}. \tag{5.2}$$

We note that β is positive when component 1 (the one employed to define the concentration c) is the heavier component and β is negative when component 1 is the lighter component. In Eq. (5.1), $\bar{T}_0$ and $\bar{c}_0$ are some reference temperature and concentration, which we will take at the center of the layer ($z = 0$). Since in the Boussinesq approximation all thermophysical properties are assumed to be constants, $\bar{T}_0$ and $\bar{c}_0$ will equal the averages temperature and concentration in the layer, respectively.

When the equation of state (5.1) is substituted into the balance laws, the Boussinesq equations for a binary mixture can be obtained by a procedure similar to the one developed in Sect. 4.1. We refer the interested reader to the relevant literature for details (Schechter et al., 1972, 1974; Platten and Chavepeyer, 1975; Cross and Hohenberg, 1993). We just mention that, in addition to the small parameters considered in Sect. 4.1, the Boussinesq approximation for mixtures also requires the parameter $\epsilon_3 = \beta \Delta c$ (with Δc being the concentration difference across the fluid layer) to be small. Just as for a one-component fluid, neglecting any pressure dependence of the density causes the continuity equation simplify to a divergence-free velocity field: $\nabla \cdot \mathbf{v} = 0$. In Eq. (2.72) for the rate of change of the entropy density, contributions arise from both the random energy flow $\delta \mathbf{Q}'$ and the random diffusion flow $\delta \mathbf{J}$. However, when the thermodynamic relation (2.74) is combined with the *random* diffusion equation, the fluctuating diffusion flux does not contribute when the heat equation is formulated in terms of temperature instead of entropy (see Eq. (5.5b) below).

Even with the additional diffusion equation, the stationary functions represented by Eqs. (4.10) continue to be a solution of the Boussinesq equations for a mixture, provided that a stationary uniform gradient ∇c_0 of the solute concentration develops such that:

$$\nabla c_0 = -\bar{c}_0 \left(1 - \bar{c}_0\right) S_T \, \nabla T_0 = -\frac{k_T}{\bar{T}_0} \, \nabla T_0. \qquad (5.3)$$

Equation (5.3) defines the Soret coefficient S_T of component 1 (solute) in component 2 (solvent) in accordance with Eq. (2.68). For isotropic mixtures, S_T is a scalar quantity and the induced concentration gradient is parallel to the imposed temperature gradient, but it can have the same or the opposite direction, depending on the sign of S_T. When S_T is positive, the concentration and temperature gradients have opposite directions, with component 1 migrating to the colder region. When S_T is negative, the concentration and temperature gradients have the same direction, with component 1 migrating to the warmer region. If c were to represent the concentration of component 2, then Eq. (5.3) would define the Soret coefficient S_T of component 2 in component 1. Conservation of mass implies that when S_T of component 1 in component 2 is positive, the S_T of component 2 in component 1 must be negative and *vice versa*. In practice, it is convenient to define a dimensionless separation ratio ψ by:

$$\psi = \bar{c}_0 \left(1 - \bar{c}_0\right) S_T \, \frac{\beta}{\alpha_p} = \frac{\beta k_T}{\alpha_p \bar{T}_0} = -\frac{\beta}{\alpha_p} \frac{\nabla c_0}{\nabla T_0}. \qquad (5.4)$$

The parameter ψ represents the ratio of the density gradient $\beta \, \nabla c_0$ produced by the concentration gradient and the density gradient $-\alpha_p \, \nabla T_0$ produced by the temperature gradient. Introducing this separation ratio ψ has the advantage that its sign is independent of which of the two components is actually chosen to represent the concentration c of the mixture.

As in the case of a one-component fluid, the stationary solution represented by Eqs. (4.10) and (5.3) is referred to as the *conductive* solution. Next, we need to derive the evolution equations for fluctuations around this conductive solution on the basis of fluctuating hydrodynamics. For small fluctuations we can adopt the linearized fluctuating Boussinesq equations that for a binary fluid are given by (Ortiz de Zárate et al., 2004):

$$\partial_t \left(\nabla^2 \delta v_z \right) = \nu \, \nabla^2 \left(\nabla^2 \delta v_z \right) + g \, \left(\partial_x^2 + \partial_y^2 \right) \left[\alpha_p \, \delta T - \beta \, \delta c \right]$$

$$+ \frac{1}{\rho} \left\{ \nabla \times \left[\nabla \times \left(\nabla \cdot \delta \Pi \right) \right] \right\}_z , \quad (5.5a)$$

$$\partial_t \, \delta T = \left[a_T + \epsilon_D D \right] \nabla^2 \delta T + \frac{D \beta \epsilon_D}{\alpha_p \psi} \, \nabla^2 \delta c - \delta v_z \nabla T_0 - \frac{a_T}{\lambda} \nabla \cdot \delta Q', \quad (5.5b)$$

$$\partial_t \, \delta c = D \left[\nabla^2 \delta c + \frac{\alpha_p}{\beta} \psi \, \nabla^2 \delta T \right] - \delta v_z \, \nabla c_0 - \frac{1}{\rho} \nabla \cdot \delta J. \quad (5.5c)$$

The notation ∇T_0 and ∇c_0 for the magnitude of the stationary gradients of temperature and concentration reminds us that both ∇T_0 and ∇c_0 in Eqs. (5.5) are constants and not gradients of the locally fluctuating temperature and concentration. As in the fluctuating Boussinesq equations for a one-component fluid, we have included in Eqs. (5.5) random dissipative fluxes to represent the contributions from rapidly varying short-ranged fluctuations. Thus, $\delta \Pi(\mathbf{r}, t)$ represents the random (symmetric) deviatoric stress tensor, $\delta Q'(\mathbf{r}, t)$ represents the random heat (energy) flux and $\delta J(\mathbf{r}, t)$ represents the random mass-diffusion flux. Again, as in Eq. (4.12), the subscript z in Eq. (5.5a) indicates that the random-noise term in this equation is to be identified with the z-component of the vector between curly brackets. The linearized random Boussinesq equations (5.5) assume that the thermophysical properties of the binary fluid depend sufficiently weakly on temperature or concentration, so that the variation of these properties as a function of z is negligible small. Thus the density ρ and all thermophysical properties in Eqs. (5.5) are identified with their average value.

Notice the presence in Eq. (5.5b) of the Dufour effect ratio ϵ_D, defined by Eq. (2.76). In most presentations of the binary Boussinesq equations the Dufour effect is neglected and $\epsilon_D = 0$ is substituted in the heat equation. In practice, the Dufour effect is only relevant in gas mixtures and in mixtures near the vapor-liquid critical point (Hollinger and Lücke, 1998; Hort et al., 1992; Bodenschatz et al., 2000). We have retained the contribution from the Dufour effect for later comparison with the work of Law and Nieuwoudt (1989).

Again, as in Sect. 4.2 for a one-component fluid, following Chandrasekhar (1961), we have found it convenient to eliminate the hydrostatic pressure gradient by taking a double curl in the equation for the fluctuating fluid velocity $\delta\mathbf{v}$, yielding Eq. (5.5a) for $\nabla^2 v_z$.

5.1.1 The linear stability problem

As already discussed in Chapter 4 for a one-component fluid, the linearized equations of fluctuating hydrodynamics are adequate as long as the nonequilibrium steady state around which fluctuations are analyzed is stable. Before proceeding any further, we shall briefly discuss the stability of the conductive solution for a binary fluid layer between two plates. This will provide us with an assessment of the range of validity of Eqs. (5.5) for fluctuations around the conductive state.

A considerable amount of information about the stability of the conductive solution for a binary fluid mixture is available in the literature (Schechter et al., 1972, 1974; Platten and Chavepeyer, 1975; Platten and Legros, 1984; Cross and Hohenberg, 1993; Hollinger and Lücke, 1998; Hollinger et al., 1998). An excellent and updated review by Lücke et al. (1998) includes a nonlinear stability analysis. The conclusion is that the stability of the conductive solution depends on whether the temperature and concentration gradients are stabilizing or not. That is, stability will depend on the sign of the Rayleigh number Ra and on the sign of the separation ratio ψ. The Rayleigh number has the same definition, Eq. (4.8), as for a one-component fluid; thus Ra is negative when the fluid is heated from above, *i.e.*, when the temperature gradient is opposite to the direction of the gravitational force $\mathbf{g}$. Other relevant dimensionless parameters for binary fluids are the Schmidt number Sc and the Lewis number Le:

$$Sc = v/D, \qquad\qquad Le = a_T/D. \tag{5.6}$$

The discussion in this section refers to the Boussinesq equations without the Dufour effect, thus $\epsilon_D \simeq 0$ in Eq. (5.5b).

To summarize the situation we show in Fig. 5.1 a typical linear stability diagram in the $\{Ra, \psi\}$ plane, plotted after Schechter et al. (1974). The stability diagram depends on the values of the Schmidt and Prandtl numbers. The diagram in Fig. 5.1 corresponds to $Sc=1000$ and $Pr =10$, which are typical values for an ordinary liquid mixture. Depending on the signs of Ra and ψ one can identify four regions in the diagram of Fig. 5.1:

1. If $Ra > 0$ and $\psi < 0$, the temperature gradient is destabilizing whereas the Soret-induced concentration gradient is weakly stabilizing. Note that if $\beta > 0$, S_T has to be negative to have an overall $\psi < 0$, see Eq. (5.4), so that the denser component 1 migrates to the warm lower plate. If $\beta < 0$, the lighter component 1 migrates to the cold upper plate. In both cases the Soret effect is weakly stabilizing. In this situation, the instability appears at a nonzero horizontal wave

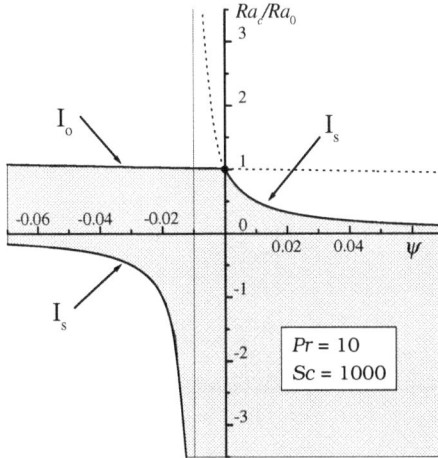

Figure 5.1: Linear stability diagram deduced from the Boussinesq equations with two free and permeable boundaries for a binary liquid with typical values of the Schmidt and Prandtl numbers, according to Schechter et al. (1974). The region of stability corresponds to the shadowed area.

number and at a nonzero frequency. The instability mechanism is an oscillatory periodic Hopf bifurcation or, in the language of Cross and Hohenberg (1993), an I_o-type instability. Above the threshold, the instability leads to a pattern of travelling waves (Schöpf and Zimmermann, 1993).

2. When $Ra > 0$ and $\psi > 0$, both the temperature and the concentration gradient are destabilizing. At given ψ the instability appears at a lower Ra than in the case of a one-component fluid ($\psi = 0$). In this situation the instability appears at a finite wave number but at zero frequency, as in a one-component fluid: it is an I_s-type instability. Above the threshold, the instability leads to a square pattern.

3. When $Ra < 0$ and $\psi < 0$, the temperature gradient is stabilizing, but the Soret effect is destabilizing. For given Ra, the separation ratio ψ could be negative enough for an instability to appear. Note that this instability develops while heating the fluid from above, a configuration in which a one-component fluid is always stable. As above, it is an I_s-type instability and it was recently observed experimentally by La Porta and Surko (1998).

4. Finally, when $Ra < 0$ and $\psi > 0$, both temperature and concentration

gradients are stabilizing and the system is always in a quiescent stable state.

5. It is interesting to note the existence of a codimension-two instability point where the I_o-instability line meets with the I_s-instability line. In the linear approximation this happens on the $\psi = 0$ axis (Cross and Kim, 1988).

The instability mechanisms described above are also referred to as bifurcations. This is because, when the nonlinear terms are taken into account, once the instability threshold is passed, new solutions to the Boussinesq equations can be identified (Cross and Hohenberg, 1993). The appearance of convection is usually explained as an exchange of stability between the conductive solution and the new (nonlinear) convective solutions.

The actual instability diagram depends on the boundary conditions. The plot in Fig. 5.1 corresponds to free and permeable boundaries, which is the simplest case mathematically. For this particular case, the critical Rayleigh number Ra_c for the I_o-instability line as a function of the separation ratio is given by (Schechter et al., 1974):

$$\left(\frac{Ra_c}{Ra_0}\right)_o = \frac{(1 + Pr)(Sc + Pr)(1 + Sc)}{Sc^2(\psi\, Pr + Pr + 1)}, \tag{5.7}$$

while the critical Rayleigh number for the I_s-instability line is given by (Schechter et al., 1974):

$$\left(\frac{Ra_c}{Ra_0}\right)_s = \frac{Pr}{Pr + \psi\,(Sc + Pr)} \tag{5.8}$$

with $Ra_0 = 27\pi^4/4$, the critical Rayleigh number for a one-component fluid with two free boundaries conditions (see Sect. 6.2). Fig. 5.1 has been constructed from Eqs. (5.7) and (5.8) The case of two permeable free interfaces is rather unrealistic. The more realistic case of fixed and impermeable boundaries has also been analyzed (Legros et al., 1972; Platten and Legros, 1984). But, as in the case of a one-component fluid, different boundary conditions cause only quantitative differences while qualitatively the stability diagram remains similar. Hence, the simpler case of two permeable walls is useful from a pedagogical point of view. Nonlinear terms, that have been neglected in Eqs. (5.5), cause also minor changes in the stability diagram, specifically shifting the position of the codimension-two instability point as discussed by Cross and Kim (1988). Nevertheless, the preceding discussion of the relationship between stability and the signs of Ra and ψ still applies.

In the present chapter we shall calculate the nonequilibrium structure factor of the fluid without considering any boundary conditions for the fluctuations, just as we did in Chapter 4 for a one-component fluid. Without any boundary conditions, the expression for the structure factor of a binary fluid to be obtained in this chapter will have a range of validity that does not extend over the entire linear-stability region, just as the validity of Eq. (4.29) for a one-component fluid at the end of Sect. 4.2 was restricted to negative Rayleigh numbers. Similarly, we restrict ourselves in the remainder of this chapter to a binary fluid with $Ra < 0$ and $\psi > 0$, for which fluctuations do not drive the system out of a stable quiescent state even in the absence of any boundary conditions. The effects of boundary conditions on the nonequilibrium fluctuations in a binary fluid will be addressed in Chapter 9.

5.2 Structure factor in a large-Lewis-number approximation

The structure factor of a binary liquid can be calculated by solving the three coupled differential equations (5.5), as was done by Law and Nieuwoudt (1989) and by Segrè and Sengers (1993). However, there are several reasons why a calculation for a binary mixture becomes more complicated than for a one-component fluid. A first complication arises from the fact that the random heat flux $\delta\mathbf{Q}'$ and the random diffusion flux $\delta\mathbf{J}$ have the same tensorial character, so that their cross-correlations are, in general, not zero: $\langle \delta Q'_i \cdot \delta J_j \rangle \neq 0$, see Eqs. (3.9) and (3.10). As discussed in formulating the FDT for a binary system in Sect. 3.2.2, these cross-correlations correspond to the Soret and Dufour effects. A second complication is that now both temperature *and* concentration fluctuations contribute to the refractive-index fluctuations. Hence, these two types of fluctuations have to be calculated, contrary to the case of a one-component liquid in Sect. 4.2, where the Rayleigh component of the structure factor results only from temperature fluctuations.

However, in practice, the diffusion coefficient D of binary liquid mixtures is much smaller than the thermometric (thermal) diffusivity a_T and the kinematic viscosity ν. Hence, to obtain the structure factor of a binary liquid system we can consider to a very good approximation the limit of large Lewis number. This is a particularly good approximation for the case of macromolecular solutions (Sengers and Ortiz de Zárate, 2002). Since $\nu > a_T$, a large value of the Lewis number Le also implies a large value of the Schmidt number, see Eq. (5.6). In the limit $Le \to \infty$ the linearized fluctuating Boussinesq equations (5.5) reduce to (Velarde and Schechter, 1972):

$$0 = \nu \, \nabla^4 \delta v_z - \beta g \, \left(\partial_x^2 + \partial_y^2 \right) \delta c + \frac{1}{\rho} \left\{ \boldsymbol{\nabla} \times \left[\boldsymbol{\nabla} \times (\boldsymbol{\nabla} \cdot \delta\Pi) \right] \right\}_z, \quad (5.9a)$$

$$\partial_t \, \delta c = D \, \nabla^2 \delta c - \delta v_z \, \nabla c_0 - \frac{1}{\rho} \, \boldsymbol{\nabla} \cdot \delta\mathbf{J}. \quad (5.9b)$$

We see that, in the limit $Le \to \infty$, any coupling between the concentration fluctuations and the temperature fluctuations vanishes and we only need to consider coupling between the concentration fluctuations and the fluctuations of the vertical velocity (viscous fluctuations). Moreover, since the viscous fluctuations decay much faster than the concentration fluctuations, we do not need to include the direct effect of the viscous mode on the dynamic structure factor, but only its indirect effect on the concentration fluctuations. That is, in the limit $Sc \to \infty$, the time derivative $\partial/\partial t(\nabla^2 \delta v_z)$ on the left-hand side of (5.9a) can be neglected.

We may mention that the large-Lewis-number approximation[‡] has been amply used in the literature to simplify the study of the stability of binary liquid mixtures (Velarde and Schechter, 1972; Schechter et al., 1974; de Giorgio, 1978; Cross and Kim, 1988). As discussed by Velarde and Schechter (1972) and by Schechter et al. (1974), the $Le \to \infty$ approximation has several shortcomings; in particular, it does not describe correctly the situation with $Ra > 0$ and $\psi < 0$, missing the interesting Hopf bifurcation and codimension-two instability point. But for $Ra < 0$ and $\psi > 0$, i.e., for a quiescent stable thermally conducting state, the $Le \to \infty$ approximation will turn out to be adequate for describing the concentration fluctuations.

Starting from Eqs. (5.9) we can readily calculate the structure factor of the binary liquid, following a procedure similar to the one adopted in previous chapters for the case of a one-component fluid. In the absence of any boundary conditions we apply a temporal and spatial Fourier transformation to Eqs. (5.9) so as to obtain:

$$\begin{pmatrix} \nu\, q^4 & \beta\, g\, q_{\parallel}^2 \\ \nabla c_0 & i\omega + D\, q^2 \end{pmatrix} \begin{pmatrix} \delta v_z(\omega, \mathbf{q}) \\ \delta c(\omega, \mathbf{q}) \end{pmatrix} = \mathbf{F}(\omega, \mathbf{q}), \tag{5.10}$$

with $q_{\parallel}$ again representing the magnitude of the wave vector $\mathbf{q}$ in the xy-plane, i.e., $q_{\parallel}^2 = q_x^2 + q_y^2$. The random-noise vector $\mathbf{F}(\omega, \mathbf{q})$ is now related to the Fourier transforms $\delta\Pi(\omega, \mathbf{q})$ of the random stress tensor and $\delta\mathbf{J}(\omega, \mathbf{q})$ of the random diffusion flux by:

$$\mathbf{F}(\omega, \mathbf{q}) = \frac{-1}{\rho_0} \begin{pmatrix} i\left[q_z\, q_i q_j \delta\Pi_{ij}(\omega, \mathbf{q}) - q^2\, q_i \delta\Pi_{iz}(\omega, \mathbf{q}) \right] \\ i\, q_i \delta J_i(\omega, \mathbf{q}) \end{pmatrix}. \tag{5.11}$$

The first component of $\mathbf{F}(\omega, \mathbf{q})$ is the same as the one previously considered in Eq. (4.15) for a one-component fluid. To deduce the autocorrelation $\langle \delta c^*(\omega, \mathbf{q}) \cdot \delta c(\omega', \mathbf{q}') \rangle$ of the concentration fluctuations from Eq. (5.10), we need the correlation functions among the components of the noise term $\mathbf{F}(\omega, \mathbf{q})$. These can be calculated from the FDT for a binary system, given

---

[‡]Notice that some authors (Velarde and Schechter, 1972; Sengers and Ortiz de Zárate, 2002) have adopted a definition of the Le number that is the inverse of the one defined in Eq. (5.6).

by Eqs. (3.9) and (3.10). Note that as a consequence of the $Le \to \infty$ approximation, the random heat flux $\delta \mathbf{Q}'$ does not appear in the random-noise term. To maintain consistency we neglect the cross-correlation between the random heat and diffusion fluxes, which, as discussed by Schöpf and Zimmermann (1993) can be neglected in most practical circumstances. A straightforward calculation then yields the usual form, Eqs. (3.22a) or (4.20), where now the correlation matrix is given by:

$$
\mathsf{C}(\mathbf{q}) = \frac{2k_B \bar{T}_0}{\rho} \begin{pmatrix} \nu\, q_\parallel^2\, q^4 & 0 \\ 0 & \left(\dfrac{\partial c}{\partial \mu}\right)_{p,T} Dq^2 \end{pmatrix}.
\tag{5.12}
$$

The autocorrelation function of the first component of the random noise $\mathbf{F}(\omega, \mathbf{q})$, i.e., the element $(0,0)$ of the correlation matrix (5.12), is the same as in Eq. (4.21) for a one-component fluid. The non-diagonal terms of (5.12) are zero because the random diffusion flux and the random stress are uncorrelated, see Sect. 3.2.2. It should be remembered that we are neglecting in this section any nonequilibrium effect that might arise from inhomogeneously correlated random noise terms $(T(\mathbf{r}) \approx \bar{T}_0)$. As discussed in Sect. 4.4, when a coupling between fluctuating fields exists, as here in Eqs. (5.9), such effects are expected to be negligible.

In the large-Lewis-number approximation, the temperature fluctuations are neglected and, according to Eq. (5.1), the Rayleigh component of the structure factor can be attributed to fluctuations of the refractive index n caused by the concentration fluctuations only. In addition, from Eq. (4.20) we note that the autocorrelation function is proportional to a delta function $\delta(\mathbf{q} - \mathbf{q}')$, unlike the autocorrelation function (4.38) for the Brillouin components when a temperature gradient is present. Hence, the relationship between the dynamic structure factor $S(\omega, \mathbf{q})$ and the autocorrelation function of the concentration fluctuations takes the simple form (see Sect. 10.1.1):

$$
\langle \delta c^*(\omega, \mathbf{q}) \cdot \delta c(\omega', \mathbf{q}') \rangle = \left(\frac{\partial c}{\partial n}\right)_T^2 S(\omega, \mathbf{q}) \, (2\pi)^4 \, \delta(\omega - \omega') \, \delta(\mathbf{q} - \mathbf{q}').
\tag{5.13}
$$

Inverting Eq. (5.10) and using Eqs. (5.12) and (5.13), one obtains for the Rayleigh component of the dynamic structure factor $S(\omega, \mathbf{q})$ of a binary liquid in the large-Le approximation:

$$
S(\omega, \mathbf{q}) = \left(\frac{\partial n}{\partial c}\right)_T^2 \frac{k_B \bar{T}_0}{\rho} \left[\left(\frac{\partial c}{\partial \mu}\right)_T + \frac{(\nabla c_0)^2\, q_\parallel^2}{\nu D\, q^6}\right] \frac{2Dq^2}{\omega^2 + [\Gamma_D(\mathbf{q})]^2}
\tag{5.14}
$$

with a single decay rate given by

$$
\Gamma_D(\mathbf{q}) = Dq^2 \left(1 - \psi RaLe\, \frac{\tilde{q}_\parallel^2}{\tilde{q}^6}\right).
\tag{5.15}
$$

In Eq. (5.15) for the decay rate $\Gamma_D(\mathbf{q})$ we have again introduced a dimensionless wave number $\tilde{q} = qL$, so as to express the effect of the gravity g through the Rayleigh number defined by Eq. (4.8). When buoyancy is neglected, or when $\tilde{q} \gg 1$, the decay rate $\Gamma_D(\mathbf{q})$ of the concentration fluctuations reduces to Dq^2, which is the same decay rate as that of equilibrium concentration fluctuations (Berne and Pecora, 1976). The quantity $\psi RaLe$ in Eq. (5.15) is commonly referred to as the concentration Rayleigh number, in terms of thermophysical properties is expressed as:

$$\psi RaLe = -\psi \frac{a_T}{D} \frac{\alpha_p g L^4 \, \nabla T_0}{\nu \, a_T} = \frac{\beta g L^4 \, \nabla c_0}{\nu \, D}. \tag{5.16}$$

A convective instability will appear when the expression (5.14) for $S(\omega, \mathbf{q})$ diverges at some finite value of q, or equivalently, when the decay rate $\Gamma_D(\mathbf{q})$ is zero. From (5.16) we conclude that for such a divergence to occur $\beta \, \nabla c_0$ must be positive, *i.e.*, the concentration of the heavier component must increase in the direction opposite to gravity. This leads to the so-called salt-fountain, double-diffusive, or salt-fingers instability (Schechter et al., 1974; Cross and Kim, 1988; Hollinger et al., 1998). Here we consider a concentration gradient induced by the Soret effect, but this instability can also be studied in the presence of an isothermal stationary concentration gradient, in which case it is referred to as the thermohaline problem (Schechter et al., 1974; Turner, 1985). However, in the present chapter we assume $Ra < 0$ (thus $\alpha_p \nabla T_0 > 0$) and $\psi > 0$, so that there are no divergences in Eq. (5.14) for any value of q.

After applying an inverse Fourier transformation in the frequency ω to Eq. (5.14), we find that the time correlation function of the scattered-light intensity (see Sect. 10.1.1) is proportional to:

$$S(\mathbf{q}, \tau) = S_{\mathrm{E},c} \{1 + A_c(\mathbf{q})\} \, \exp\left[-\Gamma_D(\mathbf{q}) \, \tau\right], \tag{5.17}$$

where $\tau = |t - t'|$ is the (absolute) difference between the two times for which the correlation is calculated, see Eq. (3.29). In Eq. (5.17)

$$S_{\mathrm{E},c} = \left(\frac{\partial n}{\partial c}\right)_T^2 \frac{k_{\mathrm{B}} \bar{T}_0}{\rho} \left(\frac{\partial c}{\partial \mu}\right)_T \tag{5.18}$$

and[‡]

$$A_c(\mathbf{q}) = \left[\left(\frac{\partial \mu}{\partial c}\right)_T \frac{(\nabla c_0)^2}{\nu \, D} + \frac{\beta g \, \nabla c_0}{\nu \, D}\right] \frac{1}{1 - \psi RaLe \, \dfrac{\tilde{q}_\parallel^2}{\tilde{q}^6}} \frac{q_\parallel^2}{q^6}. \tag{5.19}$$

[‡]Notice that there is a sign error inside the brackets in Eq. (15) of Sengers and Ortiz de Zárate (2002)

Evaluating Eq. (5.17) at $\tau = 0$, we obtain the static structure factor of the fluid. Therefore the coefficient $S_{E,c}$ represents the structure factor of a mixture in thermodynamic equilibrium, *i.e.*, for $\nabla c_0 = 0$ (Boon and Yip, 1980; Berne and Pecora, 1976). Note that gravity does not appear in the structure factor at $\nabla c_0 = 0$. Just as for a one-component fluid, gravity modifies the dynamic structure factor of a fluid in equilibrium, but not the static structure factor (Segrè et al., 1993b; Segrè and Sengers, 1993).

The dimensionless quantity A_c in Eq. (5.17) is zero in equilibrium ($\nabla c_0 = 0$), so that it represents a dimensionless nonequilibrium enhancement of the concentration fluctuations (Segrè et al., 1993a; Li et al., 1998). The term proportional to ∇c_0 inside the brackets in the LHS of Eq. (5.19) is for common mixtures and gradients negligibly small, so that the nonequilibrium enhancement of the structure factor is, in practice, proportional to the square of the concentration gradient.

Just as Eq. (5.14) for the dynamic structure factor, Eq. (5.19) will be valid only in the $\{Ra < 0, \psi > 0\}$ quadrant of the stability diagram in Fig. 5.1. In any of the other cases a Soret-driven double diffusive instability will appear as a divergence in the structure factor at some finite value of $q_{\parallel}$. To obtain a more general expression for $S(\mathbf{q})$, valid not only for negative Ra, but also for a finite interval of positive Ra, up to a certain critical value Ra_c, we need to incorporate the effects of boundary conditions, as will be done in Chapter 9. An additional problem is that Eq. (5.19) has also no divergence for $Ra > 0$ and $\psi < 0$, where a instability in the form of a Hopf bifurcation is expected (Schechter et al., 1974; Cross and Hohenberg, 1993; Lücke et al., 1998; Cross and Kim, 1988). The fact that we do not find this instability here is a consequence of the $Le \to \infty$ approximation (Velarde and Schechter, 1972) which neglects Hopf fluctuations. How the presence of a Hopf instability affects the nonequilibrium structure factor will be also addressed in Chapter 9.

For the interpretation of light-scattering experiments in liquids far away from any convective instability we may neglect the effects of gravity. In the limit $g \to 0$, the two-times structure factor $S(\mathbf{q}, \tau)$ continues to be represented by Eq. (5.17), but now Eq. (5.19) reduces to:

$$A_c(\mathbf{q}) = \left(\frac{\partial \mu}{\partial c}\right)_T \frac{(\nabla c_0)^2}{\nu D} \frac{q_{\parallel}^2}{q^6}, \tag{5.20}$$

while the decay rate of concentration fluctuations reduces to $\Gamma_D(q) = Dq^2$. Expression (5.20) for the enhancement of the nonequilibrium concentration fluctuations has been checked experimentally as will be discussed in Chapter 10. Note that Eq. (5.20), in contrast to Eq. (5.19), does not include any divergence and is valid for any positive or negative value of ∇c_0. This is to be expected since in the derivation of Eq. (5.20) buoyancy was neglected, so that the system is always in a stable quiescent state. We note that the

q^{-4} proportionality of the structure factor implied by Eq. (5.20) may also be recovered as the first term of an asymptotic expansion for $q \to \infty$ of Eq. (5.19), because the effects of gravity are only important for fluctuations with small wave number q. For $qL \gg 1$ the effects of gravity are negligible.

We finally remark that we could have obtained the nonequilibrium structure factor from Eq. (5.9) just as easily without neglecting the time derivative $\partial/\partial t(\nabla^2 w)$ in the left-hand side of Eq. (5.9a). Instead of (5.20) one then obtains:

$$A_c(\mathbf{q}) = \left(\frac{\partial \mu}{\partial c}\right)_T \frac{(\nabla c_0)^2}{(\nu^2 - D^2)} \frac{\nu}{D} \frac{q_\parallel^2}{q^6},\tag{5.21}$$

which for $\nu \gg D$ reduces to (5.20). However, we have preferred to neglect the effect of the time derivative of $\nabla^2 w$ to retain consistency with the hydrodynamic approximations originally introduced by Velarde and Schechter (1972).

5.3   Comprehensive structure factor of a non-equilibrium binary mixture

By adopting a large-Lewis-number approximation in the previous section we were able to present a simple derivation of the structure factor associated with concentration fluctuations resulting from the nonequilibrium coupling with velocity fluctuations. However, in general, temperature fluctuations (neglected in the previous section) will also contribute to the structure factor. In this section we review the more complete theoretical expressions for the Rayleigh component of the dynamic structure factor of a binary fluid including contributions from temperature and velocity fluctuations.

In this section we first review the expressions for the dynamic structure factor without including any gravity effects as originally obtained by Law and Nieuwoudt (1989). We then shall make some comments how the results are affected by gravity. If we apply a spatiotemporal Fourier transformation to the random Boussinesq equations (5.5) with $g = 0$, the resulting algebraic equations may be expressed in the typical form:

$$\mathsf{G}^{-1}(\omega, q) \begin{pmatrix} \delta v_z(\omega, q) \\ \delta T(\omega, q) \\ \delta c(\omega, q) \end{pmatrix} = \mathbf{F}(\omega, q),\tag{5.22}$$

where the inverse linear response function for our current problem is given by

$$G^{-1}(\omega, q) = \begin{pmatrix} q^2[i\omega + \nu q^2] & 0 & 0 \\ \nabla T_0 & i\omega + (a_T + D\epsilon_D)q^2 & \dfrac{D\beta\epsilon_D}{\alpha_p\psi}q^2 \\ \nabla c_0 & \dfrac{D\alpha_p\psi}{\beta}q^2 & i\omega + Dq^2 \end{pmatrix}, \qquad (5.23)$$

while the vector $\mathbf{F}(\omega, q)$ of random noise is expressed in terms of the Fourier transforms of the dissipative fluxes as:

$$\mathbf{F}(\omega, q) = \dfrac{-1}{\rho_0} \begin{pmatrix} i\left[q_z\, q_iq_j\delta\Pi_{ij}(\omega, \mathbf{q}) - q^2\, q_i\delta\Pi_{iz}(\omega, \mathbf{q})\right] \\ i\,(c_p)^{-1}q_i\delta Q_i'(\omega, q) \\ i\,q_i\delta J_i(\omega, q) \end{pmatrix}. \qquad (5.24)$$

To solve Eq. (5.22) for the fluctuating fields, we need to compute the linear response function $G(\omega, q)$ by inverting Eq. (5.23). For this purpose we need the determinant of $G^{-1}(\omega, q)$, which now is a cubic polynomial in the frequency ω. This determinant is conveniently expressed in terms of decay rates as (Law and Nieuwoudt, 1989):

$$\det[G^{-1}(\omega, q)] = q^2\,(i\omega + \Gamma_\nu)\,(i\omega + \Gamma_+)\,(i\omega + \Gamma_-)\,, \qquad (5.25)$$

where Γ_ν, Γ_+ and Γ_- are the three ω-roots of the determinant. Solving for the roots of $\det[G^{-1}]$, one finds:

$$\Gamma_\nu = \nu q^2$$

$$\Gamma_\pm = \left\{\dfrac{(a_T + \mathcal{D})}{2} \pm \dfrac{1}{2}\sqrt{(a_T + \mathcal{D})^2 - 4\mathcal{D}a_T}\right\}q^2 \qquad (5.26)$$

with

$$\mathcal{D} = D(1 + \epsilon_D). \qquad (5.27)$$

In the limit $\epsilon_D = 0$ the decay rates $\Gamma_\pm$ reduce to a thermal decay rate $\Gamma_+ = a_T q^2$ and a diffusion decay rate $\Gamma_- = Dq^2$. However, for arbitrary $\epsilon_D \neq 0$ the thermal and diffusion decay rates are generically coupled. We note that such a coupling, although dependent on the Soret effect, is also present in binary mixtures in equilibrium (Berne and Pecora, 1976; Boon and Yip, 1980). We conclude that when buoyancy is neglected, the decay rates of the nonequilibrium fluctuations are equal to the decay rates of equilibrium fluctuations, as was also the case for a one-component fluid discussed after Eq. (4.26).

The coupling between thermal and diffusion decay rates predicted by Eq. (5.26) is usually referred to as two-exponential decay, and has been recently studied by Anisimov et al. (1998). Such a coupling has been observed experimentally only in supercritical fluid mixtures not far from the consolute critical point (Ackerson and Hanley, 1980; Fröba et al., 2000); under these conditions the Dufour effect is more important and the coupling more significant. For noncritical fluid mixtures the two decay rates of the fluctuations, whether equilibrium or nonequilibrium, can be safely identified with the thermal and the diffusion decay rates.

From Eq. (5.17) in the previous section, we note that the time correlation function $S(\mathbf{q}, \tau)$ of a binary mixture for large values of Le is a single exponential in the time (difference) τ. The single decay rate was $\Gamma_D = Dq^2$. However, for arbitrary values of Le one has to account for three distinct decay rates. We note from Eqs. (5.26) that the three decay rates are always real and positive numbers for any value of the wave number q. This means two things: First, when the effects of gravity are neglected, the conductive solution is always stable. Second, there are no propagating modes in the system. These results are independent of the value of the imposed temperature gradient ∇T_0 or the sign of the separation ratio ψ.

To obtain the correlation functions between the fluctuating fields we need the correlation between the different components of the random force $\mathbf{F}(\omega, q)$ defined by Eq. (5.24). These can be derived by applying a Fourier transformation to Eq. (3.9) for the FDT of a binary mixture. Just as in Eqs. (3.22a) or (4.20), these correlation functions are conveniently expressed in terms of a noise correlation matrix $\mathsf{C}(q)$, such that:

$$\langle F_\alpha^*(\omega, \mathbf{q}) \, F_\beta(\omega', \mathbf{q}'_\|) \rangle = C_{\alpha\beta}(\mathbf{q}) \, (2\pi)^4 \, \delta(\mathbf{q} - \mathbf{q}') \, \delta(\omega - \omega'). \tag{5.28}$$

After some algebra one obtains for the correlation matrix $\mathsf{C}(\mathbf{q})$ (Law and Nieuwoudt, 1989):

$$\mathsf{C}(\mathbf{q}) = \frac{2k_\mathrm{B}T}{\rho} \begin{pmatrix} \nu q_\|^2 q^4 & 0 & 0 \\[2mm] 0 & \dfrac{T}{c_p}(a_T + D\epsilon_\mathrm{D})q^2 & \dfrac{Dk_T}{c_p}q^2 \\[4mm] 0 & \dfrac{Dk_T}{c_p}q^2 & D\left(\dfrac{\partial c}{\partial \mu}\right)_T q^2 \end{pmatrix}. \tag{5.29}$$

When temperature fluctuations are not negligible, they also contribute to the index-of-refraction fluctuations. Then Eq. (5.13) needs to be replaced by a relationship between the dynamic structure factor and both temperature and concentration fluctuations. It is customary to decompose the structure factor $S(\omega, \mathbf{q})$ into three contributions:

$$S(\omega, \mathbf{q}) = \left(\frac{\partial n}{\partial c}\right)_T^2 S_{cc}(\omega, \mathbf{q})$$

$$+ \left(\frac{\partial n}{\partial T}\right)_c^2 S_{TT}(\omega, \mathbf{q}) + \left(\frac{\partial n}{\partial c}\right)_T \left(\frac{\partial n}{\partial T}\right)_c S_{cT}(\omega, \mathbf{q}), \quad (5.30)$$

where $S_{cc}(\omega, \mathbf{q})$, $S_{TT}(\omega, \mathbf{q})$ and $S_{cT}(\omega, \mathbf{q})$ are referred to as *partial structure factors*. They represent, respectively, the contributions to $S(\omega, \mathbf{q})$ from concentration fluctuations, from temperature fluctuations, and from cross-fluctuations. The partial structure factors are readily expressed in terms of the linear response function G and the correlation matrix C as:

$$S_{cc}(\omega, \mathbf{q}) = \sum_{i,j=0}^{2} G_{2i}^*(\omega, q) \, C_{ij}(\mathbf{q}) \, G_{2j}(\omega, q) \qquad (5.31a)$$

$$S_{TT}(\omega, \mathbf{q}) = \sum_{i,j=0}^{2} G_{1i}^*(\omega, q) \, C_{ij}(\mathbf{q}) \, G_{1j}(\omega, q) \qquad (5.31b)$$

$$S_{Tc}(\omega, \mathbf{q}) = \sum_{i,j=0}^{2} [G_{1i}^*(\omega, q) \, C_{ij}(\mathbf{q}) \, G_{2j}(\omega, q) + G_{2i}^*(\omega, q) \, C_{ij}(\mathbf{q}) \, G_{1j}(\omega, q)],$$

where the indices 1 and 2 appear because temperature is the second element and concentration the third element of the vector of fluctuating fields defined in Eq. (5.22). It is worth mentioning that, because of the Soret and Dufour effects, the cross-correlation is not zero. Again we note that the representation of the structure factor by Eqs. (5.30) and (5.31), with a nonvanishing cross correlation between the concentration and temperature fluctuations is generic for binary mixtures in equilibrium (Berne and Pecora, 1976; Boon and Yip, 1980), as well as in nonequilibrium.

We can write the dynamical structure factor as the sum of an equilibrium and a nonequilibrium contribution:

$$S(\omega, \mathbf{q}) = S_{\mathrm{E}}(\omega, \mathbf{q}) + S_{\mathrm{NE}}(\omega, \mathbf{q}), \qquad (5.32)$$

with a similar decomposition for any of the three partial structure factors introduced in Eq. (5.30). It should be noticed that the additive decomposition (5.32) of the dynamic structure factor is possible here because we are neglecting buoyancy. As shown in Eq. (5.14) for the $Le \rightarrow \infty$ limit, when gravity effects are included the only decay rate in Eq. (5.14) for the nonequilibrium structure factor depends on the temperature gradient ∇T_0, and an additive decomposition like (5.32) is not possible. However, upon integration over the frequency, the nonequilibrium contribution to the *static* structure factor show up as an additive contribution, even in the presence of gravity, see Eq. (5.17).

Upon substituting Eqs. (5.23) and (5.29) into Eqs. (5.31), one recovers for the three equilibrium partial structure factors the traditional result for the components of the equilibrium Rayleigh line (Berne and Pecora, 1976; Boon and Yip, 1980):

$$S_{cc}^{(E)}(\omega, q) = \frac{k_B \bar{T}_0}{\rho} \left(\frac{\partial c}{\partial \mu}\right)_T \frac{2Dq^2 \left[\omega^2 + a_T(a_T + \epsilon_D D)q^4\right]}{\left(\omega^2 + \Gamma_+^2\right)\left(\omega^2 + \Gamma_-^2\right)}, \tag{5.33a}$$

$$S_{TT}^{(E)}(\omega, q) = \frac{k_B \bar{T}_0^2}{\rho c_P} \frac{2q^2 \left[\omega^2(a_T + \epsilon_D D) + a_T D^2 q^4\right]}{\left(\omega^2 + \Gamma_+^2\right)\left(\omega^2 + \Gamma_-^2\right)}, \tag{5.33b}$$

$$S_{cT}^{(E)}(\omega, q) = \frac{k_B \bar{T}_0 k_T}{\rho c_P} \frac{4Dq^2 \left[\omega^2 - a_T D q^4\right]}{\left(\omega^2 + \Gamma_+^2\right)\left(\omega^2 + \Gamma_-^2\right)}. \tag{5.33c}$$

In deriving Eqs. (5.33), we have made use of the definition (2.76) of the Dufour effect ratio. After substitution of Eqs. (5.33) into Eq. (5.30), the expression for the structure factor may be rewritten as a complicated superposition of two lorentzians (Berne and Pecora, 1976; Boon and Yip, 1980). When Eqs. (3.25) are integrated over the frequency ω, see Eq. (3.24), one obtains the contributions from the concentration fluctuations and the temperature fluctuations to the static structure factor $S_E(\mathbf{q})$ of a fluid mixture in equilibrium:

$$S_{cc}^{(E)}(q) = \frac{k_B \bar{T}_0}{\rho} \left(\frac{\partial c}{\partial \mu}\right)_T, \qquad S_{TT}^{(E)}(q) = \frac{k_B \bar{T}_0^2}{\rho c_P}, \tag{5.34}$$

respectively. The cross term (5.33c) vanishes upon integration over ω and does not contribute to the equilibrium static structure factor. As we see from Eqs. (5.34), none of the equilibrium partial static structure factors depends on q, which confirms the short-ranged nature of the corresponding correlation functions as was elucidated in Sect. 3.4 for a one-component fluid.

By substituting Eq. (5.23) and (5.29) into Eqs. (5.31), one can derive the nonequilibrium partial structure factors. We then recover the results originally obtained by Law and Nieuwoudt (1989):

$$S_{cc}^{(NE)}(\omega, q) = \frac{k_B \bar{T}_0}{\rho} (\nabla c_0)^2 \frac{2\nu q^2 \left[\omega^2 + (a_T + D + \epsilon_D D)^2 q^4\right]}{\left(\omega^2 + \Gamma_\nu^2\right)\left(\omega^2 + \Gamma_+^2\right)\left(\omega^2 + \Gamma_-^2\right)}, \tag{5.35a}$$

$$S_{TT}^{(NE)}(\omega, q) = \frac{k_B \bar{T}_0}{\rho} (\nabla T_0)^2 \frac{2\nu q^2 \left[\omega^2 + D^2(1 + \epsilon_D)^2 q^4\right]}{\left(\omega^2 + \Gamma_\nu^2\right)\left(\omega^2 + \Gamma_+^2\right)\left(\omega^2 + \Gamma_-^2\right)}, \tag{5.35b}$$

$$S_{cT}^{(NE)}(\omega, q) = \frac{k_B \bar{T}_0}{\rho} \nabla c_0 \nabla T_0 \frac{4\nu q^2 \left[\omega^2 + D(1 + \epsilon_D)[a_T + D(1 + \epsilon_D)]q^4\right]}{\left(\omega^2 + \Gamma_\nu^2\right)\left(\omega^2 + \Gamma_+^2\right)\left(\omega^2 + \Gamma_-^2\right)}. \tag{5.35c}$$

For simplicity we have followed Law and Nieuwoudt (1989) by assuming that $\mathbf{q} \perp \nabla T_0$, in which case $q_{\parallel} \simeq q$ (see Sect. 10.1.1), and the resulting expression depends only on the magnitude q of the vector $\mathbf{q}$. We note that the nonequilibrium partial structure factors not only contain terms with decay rates Γ_+ and Γ_-, as do the equilibrium ones, but they also contain contributions from viscous fluctuations with decay rate Γ_{ν}.

Equations (5.33) and (5.35) are long and complicated, so that for practical applications it is useful to introduce some approximations. Of course, the first possible approximation is to take the large-Lewis-number limit. It can be verified that in the limit $Le \rightarrow \infty$ (which implies $\Gamma_- \rightarrow Dq^2$ and $\Gamma_+ \rightarrow \infty$), Eqs. (5.33) and (5.35) reduce to Eq. (5.14) with $g = 0$ and $q_{\parallel} = q$. Equation (5.14) was obtained in the previous section directly from the Boussinesq equations in the limit of large Lewis number. The advantage of having adopted the $Le \rightarrow \infty$ limit a priori, when establishing the relevant hydrodynamic equations, is now evident.

A second possible simplification is to treat the Dufour effect ratio ϵ_D as a small parameter. An expansion of Eq. (5.26) up to linear order in ϵ_D yields for the decay rates Γ_+ and Γ_-:

$$\Gamma_+ \simeq a_T q^2 \left[1 + \frac{D\epsilon_D}{a_T - D}\right], \qquad \Gamma_- \simeq Dq^2 \left[1 - \frac{D\epsilon_D}{a_T - D}\right]. \qquad (5.36)$$

Thus in zeroth approximation $\Gamma_+ = a_T q^2$ and $\Gamma_- = Dq^2$, to be substituted into Eqs. (5.33) and (5.35). By substituting Eq. (5.36) into Eqs. (5.33) for the equilibrium partial structure factors, and then using Eq. (5.30), we obtain an expression for $S_E(\omega, q)$. If we retain in this expression only terms of zeroth order in ϵ_D, we recover the approximate classical result for the Rayleigh line of a newtonian binary fluid in equilibrium (Berne and Pecora, 1976; Boon and Yip, 1980):

$$S_E(\omega, q) \simeq S_{E,c} \left\{\frac{2Dq^2}{\omega^2 + D^2q^4} + \frac{1}{\mathcal{R}} \frac{2a_T q^2}{\omega^2 + a_T^2 q^4}\right\}, \qquad (5.37)$$

where $S_{E,c}$ is the intensity of concentration fluctuations in equilibrium, as given by Eq. (5.18), while the dimensionless quantity $\mathcal{R}$ is the so-called Rayleigh factor ratio,

$$\mathcal{R} = \left(\frac{\partial c}{\partial \mu}\right)_T \frac{c_P}{\bar{T}_0} \frac{\left(\frac{\partial n}{\partial c}\right)_T^2}{\left(\frac{\partial n}{\partial T}\right)_c^2}, \qquad (5.38)$$

introduced by Dubois and Bergé (1971) to represent the ratio of the intensity of scattering from concentration and thermal fluctuations in equilibrium. For many liquid mixtures $\mathcal{R}$ is quite large ($\mathcal{R} \simeq 10^4$) so that it is often difficult to observe experimentally the presence of two modes in equilibrium binary liquid mixtures (Leipertz, 1988). Hence, one usually observes concentration fluctuations, while measuring temperature fluctuations requires use of liquid mixtures whose components have a small refractive-index difference, so that $(\partial n/\partial c)_T$ is small and $\mathcal{R}$ is not too large (Wu et al., 1988). It is interesting to note that, upon integrating Eq. (5.37) over ω, we obtain for the static equilibrium structure factor the same result, Eq. (5.34), as before performing the $\epsilon_D \to 0$ approximation. While the Dufour effect, does affect the equilibrium dynamic structure factor, it does not affect the equilibrium static structure factor (Boon and Yip, 1980; Berne and Pecora, 1976).

To perform a small-ϵ_D expansion of the nonequilibrium contribution $S_{NE}(\omega, q)$ obtained by adding the three partial structure factors (5.35) in accordance with Eq. (5.30), one has to be careful. The products of the concentration and temperature gradients and the refractive-index derivatives, appearing as prefactors in Eqs. (5.35) are related through ϵ_D, since:

$$2\sqrt{\epsilon_D \mathcal{R}} = 2 \left| \frac{k_T}{\bar{T}_0} \frac{\left(\dfrac{\partial n}{\partial c}\right)_T}{\left(\dfrac{\partial n}{\partial T}\right)_c} \right| = \pm 2 \frac{\nabla c_0}{\nabla T_0} \frac{\left(\dfrac{\partial n}{\partial c}\right)_T}{\left(\dfrac{\partial n}{\partial T}\right)_c}, \tag{5.39}$$

where the plus signs applies for $(\partial n/\partial T)_c \, k_T < 0$ and the minus sign in the opposite case. However, as mentioned before, although ϵ_D might be small, the product $\epsilon_D \mathcal{R}$ is of $\mathcal{O}(1)$ for many liquid mixtures. Thus, to apply a small-ϵ_D expansion to the nonequilibrium contribution we follow Segrè et al. (1993a) and Li et al. (1994a) and substitute Eq. (5.36) into the decay rates of the various modes, but do not substitute Eq. (5.39) into the corresponding prefactors. Then, to zeroth order in ϵ_D, we obtain:

$$\frac{\rho \, S_{NE}(\omega, q)}{k_B \bar{T}_0} = \left(\frac{\partial n}{\partial c}\right)_T^2 \frac{2\nu q^2 \left[\omega^2 + (a_T + D)^2 q^4\right] (\nabla c_0)^2}{(\omega^2 + D^2 q^4)(\omega^2 + a_T^2 q^4)(\omega^2 + \nu^2 q^4)}$$

$$+ \left(\frac{\partial n}{\partial c}\right)_T \left(\frac{\partial n}{\partial T}\right)_c \frac{4\nu q^2 \left[\omega^2 + D(a_T + D) q^4\right] \nabla c_0 \nabla T_0}{(\omega^2 + D^2 q^4)(\omega^2 + a_T^2 q^4)(\omega^2 + \nu^2 q^4)}$$

$$+ \left(\frac{\partial n}{\partial T}\right)_c^2 \frac{2\nu q^2 \, (\nabla T_0)^2}{(\omega^2 + a_T^2 q^4)(\omega^2 + \nu^2 q^4)}. \tag{5.40}$$

After a little bit of algebra and use of the dimensionless numbers $\mathcal{R}$ and ϵ_D, Eq. (5.40) can be rewritten as a sum of three lorentzians (Segrè et al., 1993a; Li et al., 1994a):

$$\frac{S_{NE}(\omega, q)}{S_{E,c}} = A_c \frac{2Dq^2}{\omega^2 + D^2 q^4} + A_T \frac{2a_T q^2}{\omega^2 + a_T^2 q^4} + A_\nu \frac{2\nu q^2}{\omega^2 + \nu^2 q^4}, \tag{5.41}$$

with the dimensionless amplitudes of the three modes, A_c, A_T and A_ν given by:

$$A_c = A_0(q) \frac{\nu a_T (\nu^2 - a_T^2)}{(\nu^2 - D^2)(a_T^2 - D^2)} \left\{ \pm 2\sqrt{\frac{\epsilon_D}{\mathcal{R}}} + \epsilon_D \frac{a_T + 2D}{D} \right\}, \qquad (5.42)$$

$$A_T = A_0(q) \frac{\nu}{a_T} \left\{ \frac{1}{\mathcal{R}} \pm \frac{2(a_T^2 - Da_T - D^2)}{a_T^2 - D^2} \sqrt{\frac{\epsilon_D}{\mathcal{R}}} + \frac{D(D + 2a_T)}{a_T^2 - D^2} \epsilon_D \right\},$$

$$A_\nu = -A_0(q) \left\{ \frac{1}{\mathcal{R}} \pm \frac{2(\nu^2 - D^2 - a_T D)}{\nu^2 - D^2} \sqrt{\frac{\epsilon_D}{\mathcal{R}}} + \frac{\nu^2 - (a_T + D)^2}{\nu^2 - D^2} \epsilon_D \right\},$$

where the temperature gradient and wave number dependence have been collected in the dimensionless parameter:

$$A_0(q) = \frac{c_p (\nabla T_0)^2}{(\nu^2 - a_T^2) T q^4}. \qquad (5.43)$$

In Eqs. (5.42), as in Eq. (5.39), the plus signs applies for $(\partial n / \partial T)_c \, k_T < 0$ and the minus sign in the opposite case. When $\epsilon_D \mathcal{R} = \mathcal{O}(1)$, all components of the amplitudes A_c, A_T and A_ν, are expected to be of the same order of magnitude and no further simplifications can be made.

From Eq. (5.41) we see that for small ϵ_D the nonequilibrium contribution to the dynamic structure factor consists of three independent lorentzian components associated with viscous, temperature and concentration fluctuations. The decay rates of these components can be directly identified with νq^2, $a_T q^2$, and $D q^2$, respectively. For actual liquid mixtures, the diffusion coefficient D is much smaller than the kinematic viscosity ν and the thermal diffusivity a_T. Thus, the viscous and thermal fluctuations have in practice decayed at times when the concentration fluctuations are still important. Hence, it is possible to determine the concentration fluctuations, the viscous fluctuations and the thermal fluctuations in an independent way (Segrè et al., 1993a; Li et al., 1994a, 1998, 2000). In usual liquid mixtures for which $\nu > a_T \gg D$ and $(\partial n / \partial T)_c \, k_T < 0$, we have from Eq. (5.42) that $A_T > 0$, $A_c > 0$ while $A_\nu < 0$ (Segrè et al., 1993a; Li et al., 1994a). Thus, the presence of a temperature gradient actually enhances the temperature and concentration fluctuations, while the contribution from viscous fluctuations, that does not exist in equilibrium, Eq. (5.37), is negative. Hence, a reduction of the amplitude of the time correlation function of the fluctuations is expected at very short times.

If we consider Eqs. (5.42) in the $D \to 0$ limit (large Lewis number), we observe that the most important term is the second one inside the curly brackets in the first line for A_c. Indeed, for very small D, such a term is $\mathcal{O}(D^{-1})$ while all other terms are at most $\mathcal{O}(1)$. If we neglect any other term, we then have $A_T \simeq A_\nu \simeq 0$, while for A_c, by using the definition of ϵ_D, we recover Eq. (5.20). We observe that, when diffusion is very slow,

contributions from temperature and viscous fluctuations to the nonequilibrium structure factor can be neglected, and we reproduce the results of the
previous section where a large-Le approximation was performed *a priori*, in
the fluctuating hydrodynamics equations.

The expressions for the nonequilibrium dynamic structure factor of liquid mixtures, as originally obtained by Law and Nieuwoudt (1989) do not
account for any gravity effects. In Sect. 5.2 we have seen that, in the
$Le \rightarrow \infty$ limit, buoyancy causes the intensity of the nonequilibrium fluctuations to become independent of q for small q (as is also the case for
one-component fluids). The presence of gravity has similar consequences
for the nonequilibrium fluctuations in mixtures for general finite Le, as investigated by Segrè and Sengers (1993). A more detailed investigation of the
buoyancy effect on nonequilibrium fluctuations is presented in Chapter 9,
where finite-size effects are also accounted for.

5.4 Nonequilibrium fluctuations in isother-
mal free-diffusion processes

We conclude this chapter by reviewing the theory of nonequilibrium fluctuations in isothermal free-diffusion processes, *i.e.*, fluctuations in the presence
of a concentration gradient, ∇c, not driven by the Soret effect, but imposed by initial or boundary conditions. This problem has been studied
by Giglio and coworkers (Vailati and Giglio, 1998; Brogioli et al., 2000b).
In this situation, temperature fluctuations can be neglected (since the system is isothermal), but there still exist a hydrodynamic coupling between
concentration and velocity fluctuations through the concentration gradient
∇c. In free-diffusion processes the concentration gradient $\nabla c(\mathbf{r}, t)$ evolves in
space and time. In typical experiments a nonuniform concentration profile is
imposed initially, which then evolves in time so that, eventually, an equilib-
rium state with uniform concentration is reached at $t \rightarrow \infty$: $\nabla c(\mathbf{r}, \infty) = 0$.
Vailati and Giglio (1998) studied this problem by considering concentration
and velocity fluctuations, $\delta c(\mathbf{r}, t)$, $\delta \mathbf{v}(\mathbf{r}, t)$, around the solution of the deterministic free-diffusion equation: $c(\mathbf{r}, t)$ and $\mathbf{v} = 0$, where $c(\mathbf{r}, t)$ will depend
on the initial and/or boundary conditions. One may assume that the evolu-
tion of $c(\mathbf{r}, t)$ occurs at a time scale much slower than the one corresponding
to the random dissipative fluxes. Then, correlation functions between random dissipative fluxes can still be calculated by assuming local equilibrium
in applying the FDT at each time t between 0 and ∞. Using this procedure,
Vailati and Giglio (1998) have calculated the static structure factor of the
binary mixture, which now will be a function of the time t and the position
$\mathbf{r}$ in the fluid: $S(q; \mathbf{r}, t)$. They also employed the large Le approximation

with the result (Eq. (26) in Vailati and Giglio (1998)):

$$S(q; \mathbf{r}, t) = \left(\frac{\partial n}{\partial c} \right)_T^2 \frac{k_B \bar{T}_0}{\rho} \frac{\left\{ \left(\frac{\partial c}{\partial \mu} \right)_T + \frac{|\nabla c(\mathbf{r}, t)|^2}{\nu D \, q^4} \right\}}{1 + \frac{\beta \, [\mathbf{g} \cdot \nabla c(\mathbf{r}, t)]}{\nu D q^4}}, \tag{5.44}$$

where, for consistency with previous sections, $\bar{T}_0$ indicates the temperature at which the isothermal free diffusion takes place. When $t \to \infty$, $\nabla c \to 0$, and Eq. (5.44) reduces to $S_{E,c}$, the intensity of concentration fluctuations in equilibrium as given by Eq. (5.18). Strictly speaking, it is not evident at which spatial average of $c(\mathbf{r}, t)$ the thermophysical properties in Eq. (5.44) are to be evaluated, but as discussed in Sect. 4.4, effects due to a local dependence of the thermophysical properties can be expected to be negligibly small.

If we replace the space and time dependent concentration gradient $c(\mathbf{r}, t)$ in Eq. (5.44) by a stationary concentration gradient in the z-direction $(\nabla c(\mathbf{r}, t) = \nabla c_0 \, \hat{\mathbf{z}})$, we recover Eq. (5.17) obtained in Sect. 5 for the concentration fluctuations in the large-Lewis-number approximation. A minor technical difference with the result of Vailati and Giglio (1998) is that the latter have assumed small scattering angles, so that $q_\parallel \simeq q$. We recall that taking the $Le \to \infty$ limit implies neglecting temperature fluctuations, so that Eq. (5.17) for the nonequilibrium concentration fluctuations is valid whenever there exists a concentration gradient in the system, independently whether it is induced by the Soret effect or by any other cause. This is true even if the concentration gradient is not stationary, but slowly evolving in time. That the nonequilibrium Rayleigh spectrum in a binary liquid in the presence of a temperature gradient appears to be identical to that for the isothermal case with a concentration gradient is a direct consequence of the large Le approximation, which neglects the temperature fluctuations and its coupling with concentration fluctuations. As we have seen in the previous section, when both temperature and concentration fluctuations are present, the expression for the structure factor is more complicated.

Theoretical expressions for nonequilibrium concentration fluctuations in a colloidal suspension in the presence of a concentration gradient have been derived by Schmitz (1994). The equivalence of the results obtained by Schmitz (1994) with the theory presented in this chapter has been discussed in detail by Li et al. (1998).

Chapter 6

Finite-size effects in hydrodynamic fluctuations

In the previous chapters we considered fluctuations in a bulk fluid subjected to a temperature (or concentration) gradient. That is, we implicity assumed that the fluid system was sufficiently large (mathematically: infinite), so that fluctuations in fluid elements far from the walls are unaffected by the presence of boundaries. However, in practice, the fluid layer is confined between two horizontal plates separated by a (small) distance L, so that the nonequilibrium structure factor will be affected by confinement effects. Such confinement effects enter into the theory through corresponding boundary conditions (BC) for the fluctuating fields, that were specified in Sect. 2.5.

A number of investigations concerning finite-size effects on nonequilibrium structure factors have been reported in the literature. Schmitz and Cohen (1985b) first considered the problem and gave a general theoretical framework for incorporating boundary conditions. Ortiz de Zárate et al. (2001) derived an analytic solution for the case of free boundaries when buoyancy can be neglected. Subsequently, Ortiz de Zárate and Sengers (2001) extended the previous calculation incorporating also buoyancy effects. Ortiz de Zárate and Muñoz Redondo (2001) proposed a Galerkin approximation for the more realistic case of rigid boundaries, but in the absence of gravity effects. Next, Sengers and Ortiz de Zárate (2001) considered the case of a binary mixture in the presence of a Soret-induced stationary concentration gradient incorporating gravity effects but for somewhat unrealistic BC corresponding to free and permeable walls. Ortiz de Zárate and Sengers (2002) adopted a Galerkin approximation to evaluate the finite-size effects on the nonequilibrium fluctuations in a one-component fluid layer subjected to a stationary temperature gradient confined between two rigid boundaries. Finally, Ortiz de Zárate et al. (2004) have considered a binary mixture with realistic rigid and impermeable walls. Confinement effects on nonequilibrium fluctuations have been also studied for the plane-Couette problem as further discussed in Sect. 11.1.

A review of the literature mentioned above shows that confinement effects become important in nonequilibrium fluctuations, even for fluid elements far away from the boundaries, when the wave number of the fluctuations is of the order of the inverse of the height L of the layer. Furthermore, incorporating BC is necessary to extend the range of applicability of fluctuating hydrodynamics for calculating nonequilibrium fluctuations. As we mentioned after Eq. (4.30), the structure factor obtained from a "bulk" calculation is only valid for negative Rayleigh numbers. Incorporating BC will enable us to calculate nonequilibrium structure factors for positive Ra, up to the value of Ra_c corresponding to the onset of convection.

In the present chapter we show how the stochastic Boussinesq equations in the presence of BC can be solved by considering an eigenvalue problem associated with the hydrodynamic operator. In subsequent chapters we shall make use of the eigenvalues (decay rates) and eigenfunctions of the hydrodynamic operator to derive the nonequilibrium structure factor of fluid layers subjected to a stationary temperature gradient. Chapters 6-8 will be concerned with fluctuations in one-component-fluid layers and Chapter 9 with binary-fluid layers.

We shall proceed as follows. In Sect. 6.1 we show how BC can be incorporated in the calculation of the structure factor by introducing the eigenvalues and eigenfunctions of an hydrodynamic operator $\mathcal{H}$. Eigenfunctions are usually referred to as hydrodynamic modes, while the corresponding eigenvalues are the decay rates of these modes. The subsequent sections of this chapter are devoted to a study of the decay rates and hydrodynamic modes for some particular cases, depending on the BC considered. As examples of BC, we shall consider perfectly conducting walls for the temperature fluctuations, and both rigid and free walls for the velocity fluctuations. It is advantageous to include the unrealistic case of two free boundaries, because for free boundaries we are able to obtain exact analytic solutions for the hydrodynamic modes and the corresponding decay rates, as shown in Sect. 6.2. The realistic case of two rigid boundaries is considered in Sect 6.3; only numerical solutions can be obtained unless some additional approximations are introduced. The remainder of the chapter is devoted to a study of some limiting cases for which analytical expressions for the hydrodynamic modes and the decay rates for rigid boundaries may be obtained. Specifically, in Sect. 6.3.1 we shall study the case $q_\parallel \to 0$, in Sect. 6.3.2 the case $q_\parallel \to \infty$. Finally in Sect. 6.4 we study the case close to the onset of convection considering again slip-free and no-slip boundary conditions.

6.1 The hydrodynamic operator

The goal of this and the next chapter is to calculate the structure factor of a nonequilibrium fluid from the linearized random Boussinesq equations (4.11). The difference with the development in Chapter 4 is that we include here BC in the vertical z-direction for the fluctuating fields. As usual, we again apply a Fourier transformation in space and in time to Eqs. (4.11), but to accommodate the BC, we now restrict the spatial Fourier transformation to the horizontal xy-plane. Thus, instead of the algebraic Eqs. (4.14), we consider now the following set of linear stochastic differential equations:

$$\begin{pmatrix} i\,\omega\,\mathsf{D} + \nu\,\mathsf{D}^2 & -\alpha_p\,g\,q_\parallel^2 \\ \nabla T_0 & i\,\omega + a_T\,\mathsf{D} \end{pmatrix} \begin{pmatrix} \delta v_z(\omega, \mathbf{q}_\parallel, z) \\ \delta T(\omega, \mathbf{q}_\parallel, z) \end{pmatrix} = \mathbf{F}(\omega, \mathbf{q}_\parallel, z), \tag{6.1}$$

with the differential operator

$$\mathsf{D} = \left[q_\parallel^2 - \frac{d^2}{dz^2} \right]. \tag{6.2}$$

The noise term $\mathbf{F}(\omega, \mathbf{q}_\parallel, z)$ in Eq. (6.1) is the Fourier transform in the horizontal xy-plane and in time of the same noise vector defined by Eq. (4.12). The vector $\mathbf{q}_\parallel = \{q_x, q_y\}$ is the horizontal component of the wave vector of the fluctuations and $q_\parallel^2 = q_x^2 + q_y^2$ its magnitude. Note that, if we had performed a full spatial Fourier transformation, then $\mathsf{D} = q^2$, and Eq. (6.1) reduces to Eq. (4.14). The LHS of (6.1) contains the inverse linear response operator $\mathcal{G}^{-1}$ of our problem. To invert (6.1) we follow Schmitz and Cohen (1985b) and define a vector $\mathbf{U}$ of fluctuating fields as:

$$\mathbf{U} = \begin{pmatrix} \delta v_z(z) \\ \delta T(z) \end{pmatrix}. \tag{6.3}$$

Then Eq. (6.1) may be written as:

$$\mathcal{G}^{-1}\,\mathbf{U} = \{i\omega\,\mathcal{D} + \mathcal{H}\}\,\mathbf{U} = \mathbf{F}, \tag{6.4}$$

where we have decomposed the inverse linear response operator $\mathcal{G}^{-1}$ into two parts:

$$\mathcal{H} = \begin{pmatrix} \nu \left[q_\parallel^2 - \dfrac{d^2}{dz^2} \right]^2 & -\alpha_p g q_\parallel^2 \\ \nabla T_0 & a_T \left[q_\parallel^2 - \dfrac{d^2}{dz^2} \right] \end{pmatrix}, \quad \mathcal{D} = \begin{pmatrix} \left[q_\parallel^2 - \dfrac{d^2}{dz^2} \right] & 0 \\ 0 & 1 \end{pmatrix}. \tag{6.5}$$

The operator $\mathcal{H}$ is called the hydrodynamic operator (Schmitz and Cohen, 1985b).

Our purpose is to calculate the linear response operator by inverting Eq. (6.1) while accounting for boundary conditions. For the temperature, we assume perfectly conducting walls, so that the only relevant boundary condition will be given by Eq. (2.81), while the value of the temperature gradient at the walls is undetermined. For the vertical component of the velocity, we consider the possibility of a fluid bounded by two flat free surfaces, as well as a fluid bounded by two rigid plates. In both cases Eq. (2.86) implies that fluctuations in the vertical velocity vanish at the two boundaries. The Boussinesq approximation includes divergence-free fluid velocity, *i.e.*, $\nabla \cdot \delta \mathbf{v} = 0$, so that for two rigid surfaces we adopt Eq. (2.88), while for two free surfaces we adopt Eq. (2.90). Hence, the relevant BC are (Chandrasekhar, 1961; Manneville, 1990):

$$
\begin{aligned}
\delta T(\omega, \mathbf{q}_\parallel, z) &= 0 & \text{at} \quad z &= \pm L/2, \\
\delta v_z(\omega, \mathbf{q}_\parallel, z) &= 0 & \text{at} \quad z &= \pm L/2, \\
\partial_z \, \delta v_z(\omega, \mathbf{q}_\parallel, z) &= 0 & \text{at} \quad z &= \pm L/2,
\end{aligned}
\tag{6.6}
$$

for two rigid boundaries, and:

$$
\begin{aligned}
\delta T(\omega, \mathbf{q}_\parallel, z) &= 0 & \text{at} \quad z &= \pm L/2, \\
\delta v_z(\omega, \mathbf{q}_\parallel, z) &= 0 & \text{at} \quad z &= \pm L/2, \\
\partial_z^2 \, \delta v_z(\omega, \mathbf{q}_\parallel, z) &= 0 & \text{at} \quad z &= \pm L/2,
\end{aligned}
\tag{6.7}
$$

for two free boundaries.

The set of BC given by Eqs. (6.6) is often referred to as *stick* or *no-slip* BC, while the set given by Eqs. (6.7) is referred to as *free-slip* BC. Note that both sets of BC imply the absence of any possible fluctuations in the temperature and velocity of the fluid adjacent to the walls. The free-slip BC (6.7) are rather unrealistic, but have the advantage of mathematical simplicity, which will enable us to obtain an exact and simple expression for the nonequilibrium static structure factor. Moreover, as in the study of the linear stability of Boussinesq problem (Chandrasekhar, 1961; Manneville, 1990), the differences resulting from free-slip BC and from stick BC on the nonequilibrium structure factor are only quantitative, while qualitatively the expressions are quite similar. While the stick BC (6.6) are more appropriate for the interpretation of experimental data, free-slip BC (6.7) are more advantageous mathematically. Considering free-slip BC makes it easier to elucidate the physical origin of the various terms entering into the expressions for nonequilibrium structure factors. For this reason we consider both boundary conditions.

In order to solve Eq. (6.4) with either the BC (6.6) or (6.7), we proceed by expanding the solution $\mathbf{U}$ in terms of a set of eigenfunctions of the hydrodynamic operator $\mathcal{H}$ that satisfy the corresponding BC. Following Schmitz and Cohen (1985b), we consider the eigenvalue problem:

$$
\mathcal{H} \, \mathbf{U}_N^{\mathrm{R}}(q_\parallel, z) = \Gamma_N(q_\parallel) \left\{ \mathcal{D} \, \mathbf{U}_N^{\mathrm{R}}(q_\parallel, z) \right\},
\tag{6.8}
$$

where the right eigenfunctions

$$\mathbf{U}_N^R(q_\parallel, z) = \begin{pmatrix} W_N(q_\parallel, z) \\ \Theta_N(q_\parallel, z) \end{pmatrix} \tag{6.9}$$

must satisfy the boundary conditions:

$$W_N(q_\parallel, z) = 0, \qquad \text{at} \qquad z = \pm\tfrac{1}{2}L \, , \tag{6.10a}$$
$$\Theta_N(q_\parallel, z) = 0, \qquad \text{at} \qquad z = \pm\tfrac{1}{2}L \tag{6.10b}$$

and

$$\partial_z \, W_N(q_\parallel, z) = 0, \qquad \text{at} \qquad z = \pm\tfrac{1}{2}L \tag{6.10c}$$

for two rigid boundaries, or

$$\partial_z^2 \, W_N(q_\parallel, z) = 0, \qquad \text{at} \qquad z = \pm\tfrac{1}{2}L \tag{6.10d}$$

for two free boundaries. In Eqs. (6.8)-(6.10) we use the index N to enumerate the discrete set of eigenvalues and eigenfunctions of the problem, anticipating a detailed discussion of the solution of Eq. (6.8) in a subsequent section. Notice that, due to the structure of the problem (6.8), the resulting eigenvalues $\Gamma_N(q_\parallel)$ and corresponding right eigenfunctions $\mathbf{U}_N^R(q_\parallel, z)$ do not depend on the frequency ω, but only on the magnitude $q_\parallel$ of the two-dimensional vector $\mathbf{q}_\parallel$.

The hydrodynamic operator $\mathcal{H}$, defined by Eq. (6.5), has real coefficients, but it is not self-adjoint. Therefore, in addition to $\mathcal{H}$ we have to consider its adjoint operator $\mathcal{H}^\dagger$. According to the usual definition of an adjoint operator (Courant and Hilbert, 1953), for any pair of two-dimensional functions $\mathbf{U}_1$ and $\mathbf{U}_2$ in the interval $[-\tfrac{1}{2}L, \tfrac{1}{2}L]$ and that satisfy the boundary conditions (6.10), $\mathcal{H}^\dagger$ is such that:

$$\int_{-\frac{1}{2}L}^{\frac{1}{2}L} \mathbf{U}_1^* \cdot (\mathcal{H} \, \mathbf{U}_2) \, dz = \int_{-\frac{1}{2}L}^{\frac{1}{2}L} (\mathcal{H}^\dagger \, \mathbf{U}_1)^* \cdot \mathbf{U}_2 \, dz. \tag{6.11}$$

We thus find that the adjoint of the hydrodynamic operator is given by:

$$\mathcal{H}^\dagger = \begin{pmatrix} \nu\mathsf{D}^2 & \nabla T_0 \\ -\alpha_p g q_\parallel^2 & a_T \mathsf{D} \end{pmatrix}, \tag{6.12}$$

where D the same (linear) differential operator defined by Eq. (6.2). Indeed, by substituting Eq. (6.12) for $\mathcal{H}^\dagger$ into Eq. (6.11), one can readily verify that Eq. (6.11) holds for any pair $\mathbf{U}_1$ and $\mathbf{U}_2$ in the usual way, namely, upon integrating by parts and using the boundary conditions (6.10b), and (6.10c) or (6.10d). It is worth mentioning that the same formula (6.12) for the adjoint hydrodynamic operator is valid for both free-slip and stick boundary conditions. We note that, for the given BC, the adjoint operator equals the transpose of $\mathcal{H}$. In addition to the eigenvalue problem defined in (6.8), we need to consider the adjoint problem:

$$\mathcal{H}^\dagger \cdot \mathbf{U}_N^L = \Gamma_N^*(\tilde{q}_\parallel) \, \mathcal{D} \cdot \mathbf{U}_N^L, \tag{6.13}$$

where the set of left eigenfunctions $\mathbf{U}_N^L$ must satisfy the same boundary conditions (6.10) as the right eigenfunctions. In Eq. (6.13) we have anticipated that the eigenvalues of the adjoint problem are the complex conjugates of the eigenvalues of the direct problem. Actually, due to the similar structure of $\mathcal{H}$ and $\mathcal{H}^\dagger$, we note that for any right eigenfunction with components $\{W_N, \Theta_N\}$, a left eigenfunction can be simply constructed as:

$$\mathbf{U}_N^L(q_\parallel, z) = \begin{pmatrix} \nabla T_0 \ W_N^*(q_\parallel, z) \\ -\alpha_p g q_\parallel^2 \ \Theta_N^*(q_\parallel, z) \end{pmatrix}. \tag{6.14}$$

Indeed, since $\mathbf{U}_N^R = \{W_N, \Theta_N\}$ satisfies Eq. (6.8) by hypothesis, it can be readily shown with the definitions of $\mathcal{H}$ and $\mathcal{H}^\dagger$ that the set of functions defined by Eq. (6.14) does indeed satisfy Eq. (6.13) (we recall that the BC are the same for the direct and the adjoint problem). Formula (6.14) for the left eigenfunctions also demonstrates that the eigenvalues of the adjoint problem (6.13) are the complex conjugates of the eigenvalues of the direct problem (6.8). This is to be expected, since the two differential operators $\mathcal{H}$ and $\mathcal{D}$ have real coefficients, so that if Γ_N is a solution of Eq. (6.8), then Γ_N^* is also an eigenvalue. Thus, there are two possibilities for the eigenvalues of both the direct and the adjoint problem: they are real or they form complex conjugate pairs. We shall come back to this feature in Sects. 6.2 and 6.3, where we shall derive explicit expressions for the eigenvalues for each one of the two sets of BC under consideration.

Next, on comparing the right and the left problem, Eqs. (6.8) and (6.13), respectively, and by using that the differential operator $\mathcal{D}$ is self-adjoint ($\mathcal{D}^\dagger = \mathcal{D}$ for both sets of BC), it can be readily demonstrated that (Courant and Hilbert, 1953):

$$\left[\Gamma_M(q_\parallel) - \Gamma_N(q_\parallel)\right] \int_{-\frac{1}{2}L}^{\frac{1}{2}L} \mathbf{U}_N^{L*}(q_\parallel, z) \cdot \left\{\mathcal{D} \ \mathbf{U}_M^R(q_\parallel, z)\right\} \ dz = 0. \tag{6.15}$$

Equation (6.15) implies that the integral has to be zero for $N \neq M$. Consequently, the set of right eigenfunctions has the important property of being "orthogonal" to the set of left eigenfunctions (Courant and Hilbert, 1953), in the sense that:

$$\int_{-\frac{1}{2}L}^{\frac{1}{2}L} \mathbf{U}_N^{L*}(q_\parallel, z) \cdot \left\{\mathcal{D} \ \mathbf{U}_M^R(q_\parallel, z)\right\} \ dz = B_N(q_\parallel) \, \delta_{NM}, \tag{6.16}$$

where B_N is to be interpreted as the "norm" of the (right) eigenfunction $\{W_N, \Theta_N\}$, or:

$$B_N(q_\parallel) = \int_{-\frac{1}{2}L}^{\frac{1}{2}L} \left\{ \nabla T_0 \left[q_\parallel^2 \, W_N^2(q_\parallel, z) + \left(\partial_z \, W_N(q_\parallel, z) \right)^2 \right] \right.$$
$$\left. - \alpha_p g q_\parallel^2 \, \Theta_N^2(q_\parallel, z) \right\} \; dz. \quad (6.17)$$

In deriving the expression for $B_N(q_\parallel)$ we have made use of Eqs. (6.2), (6.9) and (6.14) and we have performed an integration by parts (using any of the BC (6.10)) in the RHS. We have put "orthogonal" and "norm" between quotation marks, because Eq. (6.16) shows that both sets of functions are bi-orthogonal not with respect to the usual scalar product, but with respect to a new scalar product, which can be defined for functions $\mathbf{U}_1$ and $\mathbf{U}_2$, that obey the boundary conditions (6.10), by:

$$(\mathbf{U}_1, \mathbf{U}_2) \;\; = \;\; \int_{-\frac{1}{2}L}^{\frac{1}{2}L} \mathbf{U}_1^* \cdot (\mathcal{D} \, \mathbf{U}_2) \; dz = \quad (6.18)$$
$$= \;\; \int_{-\frac{1}{2}L}^{\frac{1}{2}L} \left\{ q_\parallel^2 \, W_1^* \, W_2 + \frac{dW_1^*}{dz} \frac{dW_2}{dz} + \Theta_1^* \, \Theta_2 \right\} \, dz,$$

where the last equality is obtained upon integrating by parts and by using the boundary conditions (6.10). Equation (6.18) shows that the scalar product used in the bi-orthogonality relationship (6.16) is a perfectly valid scalar product for the set of functions that satisfy the boundary conditions (6.10). When $\mathbf{U}_1 = \mathbf{U}_2$, one obtains real and positive numbers.

There may be a problem in the "orthogonality" relationship (6.16), when some of the eigenvalues are degenerate, *i.e.*, when there are different eigenfunctions corresponding to the same eigenvalue Γ_N. This problem can always be overcome by appropriate techniques (Courant and Hilbert, 1953). However, for our particular problem, given by Eq. (6.8) with BC (6.10), there are no degenerate eigenvalues except for particular values of $q_\parallel$. This issue will be discussed in more detail in Sects. 6.2 and 6.3.

We now have all the ingredients needed to solve the linear stochastic differential equation (6.4) with the boundary conditions (6.10) by expanding the solution in a series of right eigenfunctions:

$$\begin{pmatrix} \delta v_z(\omega, \mathbf{q}_\parallel, z) \\ \delta T(\omega, \mathbf{q}_\parallel, z) \end{pmatrix} = \sum_{N=1}^{\infty} G_N(\omega, \mathbf{q}_\parallel) \begin{pmatrix} W_N(q_\parallel, z) \\ \Theta_N(q_\parallel, z) \end{pmatrix}. \quad (6.19)$$

Since the eigenfunctions satisfy the BC (6.10), the fluctuating fields, represented as a series of eigenfunctions, will also satisfy the same BC. To obtain the coefficients $G_N(\omega, \mathbf{q}_\parallel)$, we substitute Eq. (6.19) into Eq. (6.4) and then project (with the usual scalar product) the result onto the set of left eigen-functions $\mathbf{U}_M^L$. Using the orthogonality relationship, Eq. (6.16), we readily solve for the amplitudes of the linear response operator:

$$G_N(\omega, \mathbf{q}_\|) = \frac{F_N(\omega, \mathbf{q}_\|)}{B_N(q_\|)\left[\mathrm{i}\,\omega + \Gamma_N(q_\|)\right]}, \tag{6.20}$$

where the parameters $F_N(\omega, \mathbf{q}_\|)$ are the projections (with the usual scalar product) of the random noise vector $\mathbf{F}$ over the left eigenfunctions:

$$
\begin{aligned}
F_N(\omega, \mathbf{q}_\|) &= \int_{-\frac{1}{2}L}^{\frac{1}{2}L} \mathbf{U}_N^{L*}(q_\|, z) \cdot \mathbf{F}(\omega, \mathbf{q}_\|, z)\,dz, \\[2mm]
&= \int_{-\frac{1}{2}L}^{\frac{1}{2}L} \left[\nabla T_0\,W_N(q_\|, z)\,F_0(\omega, \mathbf{q}_\|, z) - \alpha_p g q_\|^2\,\Theta_N(q_\|, z)\,F_1(\omega, \mathbf{q}_\|, z)\right] dz
\end{aligned}
\tag{6.21}
$$

Here, $F_i(\omega, \mathbf{q}_\|, z)$ are the components of the vector of random forces (6.3). We see from Eq. (6.20) that the eigenvalues Γ_N give the decay rate of the corresponding mode (eigenfunction). From here on, we shall refer to the eigenvalues as decay rates and to the eigenfunctions as hydrodynamic modes.

In deriving Eq. (6.20) we have implicitly assumed that $B_N(q_\|) \neq 0$ and that $\Gamma_N(q_\|) \neq 0$ for any value of $q_\|$. As will be discussed later, below the convective instability $B_N(q_\|)$ is indeed different from zero. However, from the eigenvalue problem Eq. (6.8) and the definition of the operator $\mathcal{H}$, given by Eq. (6.5), we observe that the problem of whether $\Gamma_N(q_\|)$ can be zero or not coincides with the linear stability problem as, for instance, discussed by Chandrasekhar (1961). Since the original calculation of Pellew and South-well (1940), it is well known that for the case of two rigid boundaries the operator $\mathcal{H}$ can have the decay rate $\Gamma_N(q_\|) = 0$ at some value of $q_\|$ only for Ra values larger than a critical Rayleigh number given by $Ra_c \approx 1708$. For $Ra = Ra_c$ the problem (6.8) admits the solution $\lambda_N(q_\|) = 0$ just for a single value $q_\|L = \tilde{q}_{\|c} \approx 3.11$. For $Ra > Ra_c$, there will be a number of isolated values of $q_\|$ where one or more of the decay rates will be zero (Chandrasekhar, 1961; Manneville, 1990). On the other hand, if $Ra < Ra_c$ none of the decay rates for which there are solutions to Eq. (6.8) can be zero, independently of the value of $q_\|$. Since the original calculations of Lord Rayleigh (1916) and Sir Harald Jeffreys (1926), it has been known that for the case of two free boundaries the same is true for $Ra_c = 27\pi^4/4$ and $q_{\|c}L = \pi/\sqrt{2}$. In the remainder of this chapter we shall assume that $Ra < Ra_c$ for the corresponding BC, so that the inversion procedure implicit in Eq. (6.20) is correct. As expected, for $Ra \geq Ra_c$ the present calculation breaks down because the fluctuations around the conductive state are no longer small (see Chapter 8) and the contribution from nonlinear terms in the Boussinesq equations can no longer be neglected.

To obtain an expression for the nonequilibrium structure factor, we thus proceed with the assumption that Eqs. (6.19) and (6.20) will yield a valid solution of the fluctuating Boussinesq equations with the BC incorporated. Then, the autocorrelation function $\langle \delta T^*(\omega, \mathbf{q}_\|, z) \, \delta T(\omega', \mathbf{q}'_\|, z') \rangle$ of the temperature fluctuations can be evaluated as a double series of eigenfunctions (for the corresponding BC) from Eqs. (6.19) and (6.20). For this calculation we need the various autocorrelation functions $\langle F_N^*(\omega, \mathbf{q}_\|) \, F_M(\omega', \mathbf{q}'_\|) \rangle$ among the random noises defined by Eq. (6.21). As was done previously in the absence of any boundary conditions (see Chapter 4), we continue to assume that the correlation functions between the different components of the random stress tensor and the random heat flux retain their equilibrium values given by the FDT (3.7) for a divergence-free (incompressible) one-component fluid. This assumption remains valid as long as L is a macroscopic distance, much larger than the molecular distances in the fluid. Using the relationship between the components $F_i(\omega, \mathbf{q}_\|, z)$ of the thermal noise and the random dissipative fluxes, as obtained by applying a temporal and horizontal Fourier transformation to Eq. (4.12), the definition (6.21) of $F_N(\omega, \mathbf{q}_\|)$, and the FDT (3.7c) of a divergence-free one-component fluid, we obtain the correlation functions between the projections of the thermal noise. As in previous calculations, they can be conveniently expressed in terms of a correlation matrix:

$$\langle F_N^*(\omega, \mathbf{q}_\|) \cdot F_M(\omega', \mathbf{q}'_\|) \rangle = \frac{m_0}{\alpha_p^2 \rho} \, C_{NM}(q_\|) \, (2\pi)^3 \, \delta(\omega - \omega') \, \delta(\mathbf{q}_\| - \mathbf{q}'_\|), \quad (6.22)$$

where the components $C_{NM}(q_\|)$ of the noise correlation matrix are also referred to as mode-coupling coefficients (Schmitz and Cohen, 1985b). They are given by (Ortiz de Zárate et al., 2001; Ortiz de Zárate and Muñoz Redondo, 2001):

$$C_{NM}(q_\|) = \frac{\alpha_p^2 \rho}{m_0} 2k_B T q_\|^2 \left\{ \frac{(\nabla T_0)^2 \nu}{\rho} \int\!\!\!\int_{-L/2}^{L/2} dz \; dz' \; W_N^*(q_\|, z) W_M(q_\|, z') \right.$$

$$\times \left[q_\|^4 + q_\|^2 \left(\frac{d^2}{dz^2} + \frac{d^2}{dz'^2} + 4 \frac{d}{dz} \frac{d}{dz'} \right) + \frac{d^2}{dz^2} \frac{d^2}{dz'^2} \right] \cdot \delta(z - z') + \frac{\alpha_p^2 g^2 T \lambda}{\rho^2 c_p^2} q_\|^2$$

$$\times \int\!\!\!\int_{-L/2}^{L/2} \Theta_N^*(q_\|, z) \Theta_M(q_\|, z') \left[q_\|^2 + \frac{d}{dz} \frac{d}{dz'} \right] \cdot \delta(z - z') \, dz \, dz' \left. \vphantom{\int} \right\}, \quad (6.23)$$

where in the double integrals both variables z and z' vary over the interval $[-L/2, L/2]$. In Eq. (6.22) we have entered the factor $\alpha_p^2 \rho / m_0$ to obtain more compact expressions later. In the derivation of Eq. (6.23) use has been made of the property that the cross correlations between the components of the random current and the random heat flux vanish.

Next, we integrate by parts the different terms of Eq. (6.23), so as to move the differential operator inside the double integral from the delta function to the components of the eigenfunctions preceding it. Note that since in all cases an even number of integrations are required, there will not be any change of sign as a result of this process. After this procedure the differential operators inside the integrals apply to the W_N and Θ_N functions, and the delta functions will be isolated. Thus the integration in the variable z' can be readily performed. We then continue to integrate by parts, but now using the boundary conditions (6.10b), so as to finally obtain:

$$
C_{NM}(q_\parallel) = S_{\mathrm{E}} \frac{\gamma - 1}{\gamma} \, 2q_\parallel^2 \left\{ \frac{(\nabla T_0)^2 \nu c_p}{T} \int_{-\frac{L}{2}}^{\frac{L}{2}} W_N^*(q_\parallel, z) \left[\mathsf{D}^2 \, W_M(q_\parallel, z) \right] \, dz \right.
$$

$$
\left. + \alpha_p^2 g^2 a_T \, q_\parallel^2 \int_{-\frac{L}{2}}^{\frac{L}{2}} \Theta_N^*(q_\parallel, z) \left[\mathsf{D} \, \Theta_M(q_\parallel, z) \right] \, dz \right\}, \quad (6.24)
$$

with the (linear) operator D defined in Eq. (6.2). The presence of the factor $\alpha_p^2 \rho / m_0$ in (6.22) allows us to express $C_{NM}(q_\parallel)$ in terms of the equilibrium static structure factor of the fluid S_{E} as defined by (3.28). Note that from Eq. (6.24) it follows that $C_{NM}(q_\parallel) = C_{MN}^*(q_\parallel)$, which means that the diagonal mode-coupling coefficients, C_{NN}, are real numbers, independently of whether the decay rates are complex or not.

6.1.1   Structure factors

In analogy to Eq. (4.19) for the "bulk" structure factor, the relationship between the dynamic structure factor $S(\omega, q_\parallel, z, z')$ and the autocorrelation function of the temperature fluctuations is now given by:

$$
\langle \delta T^*(\omega, \mathbf{q}_\parallel, z) \, \delta T(\omega', \mathbf{q}_\parallel', z') \rangle = \rho m_0 \frac{(2\pi)^3}{\alpha_p^2 \rho^2}
$$

$$
\times \; S(\omega, q_\parallel, z, z') \, \delta(\omega - \omega') \, \delta(\mathbf{q}_\parallel - \mathbf{q}_\parallel'). \quad (6.25)
$$

Notice that (6.25) is exactly the same as Eq. (4.56) for fluctuations in the temperature: In both cases contributions from pressure fluctuations are neglected and only Fourier transformations in the horizontal xy-plane were applied to the fluctuating hydrodynamics equations. Next, substituting Eqs. (6.19), (6.20), and (6.22) into Eq. (6.25), we obtain an expression for the dynamic structure factor of the nonequilibrium fluid which takes into account both finite-size effects and the effect of gravity, namely:

$$
S(\omega, q_\parallel, z, z') = \sum_{N,M=0}^{\infty} \frac{C_{NM}(q_\parallel) \, \Theta_N^*(q_\parallel, z) \, \Theta_M(q_\parallel, z')}{B_N^*(q_\parallel) B_M(q_\parallel) [-i\omega + \Gamma_N^*(q_\parallel)][i\omega + \Gamma_M(q_\parallel)]}. \quad (6.26)
$$

When $\Gamma_N(q_\parallel) > 0$ (stability) Eq. (6.26) may be rewritten as a sum of lorentzians:

$$S(\omega, q_\parallel, z, z') = \sum_{N,M=0}^{\infty} \frac{C_{NM}\, \Theta_N^*(z)\, \Theta_M(z')}{B_N^* B_M [\Gamma_N^* + \Gamma_M]} \left\{ \frac{\mathrm{Re}(\Gamma_N)}{|i\omega + \Gamma_N|^2} + \frac{\mathrm{Re}(\Gamma_M)}{|i\omega + \Gamma_M|^2} \right\}$$

$$+ \sum_{N,M=0}^{\infty} \frac{C_{NM}\, \Theta_N^*(z)\, \Theta_M(z')}{B_N^* B_M [\Gamma_N^* + \Gamma_M]} \left\{ \frac{i[\omega + \mathrm{Im}(\Gamma_N)]}{|i\omega + \Gamma_N|^2} - \frac{i[\omega + \mathrm{Im}(\Gamma_M)]}{|i\omega + \Gamma_M|^2} \right\}, \quad (6.27)$$

where, to simplify the notation, we have dropped the $q_\parallel$ dependence in the RHS. Upon applying a temporal Fourier transformation to Eq. (6.27) we observe, as already noted previously, that the time correlation function of nonequilibrium fluctuations can be expressed as a series of exponentials[‡]. This is a consequence of the linear approximation we have adopted to fluctuating hydrodynamics. In addition, since $C_{NM}(q_\parallel) = C_{MN}^*(q_\parallel)$, it follows that $S(\omega, q_\parallel, z, z') = S^*(\omega, q_\parallel, z', z)$.

As will be discussed in more detail in Sect. 10.1.1, it turns out that the structure factor actually measured in the experiments is obtained by integrating $S(\omega, q_\parallel, z, z')$ over the variables z and z'. Specifically, it can be shown that the spectrum of light scattered with scattering vector $\mathbf{q} = \{q_\parallel, q_\perp\}$ is proportional to:

$$S(\omega, \mathbf{q}) = \frac{1}{L} \int_{-L/2}^{L/2} \int_{-L/2}^{L/2} e^{-iq_\perp(z - z')}\, S(\omega, q_\parallel, z, z')\, dz\, dz'. \quad (6.28)$$

Exchange of the integration variables in the RHS of (6.28) shows that $S(\omega, \mathbf{q}) = S^*(\omega, \mathbf{q})$, *i.e.*, the result of the integral is a real number. When all decay rates and hydrodynamic modes are real numbers (as happens for positive Ra), the second line of Eq. (6.27) is purely imaginary and, consequently, has to cancel upon vertical integration and does not contribute to the experimental structure factor (6.28). This fact will be used later in Eq. (8.1). It is worth noticing that in many experiments $S(\omega, \mathbf{q})$ is obtained in a small-angle approximation, which means: $q_\parallel \simeq q$ and $q_\perp \simeq 0$, see Eq. (10.21). This small-angle approximation will be used quite often in this book.

In addition to the dynamic structure factors, defined by (6.25) and (6.28), we shall also need static structure factors, obtained by integrating over the frequency ω in accordance with Eq. (3.24). For instance, from the dynamic structure factor $S(\omega, q_\parallel, z, z')$, we deduce for the static structure factor:

$$S(q_\parallel, z, z') = \frac{1}{2\pi} \int_{-\infty}^{\infty} S(\omega, q_\parallel, z, z')\, d\omega. \quad (6.29)$$

[‡]Possibly multiplied by trigonometric functions, if the imaginary part of the decay rates is not zero

By applying a double inverse Fourier transformation in the two frequencies to Eq. (6.25), we notice that the static structure factor (6.29) will be proportional to the equal-time temperature fluctuations autocorrelation, namely:

$$\langle \delta T^*(\mathbf{q}_{\parallel}, z, t) \, \delta T(\mathbf{q}'_{\parallel}, z', t) \rangle = \frac{m_0}{\rho \alpha_p^2} \, S(q_{\parallel}, z, z') \, (2\pi)^2 \, \delta(\mathbf{q}_{\parallel} - \mathbf{q}'_{\parallel}). \quad (6.30)$$

As was the case for the dynamic structure factor, the dimensionless static structure factor actually measured in light-scattering experiments is obtained by a double vertical integration of $S(q_{\parallel}, z, z')$, namely:

$$S(\mathbf{q}) = \frac{1}{L} \int_{-L/2}^{L/2} \int_{-L/2}^{L/2} e^{-iq_{\perp}(z - z')} \, S(q_{\parallel}, z, z') \, dz \, dz', \quad (6.31)$$

where, again, the small-angle approximation: $q_{\parallel} \simeq q$ and $q_{\perp} \simeq 0$, is often employed. Notice that the experimental structure factors $S(\omega, \mathbf{q})$ and $S(\mathbf{q})$ are related by exactly the same expression (3.24) as for a fluid in equilibrium. Because the static structure factor is the integral of the spectrum $S(\omega, \mathbf{q})$ of light scattered by the fluid, it will be proportional to the total intensity of scattered light. As was the case for $S(\omega, \mathbf{q})$, the static $S(\mathbf{q})$ is a real quantity independent of whether there are propagating modes (complex decay rates) in the fluid.

Before continuing with the explicit calculation of the hydrodynamic structure factor $S(\omega, q)$, we first need to discuss the behavior of the decay rates $\Gamma_N(q_{\parallel})$ and the corresponding right and left eigenfunctions, defined by Eqs. (6.8) and (6.13), respectively. This problem has been discussed by Schmitz and Cohen (1985b) who studied the eigenvalues at a constant wave number $q_{\parallel}$. Since, from a practical point of view, a study of the dependence of the eigenvalues on the wave vector $q_{\parallel}$ turns out to be most interesting, we extend here the treatment of Schmitz and Cohen (1985b). The equations derived in the present section apply to both free-slip BC and stick BC. To discuss decay rates and hydrodynamic modes we need to consider the two types of BC separately. Specifically, for the case of two free boundaries, we shall give explicit analytic expressions for the decay rates and the hydrodynamic modes. For two rigid boundaries, we shall discuss how the decay rates can be computed numerically, and shall give expressions that, upon substitution of the numerically computed decay rates, shall represent the hydrodynamic modes. For this case, we shall also evaluate the asymptotic behavior of the decay rates in the limits $q_{\parallel} \to 0$ and $q_{\parallel} \to \infty$.

6.2 Hydrodynamic modes and decay rates for two free boundaries

The decay rates and corresponding hydrodynamic modes of the operator $\mathcal{H}$ for slip-free BC can be obtained by simple inspection. First of all, due to the symmetry of the boundary conditions, it is evident that the eigenfunctions can be classified into two types, depending on the parity of the vertical dependence: odd eigenfunctions and even eigenfunctions (Schmitz and Cohen, 1985a). This is also true for the case of two rigid boundaries, to be discussed later. Furthermore, for free boundaries, due to the simplicity of the boundary conditions (6.7), it is clear that the two components of the hydrodynamic modes W_N and Θ_N are proportional to trigonometric functions. Thus, even eigenfunctions will be proportional to $\cos{(N\pi z/L)}$ for odd N and odd eigenfunctions will be proportional to $\sin{(N\pi z/L)}$ for even N. This Ansatz guarantees that the boundary conditions (6.7) are satisfied. Substituting this form of the eigenfunctions into the eigenvalue problem (6.8), we find that for each N (even or odd) there are two decay rates which we denote by $\Gamma_N^{(+)}(q_\|)$ and by $\Gamma_N^{(-)}(q_\|)$. Introducing dimensionless decay rates, $\tilde{\Gamma}= \Gamma\, L^2/\nu$, they can be written as:

$$\tilde{\Gamma}_N^{(\pm)}(\tilde{q}_\|) = \frac{1}{2}\,(N^2\pi^2 + \tilde{q}_\|^2)$$

$$\times \left\{ \frac{Pr+1}{Pr} \pm \frac{Pr-1}{Pr}\sqrt{1 + \frac{4Pr\,Ra\,\tilde{q}_\|^2}{(Pr-1)^2(N^2\pi^2 + \tilde{q}_\|^2)^3}}\,\right\}, \quad (6.32)$$

where $\tilde{q}_\| = q_\| L$, $Pr = \nu/a$ is again the Prandtl number (4.7) and Ra the same Rayleigh number defined by Eq. (4.8). Note that the formula for the decay rates is the same independent of whether N is odd or even.

The decay rates $\tilde{\Gamma}_N^{(\pm)}(\tilde{q}_\|)$, given by Eq. (6.32), are obtained as the roots of a second-order algebraic equation with real coefficients. Hence, they are either both real numbers or they form a complex conjugate pair, depending on the value of Ra. This fact was anticipated above, when we introduced the Boussinesq's adjoint problem, Eq. (6.13). It is evident form Eq. (6.32) that for positive Ra the decay rates are always real, while for negative Rayleigh numbers the decay rates for a given N may form a complex conjugate pair. Hence, for negative Ra propagating modes may appear, *i.e.*, a nonzero imaginary part of $\Gamma_N^{(\pm)}(q_\|)$. From Eq. (6.32) we can deduce that propagating modes exist for negative Rayleigh numbers such that:

$$Ra < -\frac{27\pi^4}{16}\frac{(Pr-1)^2}{Pr}. \quad (6.33)$$

The existence of propagating modes adds some mathematical difficulties to the solution of the random Boussinesq equations (6.4). It turns out that the

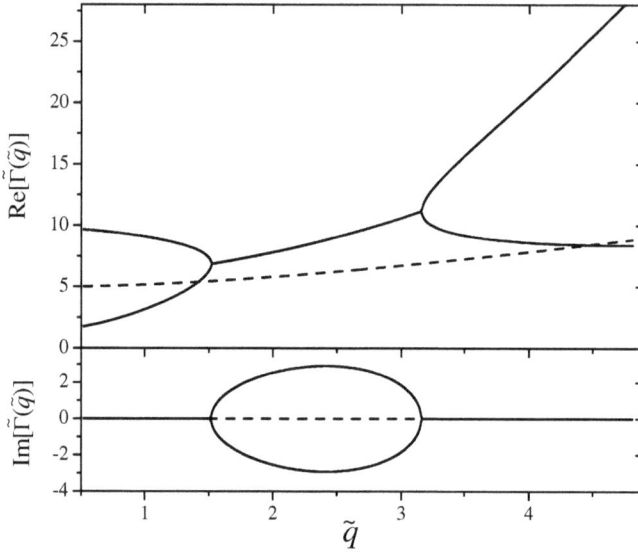

Figure 6.1: Real (upper panel) and imaginary (lower panel) parts of the decay rates $\tilde{\Gamma}_1^{(\pm)}$ (solid curves) and $\tilde{\Gamma}_3^{(-)}$ (dotted curve) from (6.32) for free boundaries, $Ra = -1200$ and $Pr = 8$. The presence of propagating modes, branch points and crossing points is evident.

appearance of propagating modes corresponds to branch points in $\Gamma_N^{(\pm)}(\tilde{q}_{\parallel})$, where the (+) and the (-) decay rates merge resulting in a degeneracy in the eigenvalues. To illustrate this behavior we show in Fig. 6.1 the three eigenvalues $\tilde{\Gamma}_1^{(\pm)}(\tilde{q})$ and $\tilde{\Gamma}_3^{(-)}(\tilde{q})$ for $Ra = -1825$ and $Pr = 10$. From the behavior displayed in Fig. 6.1 we first observe the presence of propagating modes in $\tilde{\Gamma}_1^{(\pm)}$ for a certain region of $\tilde{q}$, as is evident from the merging of the real parts in the upper panel of the figure and imaginary parts different from zero in the lower panel. Most importantly, we note that the decay rate $\tilde{\Gamma}_3^{(-)}$ crosses $\tilde{\Gamma}_1^{(\pm)}$ leading to additional degenerate points. Obviously, near branch or crossing points, like the ones shown in Fig. 6.1, the expansion (6.19) is not possible in its present form. However, these technical difficulties can be overcome by using the techniques of Kato (1980); implementation of these techniques has been discussed in detail by Schmitz and Cohen (1985b) and they are not considered here. Furthermore, as shown in detail in Sect. 7, the calculation of the nonequilibrium structure factor (which is our actual goal) is not affected if we continue to assume Eq. (6.19) to hold even for values of the Rayleigh number Ra for which propagating modes are present.

The existence of propagating modes is a new property of fluctuations in inhomogeneous fluids, that does not exist in homogeneous equilibrium fluids, where temperature fluctuations are always purely diffusive. This topic has been theoretically discussed in more detail by Segrè et al. (1993b), and, to some extent, experimentally verified as further discussed in Sect. 10.2.1. A similar nonequilibrium effect has been investigated by Ko and Cohen (1987) for the Taylor-Couette system, for which the viscous mode becomes propagating, also as a consequence of the nonequilibrium forcing.

The right eigenfunctions (hydrodynamic modes) corresponding to the decay rates $\Gamma_N^{(\pm)}(\tilde{q}_\parallel)$ can be formally written as:

$$
\mathbf{U}_N^{R,(\pm)}(\tilde{q}_\parallel, z) = \begin{pmatrix} \dfrac{\alpha_p g \, \tilde{q}_\parallel^2}{(N^2\pi^2 + \tilde{q}_\parallel^2)} \\[2ex] \dfrac{\nu}{L^2}\left[N^2\pi^2 + \tilde{q}_\parallel^2 - \tilde{\Gamma}_N^{(\pm)}(\tilde{q}_\parallel)\right] \end{pmatrix} H_N(z),
\tag{6.34}
$$

where the functions $H_N(z)$ are defined as:

$$
H_N(z) = \begin{cases} \cos\left(\dfrac{N\pi}{L}z\right), & \text{for } N = 1, 3, 5, \ldots \\[2ex] \sin\left(\dfrac{N\pi}{L}z\right), & \text{for } N = 2, 4, 6, \ldots. \end{cases}
\tag{6.35}
$$

The eigenfunctions (6.34) are defined except for a constant normalization coefficient (independent of z). The normalization $B_N(q_\parallel)$ may be computed by substitution of Eq. (6.34) into Eq. (6.17). However, they are more easily calculated by comparing the vectors $\mathbf{U}_N^{R,(+)}$ and $\mathbf{U}_N^{R,(-)}$. Since they correspond to different eigenvalues, they form an orthogonal pair, see Eq. (6.16), so that:

$$
0 = \int_{-\frac{1}{2}L}^{\frac{1}{2}L} \left\{ \nabla T_0 \left[q_\parallel^2 \, W_N^{(+)}(z) \, W_N^{(-)}(z) + \partial_z W_N^{(+)}(z) \, \partial_z W_N^{(-)}(z) \right] \right.
$$
$$
\left. - \alpha_p g q_\parallel^2 \, \Theta_N^{(+)}(z) \, \Theta_N^{(-)}(z) \right\} \, dz.
\tag{6.36}
$$

Next, we observe that the components of the $(\pm)$ eigenvectors in Eq. (6.34) are related in such a way that: $W_N^{(+)} = W_N^{(-)}$ and $\Theta_N^{(+)} = \Theta_N^{(-)} + (\Gamma^{(-)} - \Gamma^{(+)}) H_N$. Substituting these relationships into Eq. (6.36), we obtain for the coefficients $B_N(q_\parallel)$:

$$
B_N^{(\pm)}(\tilde{q}_\parallel) = \pm \frac{\nu}{L^2}\left[N^2\pi^2 + \tilde{q}_\parallel^2 - \tilde{\Gamma}_N^{(\pm)}(\tilde{q}_\parallel)\right] D_N(q_\parallel)
\tag{6.37}
$$

with

$$
D_N(\tilde{q}_\parallel) = \frac{\alpha_p g \nu}{2L^3} \, \tilde{q}_\parallel^2 \left[\tilde{\Gamma}_N^{(+)}(q_\parallel) - \tilde{\Gamma}_N^{(-)}(q_\parallel)\right].
\tag{6.38}
$$

The formulas for the normalization constants $B_N^{(\pm)}$ do not depend on whether N is even or odd. Notice, however, that depending on the value of Ra, the normalization constants $B_N^{(\pm)}(\tilde{q}_\parallel)$ for the hydrodynamic modes may form complex conjugate pairs. For $Ra > 0$ they are always real and negative numbers, for negative values of Ra larger than Eq. (6.33) they are real and positive numbers, while for Ra satisfying the condition (6.33) they may form complex conjugate pairs: $B_N^{(+)*} = B_N^{(-)}$, depending on the value of the horizontal wave number. The constants C_N are real when the corresponding eigenvalues are two real numbers, while they are purely imaginary when the corresponding N-eigenvalues form a complex conjugate pair (propagating modes).

We remark that the conditions for $D_N(\tilde{q}_\parallel) \neq 0$ are the same as the condition (6.33) for the appearance of propagating modes. Thus we conclude that the problem of $B_N^{(\pm)}(q_\parallel) = 0$ occurs at the branch points, and also when $Ra = 0$. For the branch points, we refer to the discussion of propagating modes after Eq. (6.33). In the present context, $Ra = 0$ corresponds to $g = 0$, *i.e.*, to a fluid layer subjected to a temperature gradient without any correction for the presence of gravity to be discussed later in Chapter 7. In any case, upon dividing (6.34) by (6.37) as required for the calculation of the temperature fluctuations, the term causing $B_N^{(\pm)} = 0$ at $Ra = 0$ cancels, so that it will not actually enter into the expression (6.19) for $\delta T(\omega, \mathbf{q}_\parallel, z)$.

The last ingredient we need for the calculation of the nonequilibrium structure factor are the mode-coupling coefficients $C_{NM}(q_\parallel)$. Since the eigenfunctions (6.34) representing the hydrodynamic modes are proportional to trigonometric functions, the mode-coupling coefficients will be zero for $N \neq M$. The only nonzero coefficients will be those mixing the $(+)$ and $(-)$ components for the same value of N. Substituting (6.34) into (6.24), the mode-coupling coefficients can be readily evaluated. The explicit expressions become simpler if one distinguishes the case when the index N corresponds to two real decay rates from the case when N gives a complex conjugate pair. If the two N-decay rates are real numbers, the (real) mode-coupling coefficients are given by the relations:

$$C_{NN}^{(+-)} = C_{NN}^{(-+)} = S_E \frac{\gamma-1}{\gamma} \frac{\alpha_p^2 g^2 \nu^3}{L^9} \tilde{q}_\parallel^6 \frac{Pr+1}{Pr^2} \left\{ \tilde{S}_{NE}^0 - Ra \right\},$$

$$\frac{C_{NN}^{(++)} \Gamma_N^{(-)} + C_{NN}^{(--)} \Gamma_N^{(+)}}{\Gamma_N^{(+)} + \Gamma_N^{(-)}} = S_E \frac{\gamma-1}{\gamma} \frac{4(N^2\pi^2 + \tilde{q}_\parallel^2)}{Pr+1} \frac{\nu D_N^2}{L^3} + C_{NN}^{(+-)}, \quad (6.39)$$

where $\tilde{S}_{NE}^0$ is the same quantity defined by Eq. (4.30). In the second line of (6.39) we have displayed a relationship between the mode-coupling coefficients that will be used for obtaining the structure factor.

If the two N-decay rates form a complex conjugate pair, the (complex) mode-coupling coefficients satisfy:

$$C_{NN}^{(++)} = C_{NN}^{(--)} = S_{\mathrm{E}} \frac{\gamma-1}{\gamma} \frac{\alpha_p^2 g^2 \nu^3}{L^9} \tilde{q}_\parallel^6 \frac{Pr+1}{Pr^2} \left\{ \tilde{S}_{\mathrm{NE}}^0 - Ra \right\},$$

$$\frac{C_{NN}^{(+-)} \Gamma_N^{(-)} + C_{NN}^{(-+)} \Gamma_N^{(+)}}{\Gamma_N^{(+)} + \Gamma_N^{(-)}} = S_{\mathrm{E}} \frac{\gamma-1}{\gamma} \frac{4(N^2 \pi^2 + \tilde{q}_\parallel^2)}{Pr+1} \frac{\nu |D_N|^2}{L^3} + C_{NN}^{(++)}. \quad (6.40)$$

All the other components of the matrix of mode-coupling coefficients C_{NM} not included in Eqs. (6.39)-(6.40) are zero.

As was already discussed after Eq. (6.24), the diagonal mode-coupling coefficients ($C_{NN}^{(++)}$ and $C_{NN}^{(--)}$) are always real numbers, while the non-diagonal terms are complex conjugate pairs. As a consequence, when the N-decay rates correspond to two real numbers we have: $C_{NN}^{(+-)} = C_{NN}^{(-+)}$, as indicated in the first line of Eq. (6.39). When the N-decay rates correspond to a pair of complex conjugate numbers, $C_{NN}^{(++)} = C_{NN}^{(--)}$, as shown in the first line of Eq. (6.40). It can be demonstrated that this last fact is a consequence of the difference $\Gamma_N^{(+)} - \Gamma_N^{(-)}$ being purely imaginary.

6.3   Hydrodynamic modes and decay rates for two rigid boundaries

Following again Schmitz and Cohen (1985b), we start by searching for solutions to the problem (6.8) that are proportional to $\exp(\tilde{\lambda}z/L)$. From the corresponding secular equation, it is found that $\tilde{\lambda}$ has to be one of the six roots of the following sixth-order algebraic equation:

$$(\tilde{q}_\parallel^2 - \tilde{\lambda}_j^2)^3 - \tilde{\Gamma}(Pr+1)(\tilde{q}_\parallel^2 - \tilde{\lambda}_j^2)^2 + \tilde{\Gamma}^2 Pr(\tilde{q}_\parallel^2 - \tilde{\lambda}_j^2) = \tilde{q}_\parallel^2 Ra, \quad (6.41)$$

where, as in the previous section, we use dimensionless decay rates $\tilde{\Gamma} = \Gamma L^2/\nu$. In Eq. (6.41), the index $j = 0, \ldots, 5$ is used to enumerate the six $\tilde{\lambda}$-roots for given values of $\tilde{\Gamma}$ and of the other dimensionless parameters. Explicit expressions for $\tilde{\lambda}_j^2(\tilde{q}_\parallel, \tilde{\Gamma})$ are given by the formulas for the roots of a cubic equation, but the resulting expressions are quite complicated and not very informative; therefore, we do not specify them here, although they have been used in some of the following calculations. Since Eq. (6.41) is quadratic in $\tilde{\lambda}_j$, there are three roots with a positive real part and three roots with a negative real part. We choose the order of the roots in such a way that for $j = 0, 1, 2$ the real part of $\tilde{\lambda}_j$ is positive. Because of the nature of the roots and the symmetry of the boundary conditions (6.10b), the hydrodynamic modes $\mathbf{U}^{\mathrm{R}}$ possess a definite parity, just as in the previous section. It is advantageous to classify them in even $\mathbf{U}^{\mathrm{R,E}}$ and odd $\mathbf{U}^{\mathrm{R,O}}$ modes or eigenfunctions, with corresponding even $\Gamma^{\mathrm{E}}(\tilde{q}_\parallel)$ and odd $\Gamma^{\mathrm{O}}(\tilde{q}_\parallel)$

decay rates. For the case of two free boundaries examined in Sect. 6.2, the odd modes were the ones proportional to $\cos(N\pi z)$, while the even modes were the ones proportional to $\sin(N\pi z)$. Now, in view of the two first BC (6.10b), we find it convenient to express the even right eigenfunctions as:

$$
\mathbf{U}_N^{\mathrm{R,E}}(\tilde{q}_\|, z) = \sum_{j=0}^{2} \frac{1}{A_j(q_\|, \Gamma)}
\begin{bmatrix}
\dfrac{\alpha_p g \tilde{q}_\|^2}{\tilde{\lambda}_j^2 - \tilde{q}_\|^2} \\[2ex]
\dfrac{\nu}{L^2}\left(\tilde{\Gamma} + \tilde{\lambda}_j^2 - \tilde{q}_\|^2\right)
\end{bmatrix}
\frac{\cosh(\tilde{\lambda}_j\, z/L)}{\cosh(\frac{1}{2}\tilde{\lambda}_j)}
\qquad (6.42)
$$

with

$$
A_j(\tilde{q}_\|, \tilde{\Gamma}) =
$$

$$
\frac{Pr\tilde{\Gamma}^2 - 2\left[\tilde{q}_\|^2 - \tilde{\lambda}_j^2(\tilde{q}_\|, \tilde{\Gamma})\right](Pr+1)\tilde{\Gamma} + 3\left[\tilde{q}_\|^2 - \tilde{\lambda}_j^2(\tilde{q}_\|, \tilde{\Gamma})\right]^2}{\left[\tilde{q}_\|^2 - \tilde{\lambda}_j^2(\tilde{q}_\|, \tilde{\Gamma})\right]\left[\tilde{q}_\|^2 - \tilde{\lambda}_j^2(\tilde{q}_\|, \tilde{\Gamma}) - Pr\tilde{\Gamma}\right]},
\qquad (6.43)
$$

where $\tilde{\lambda}_j(q_\|, \Gamma)$, $j = \{0, 1, 2\}$, are the three complex roots of Eq. (6.41) with positive real part. We note that the eigenfunctions (6.42) already satisfy the two first BC (6.10b), since from (6.41) it follows that:

$$
\sum_{j=0}^{2} \frac{1}{(\tilde{\lambda}_j^2 - q_\|^2)A_j} = \sum_{j=0}^{2} \frac{\tilde{\Gamma} + \tilde{\lambda}_j^2 - \tilde{q}_\|^2}{A_j} = 0.
\qquad (6.44)
$$

If the BC are satisfied at $z = \frac{1}{2}L$, then the BC at $z = -\frac{1}{2}L$ are satisfied automatically because of the parity of the functions in the vertical variable. As the next step we require the first component of $\mathbf{U}_N^{\mathrm{R,E}}(q_\|, z)$ to satisfy the third boundary condition (6.10c) at $z = \frac{1}{2}L$, which implies that:

$$
\sum_{j=0}^{2} \frac{\left[\tilde{\lambda}_j^2(\tilde{q}_\|, \tilde{\Gamma}) - \tilde{q}_\|^2 + Pr\tilde{\Gamma}\right]\tilde{\lambda}_j(\tilde{q}_\|, \tilde{\Gamma})\,\tanh\left[\frac{1}{2}\tilde{\lambda}_j(\tilde{q}_\|, \tilde{\Gamma})\right]}{Pr\tilde{\Gamma}^2 - 2\left[\tilde{q}_\|^2 - \tilde{\lambda}_j^2(\tilde{q}_\|, \tilde{\Gamma})\right](Pr+1)\tilde{\Gamma} + 3\left[\tilde{q}_\|^2 - \tilde{\lambda}_j^2(\tilde{q}_\|, \tilde{\Gamma})\right]^2} = 0.
\qquad (6.45)
$$

Upon substitution of the three solutions $\tilde{\lambda}_j(\tilde{q}_\|, \tilde{\Gamma})$ of (6.41) with positive real part into Eq. (6.45), we obtain a complicated algebraic equation from which the decay rates $\tilde{\Gamma}(q_\|)$ of the even eigenfunctions can be determined. In general, this equation can only be solved numerically. Due to the periodicity of the hyperbolic tangent, there is an infinite numerable set of solutions for the even eigenvalues of the differential operator $\mathcal{H}$ for two rigid boundaries that we represent by $\tilde{\Gamma}_N^{\mathrm{E}}(\tilde{q}_\|)$.

The odd eigenfunctions have a structure similar to Eq. (6.42), but with the hyperbolic cosines replaced by hyperbolic sines. They satisfy the two first BC (6.10b) also. Imposing the boundary condition (6.10d), we obtain a secular equation similar to Eq. (6.45), but with the hyperbolic tangent in the numerator replaced by a hyperbolic cotangent. Using a similar numerical procedure as used for calculating $\tilde{\Gamma}_N^{\rm E}(\tilde{q}_\parallel)$, we can compute the set of odd decay rates $\tilde{\Gamma}_N^{\rm O}(\tilde{q}_\parallel)$.

To determine the eigenfunctions completely, we need to calculate the coefficients $B_N(\tilde{q}_\parallel)$ defined by Eq. (6.17). The explicit expressions in terms of $\tilde{\lambda}_j(q_\parallel, \Gamma_N^E)$ and $\Gamma_N^E(\tilde{q}_\parallel)$, obtained upon substituting Eq. (6.42) into Eq. (6.17) and performing the vertical integrations, are again complicated and not given here. However, explicit expressions for $\tilde{\lambda}_j(q_\parallel, \Gamma_N^E)$ and $\Gamma_N^E(\tilde{q}_\parallel)$ in the small-$q_\parallel$ and the large-$q_\parallel$ limits will be presented later in Sects. 6.3.1 and 6.3.2, respectively.

For an analysis of the nonequilibrium structure factor measured in experiments in Sect. 7.2, we shall need the result of the vertical integration of the temperature component of the even hydrodynamic modes divided by the normalization constants, namely:

$$
\begin{aligned}
\Xi_N^{\rm E}(\tilde{q}_\parallel) &= \frac{1}{B_N^{\rm E}(q_\parallel)} \int_{-\frac{1}{2}L}^{\frac{1}{2}L} \Theta_N^{\rm E}(q_\parallel, z)\, dz, \\
&= \frac{\nu}{L}\, \frac{1}{B_N^{\rm E}(q_\parallel)} \sum_{j=0}^{2} \frac{\left(\tilde{\Gamma} + \tilde{\lambda}_j^2 - \tilde{q}_\parallel^2\right)}{A_j(q_\parallel, \Gamma)} \int_{-\frac{1}{2}}^{\frac{1}{2}} \frac{\cosh(\tilde{\lambda}_j\, \tilde{z})}{\cosh(\frac{1}{2}\tilde{\lambda}_j)}\, d\tilde{z}.
\end{aligned}
\tag{6.46}
$$

A numerical investigation of the dependence of the decay rates on Ra and Pr has been carried out by Schmitz and Cohen (1985b). They demonstrated that for positive Ra all decay rates are real numbers, so that the corresponding modes are diffusive. For negative Ra, as in Sect. 6.2 for free boundaries, propagating modes may exist. Schmitz and Cohen (1985b) analyzed the conditions under which such propagating modes may be present. We shall not further investigate the propagating modes for the case of rigid boundaries here. We shall be primarily interested in the dependence of the decay rates on $q_\parallel$, rather than the dependence on Ra or Pr studied by Schmitz and Cohen (1985b). Specifically, we shall consider the limiting cases $q_\parallel \to 0$ and $q_\parallel \to \infty$, for which interesting analytical expansions of the decay rates can be obtained. In these two asymptotic limits there are no propagating modes; as shown in Fig. 6.1 for free boundaries, propagating modes only appear at intermediate values of the horizontal wave numbers.

6.3.1   Decay rates for rigid boundaries in the limit of small $q_\parallel$

To simplify the notation in this and the following subsection we shall always use dimensionless wave numbers and decay rates, so that we skip the tildes above the corresponding symbols. For small values of $q_\parallel$ it is possible to find a series expansion of the decay rates in powers of $q_\parallel^2$. We start by assuming that such a series exists:

$$\Gamma_N(q_\parallel) = \Gamma_{N,0} + \Gamma_{N,2}\, q_\parallel^2 + \Gamma_{N,4}\, q_\parallel^4 + \mathcal{O}(q_\parallel^6). \tag{6.47}$$

Substituting Eq. (6.47) into Eq. (6.41), expanding in powers of $q_\parallel$, and solving systematically up to $\mathcal{O}(q_\parallel^4)$, we find the square of roots $\tilde{\lambda}_N^2(q_\parallel)$:

$$\tilde{\lambda}_N^2(q_\parallel) = \begin{bmatrix} 0 \\ -\Gamma_{N,0} \\ -Pr\,\Gamma_{N,0} \end{bmatrix} + \begin{bmatrix} 1 - \dfrac{Ra}{Pr(\Gamma_{N,0})^2} \\[2mm] 1 - \Gamma_{N,2} + \dfrac{Ra}{(Pr-1)(\Gamma_{N,0})^2} \\[2mm] 1 - Pr\,\Gamma_{N,2} - \dfrac{Ra}{Pr(Pr-1)(\Gamma_{N,0})^2} \end{bmatrix} q_\parallel^2 + \mathcal{O}(q_\parallel^4). \tag{6.48}$$

Next, we need to distinguish between the case of even and of odd hydrodynamic modes. For the even ones, we substitute Eqs. (6.47) and (6.48) into Eq. (6.45), expand the resulting expression in powers of $q_\parallel^2$. We then find that there are two kinds of solutions for the even decay rates, to be designated by $(\pm)$ as in Sect. 6.2. The series expansion (6.47) for the solution of the first kind has the coefficients:

$$\Gamma_{N,0}^{(E,-)} = \frac{N^2\pi^2}{Pr}$$

$$\Gamma_{N,2}^{(E,-)} = \frac{1}{Pr} + \frac{Ra}{N^4\pi^4(Pr-1)}\left[\frac{4\sqrt{Pr}}{N\pi(Pr-1)}\cot\left(\frac{N\pi}{2\sqrt{Pr}}\right) - 1\right], \tag{6.49}$$

for odd integers $N = 1, 3, 5\ldots$. Indeed, the coefficients (6.49) ensure the cancellation of the zeroth-order term in the $q_\parallel$-series expansion of Eq. (6.45). For the decay rates of the second kind, solving Eq. (6.45) up to zeroth order only determines the first coefficient in (6.47). To obtain the second coefficient, we need to cancel also the term of $\mathcal{O}(q_\parallel^2)$ in the $q_\parallel$-series expansion of Eq. (6.45). We then obtain:

$$\Gamma_{N,0}^{(E,+)} = N^2\pi^2,$$

$$\Gamma_{N,2}^{(E,+)} = -1 + \frac{Ra}{N^4\pi^4 Pr(Pr-1)}\left[3Pr - 2 - \frac{4\tan\left(N\frac{\pi}{2}\sqrt{Pr}\right)}{N\pi\sqrt{Pr}(Pr-1)}\right]. \tag{6.50}$$

Equation (6.50) is valid for even integers $N = 2, 4, 6\ldots$.

For the case of the odd decay rates, solving Eq. (6.45), with coth instead of tanh, in zeroth order, we again find two kind of solutions. The coefficients in the series expansion (6.47) for the odd modes of the first kind are:

$$\Gamma_{N,0}^{(O,-)} = \frac{N^2\pi^2}{Pr}, \tag{6.51}$$

$$\Gamma_{N,2}^{(O,-)} = \frac{1}{Pr} - \frac{Ra}{N^4\pi^4(Pr-1)}\left\{1 + \frac{2}{(Pr-1)\left[\frac{N\pi}{2\sqrt{Pr}}\cot\left(\frac{N\pi}{2\sqrt{Pr}}\right)-1\right]}\right\},$$

for odd integers $N = 1, 3, 5 \ldots$. As was the case for the even decay rates, to obtain the solution of the second kind for the odd ones solving Eq. (6.45) up to order zeroth-order in $q_\parallel^2$ only provides the first coefficient, which are the roots of the trascendent equation:

$$\sqrt{\Gamma_{N,0}^{(O,+)}}\,\cot\left(\frac{1}{2}\sqrt{\Gamma_{N,0}^{(O,+)}}\right) - 2 = 0. \tag{6.52}$$

Hence, for the odd decay rates of the second kind it is not possible to write down explicit expressions like Eq. (6.51)-(6.50) above. However, it is clear that Eq. (6.52) has an infinite numerable set of solutions, $\Gamma_{N,0}^{(O,+)}$. Numerically we find for the first root of Eq. (6.52): $\Gamma_{1,0}^{(O,+)} \approx 80.7629$.

For free boundaries we obtained in Sect. 6.2 analytical expressions for the decay rates in the full range of $q_\parallel$, and not just in the limit $q_\parallel \to 0$ discussed here for rigid boundaries. Nevertheless, there is a close analogy in the nature of the decay rates for free and rigid boundaries. In both cases, for each order N of the respective even/odd hydrodynamic modes, we find two decay rates which have been designated as ($\pm$). It is interesting to note that performing a small-$q_\parallel$ expansion of (6.32) for the decay rates for free boundaries, we obtain the same limits as here for rigid boundaries; *i.e.*, $N^2\pi^2/Pr$ for the ($-$) eigenvalues and $N^2\pi^2$ for the ($+$) eigenvalues. However, notice that the limit $N^2\pi^2$ with even N corresponds to even modes in the case of rigid boundaries, while it corresponds to odd modes for free boundaries. The $q_\parallel \to 0$ limit of the odd decay rates for rigid boundaries (not discussed here) has no correspondence with the small-$q_\parallel$ limit of the decay rates for free boundaries. We also emphasize that the small-$q_\parallel$ expansions of the decay rates for rigid boundaries have real coefficients (6.49)-(6.52), independent of the Ra number. This confirms that the phenomenon of propagating modes only occurs in a finite "window" of intermediate horizontal wave numbers, as discussed previously for free boundaries in Sect. 6.2

As a final remark about the decay rates, note that for typical values of the Prandtl number ($Pr > 1/4$), the lower decay rate corresponds to the first $\Gamma^{(E,-)}$ and attains the nonzero limit π^2/Pr when $q_\parallel \to 0$, see Eq. (6.49). This fact is true for both rigid and free boundaries, and in both cases the corresponding hydrodynamic mode has even vertical parity.

For the calculation of the nonequilibrium enhancement of the structure factor in the small-$q_\parallel$ limit, to be performed in Sect. 7.3, we need the normalization constants for the even modes. They can be obtained by substituting Eqs. (6.48)-(6.50) into Eq. (6.42), and then into the definition (6.17) of $B_N(q_\parallel)$. After performing the vertical integration in (6.17), we expand the result in a power series of $q_\parallel^2$ to obtain for the even hydrodynamic modes:

$$B_N^{(E,-)}(q_\parallel) = -\frac{\alpha_p g \nu^2}{L^5} \frac{(Pr-1)^2 N^4 \pi^4}{2Pr} \tan^2\left(\frac{N\pi}{2\sqrt{Pr}}\right) q_\parallel^2 + \mathcal{O}(q_\parallel^4)$$

$$B_N^{(E,+)}(q_\parallel) = -\frac{\alpha_p g \nu^2}{L^5} \frac{Ra}{2Pr\, N^2\pi^2} q_\parallel^4 + \mathcal{O}(\tilde{q}_\parallel^6), \tag{6.53}$$

where the first line is for odd integers $N = 1, 3, 5 \ldots$, while the second line is for even integers $N = 2, 4, 6 \ldots$. A similar expansion can be obtained for the normalization constants of the odd hydrodynamic modes, $\tilde{B}_N^{(O,-)}(q_\parallel)$. However, we do not present it here, since it will not be needed for our goal which is to obtain the nonequilibrium structure factor.

Finally, substitution of (6.48)-(6.50) into Eq. (6.46), enables us to evaluate the ratios $\Xi_N^E(\tilde{q}_\parallel)$ of the vertical integral of the temperature component of the modes to the normalization constants, for which we obtain:

$$\Xi_N^{(E,-)}(q_\parallel) = \frac{-L^4}{\alpha_p g \nu} \frac{4\sqrt{Pr}}{(Pr-1)N^3\pi^3} \cot\left(\frac{N\pi}{2\sqrt{Pr}}\right) \frac{1}{q_\parallel^2} + \mathcal{O}(1) \tag{6.54}$$

$$\Xi_N^{(E,+)}(q_\parallel) = \frac{L^4}{\alpha_p g \nu} \frac{2}{N^2\pi^2} \left[\frac{2\tan\left(N\frac{\pi}{2}\sqrt{Pr}\right)}{\sqrt{Pr}(Pr-1)N\pi} + 1\right] \frac{1}{q_\parallel^2} + \mathcal{O}(1),$$

where, again, the first line is for odd integers $N = 1, 3, 5 \ldots$, while the second line is for even integers $N = 2, 4, 6 \ldots$. We note that for the modes of both $(\pm)$ kinds, the ratios $\Xi_N^{(E,\pm)}$ are $\mathcal{O}(\tilde{q}_\parallel^{-2})$ and independent of Ra. The calculation of the mode-coupling coefficients $C_{NM}(q_\parallel)$ from Eq. (6.24) in the small-$q_\parallel$ limit will be addressed in Sect. 7.3.

The expressions obtained above for the coefficients of the series expansion (6.47) for the decay rates will not be strictly valid for a large number of Pr values, because of divergences in Eq. (6.50). However, as will be elucidated in Sect. 7.3, it turns out that for the calculation of the small-wavenumber expansion of the static structure factor, the divergences cancel, and we shall indeed obtain a result for the nonequilibrium static structure factor that is valid for any value of Pr.

6.3.2 Decay rates for rigid boundaries in the limit of large $q_\parallel$

It is also possible to solve equations (6.41) and (6.45) perturbatively for large values of $q_\parallel$. Since the leading term in the expansion for the decay rates is of order $q_\parallel^2$, we start by assuming that the decay rates in the large-$q_\parallel$ limit have an asymptotic expansion of the form:

$$\Gamma_N(q_\parallel) = \Gamma_{N,-2}\, q_\parallel^2 + \Gamma_{N,0} + \Gamma_{N,1}\,\frac{1}{q_\parallel} + \Gamma_{N,2}\,\frac{1}{q_\parallel^2} + \Gamma_{N,3}\,\frac{1}{q_\parallel^3} + \cdots \quad (6.55)$$

Substituting (6.55) into (6.41), we obtain the asymptotic expansion for the roots $\tilde\lambda_j(q_\parallel)$:

$$\tilde\lambda_N^2(q_\parallel) = \begin{bmatrix} 1 \\ 1 - \Gamma_{N,-2} \\ 1 - Pr\,\Gamma_{N,-2} \end{bmatrix} q_\parallel^2 - \begin{bmatrix} 0 \\ \Gamma_{N,0} \\ Pr\,\Gamma_{N,0} \end{bmatrix} - \begin{bmatrix} 0 \\ \Gamma_{N,1} \\ Pr\,\Gamma_{N,1} \end{bmatrix} \frac{1}{q_\parallel} \quad (6.56)$$

$$+ \begin{bmatrix} \dfrac{-Ra}{Pr\,\Gamma_{N,-2}^2} \\[2mm] \dfrac{Ra}{(Pr-1)\,\Gamma_{N,-2}^2} - \Gamma_{N,2} \\[2mm] \dfrac{-Ra}{Pr(Pr-1)\,\Gamma_{N,-2}^2} - Pr\,\Gamma_{N,2} \end{bmatrix} \frac{1}{q_\parallel^2} - \begin{bmatrix} 0 \\ \Gamma_{N,3} \\ Pr\,\Gamma_{N,3} \end{bmatrix} \frac{1}{q_\parallel^3} + \cdots$$

As was the case in the previous section for small $q_\parallel$, we need to distinguish between even and odd modes. To obtain the expansion of the even decay rates for large-$q_\parallel$, we substitute Eqs. (6.55) and (6.56) into Eq. (6.45), and expand the resulting expression in powers of $q_\parallel^{-1}$. We find that the leading $\mathcal{O}(q_\parallel)$ term cancels only if $\Gamma_{N,-2} = 1$ or $\Gamma_{N,-2} = Pr^{-1}$. Again, we find two solutions for the even decay rates depending on the value of the $\Gamma_{N,-2}$ term in the asymptotic expansion (6.55). We shall retain the notation $(\pm)$ of the previous sections. For the eigenvalues of the $(+)$ kind, cancellation of the leading term in the asymptotic expansion of (6.45) determines the three first terms in (6.55), namely:

$$\Gamma_{N,-2}^{(E,+)} = 1, \qquad \Gamma_{N,0}^{(E,+)} = N^2\pi^2, \qquad \Gamma_{N,1}^{(E,+)} = 4N^2\pi^2, \qquad (6.58)$$

for odd integers $N = 1,3,5\ldots$ For the decay rates of the $(-)$ kind, the cancellation of the leading term in the asymptotic expansion of (6.45) determines the five first terms in (6.55), namely:

$$\Gamma_{N,-2}^{(E,-)} = \frac{1}{Pr}, \qquad \Gamma_{N,0}^{(E,-)} = \frac{N^2\pi^2}{Pr}, \qquad \Gamma_{N,1}^{(E,-)} = 0,$$

$$\Gamma_{N,2}^{(E,-)} = \frac{-Ra}{Pr-1}, \qquad \Gamma_{N,3}^{(E,-)} = 0, \qquad \Gamma_{N,4}^{(E,-)} = \frac{2\pi^2 N^2\,Ra}{Pr-1}, \qquad (6.59)$$

$$\Gamma_{N,5}^{(E,-)} = \frac{4N^2\pi^2\sqrt{Pr}\,Ra}{(Pr-1)^2\left[\sqrt{Pr} - \sqrt{Pr-1}\right]},$$

for odd integers $N = 1, 3, 5 \ldots$

The coefficients for the expansion of the odd eigenvalues can be obtained in a similar way, but they are not actually required for the derivation of the nonequilibrium structure factor (see Sect. 7.2).

Next, we determine the normalization constants. Substituting Eqs. (6.55) and (6.56) into the definition of B_N, performing the corresponding integrals and retaining only the leading term in a large-$q_\parallel$ expansion, we obtain:

$$B_N^{(E,-)}(q_\parallel) = -\frac{\alpha_p g \nu^2}{L^5} \frac{(Pr-1)^2 \left[\sqrt{Pr} - \sqrt{Pr-1}\right]^2}{2\pi^2 N^2 Pr} q_\parallel^8 + \mathcal{O}(q_\parallel^6),$$

$$B_N^{(E,+)}(q_\parallel) = -\frac{\alpha_p g \nu^2}{L^5} \frac{Ra}{Pr} \frac{1}{2N^2\pi^2} q_\parallel^4 + \mathcal{O}(q_\parallel^3). \tag{6.60}$$

We remind the reader that in this section we are always using dimensionless wave numbers and decay rates. Finally, substitution of the asymptotic expansions (6.55) and (6.56) into Eq. (6.46) enables us to evaluate the ratios of the vertical integrals of the temperature component of the hydrodynamic modes to the normalization constants (6.60), for which we obtain:

$$\Xi_N^{(E,-)}(q_\parallel) = \frac{-L^4}{\alpha_p g \nu} \frac{4\sqrt{Pr}}{(Pr-1)\left[\sqrt{Pr} - \sqrt{Pr-1}\right]} \frac{1}{q_\parallel^5} + \mathcal{O}\left(\frac{1}{q_\parallel^7}\right),$$

$$\Xi_N^{(E,+)}(q_\parallel) = \frac{L^4}{\alpha_p g \nu} \frac{4Pr}{(Pr-1)} \frac{1}{q_\parallel^5} + \mathcal{O}\left(\frac{1}{q_\parallel^7}\right). \tag{6.61}$$

We note that for the even modes of both kinds, the ratio $\Xi_N^{(E,\pm)}$ is $\mathcal{O}(\tilde{q}_\parallel^{-5})$, independently of Ra and of the index N. The calculation of the mode-coupling coefficients $C_{NM}(q_\parallel)$ in the large-$q_\parallel$ limit will be addressed in Sect. 7.3.

6.4   The slowest decay rate

The decay rates of the hydrodynamic operator for $Ra > 0$ form a discrete set of well-separated real and positive numbers. The slowest decay rate is $\Gamma_0^E(q_\parallel)$, which corresponds to a hydrodynamic mode of even vertical parity. For the case of free boundaries, Eq. (6.32) gives an explicit expression for all the decay rates. For usual fluids ($Pr > 1$), it follows from (6.32) that the slowest decay rate for free boundaries corresponds to $N = 1$ and the minus sign: $\Gamma_1^{(-)}(q)$. Since the system will become unstable when the decay rate of at least one of the modes becomes zero, so that the critical condition is $\Gamma_1^{(-)}(q) = 0$, or:

$$Ra = \frac{(\pi^2 + \tilde{q}_\parallel^2)^3}{\tilde{q}_\parallel^2}. \tag{6.62}$$

We note that the instability condition (6.62) is independent of the Prandtl number Pr. As is well known, from condition (6.62) it follows that the system will be unstable when $Ra \geq Ra_c = 27\pi^4/4$, corresponding to the global minimum of the RHS of Eq. (6.62) attained at $\tilde{q}_{\|c} = \pi/\sqrt{2}$ (Chandrasekhar, 1961).

When $\tilde{q}_\| = \tilde{q}_{\|,c}$, the slowest decay rate is zero for $Ra = Ra_c$. If Ra differs slightly from the critical value one may write (Swift and Hohenberg, 1977; Westried et al., 1978):

$$Pr\, \tilde{\Gamma}_1^{(-)}(\tilde{q}_{\|,c}) = -\tau_0^{-1}\, \epsilon \tag{6.63}$$

where $\epsilon = (Ra - Ra_c)/Ra_c$ is the distance to the threshold. Note that the minus sign in the RHS of Eq. (6.63) arises because the coefficient τ_0 is usually defined in terms of growth rates instead of decay rates as here, while the factor Pr in the LHS of Eq. (6.63) accounts for a slightly different definition of the decay rates commonly used in the literature (Swift and Hohenberg, 1977; Westried et al., 1978). The relevance of Eq. (6.63) is that it defines a generic class of instabilities, *i.e.*, instabilities where the critical growth rate increases linearly with the distance to threshold, as opposed to other possibilities as quadratic or algebraic dependence on ϵ. For the case of RB between two free boundaries, upon replacing Ra by $Ra_c(1 + \epsilon)$ in Eq. (6.32) and expanding in powers of ϵ, we see that $\tilde{\Gamma}_1^{(-)}(\tilde{q}_{\|,c})$ indeed satisfies Eq. (6.63) with (Swift and Hohenberg, 1977; Westried et al., 1978; Hohenberg and Swift, 1992):

$$\tau_0^{-1} = \frac{3\pi^2}{2} \frac{Pr}{Pr + 1}, \tag{6.64}$$

so that in the case of Rayleigh-Bénard convection, τ_0 depends on the Prandtl number.

For the realistic case of a fluid layer confined between two rigid boundaries, the decay rates for even vertical parity modes are obtained by solving the set of algebraic equations (6.41) and (6.45) numerically. The slowest (first) eigenvalue obtained from this procedure has been studied, after the pioneering work of Pellew and Southwell (1940), by many authors (Westried et al., 1978; Domínguez Lerma et al., 1984). Most of these studies have been concerned with an instability analysis for the onset of convection, while we are here interested in the dynamics of the fluctuations for arbitrary $q_\|$. As shown by Eq. (6.26), the nonequilibrium static structure factor of a fluid subjected to a stationary temperature gradient is represented by a (double) series expansion in the hydrodynamic modes of even vertical parity. The most important terms in this series correspond to the modes with slower decay rates, so that very good numerical approximations are obtained by only retaining a few of the lowest $\tilde{\Gamma}_N^E(q_\|)$. As an illustration, we present in Table 6.1 values for the first two dimensionless even decay rates, $\tilde{\Gamma}_1^{(E,-)}$ and

Table 6.1: Slowest pair of decay rates for modes with even vertical parity and for rigid boundaries, as computed numerically from Eqs. (6.41) and (6.45). Data are for $Ra = 1650$ and $Pr = 8$.

$\tilde{q}$	0.15	0.25	0.5	1	1.5	2	2.5
$\tilde{\Gamma}_1^{(E,-)}$	1.2272	1.2157	1.1630	0.9671	0.6926	0.4068	0.1823
$\tilde{\Gamma}_1^{(E,+)}$	39.465	39.442	39.340	39.006	38.681	38.601	38.955
$\tilde{q}$	2.75	3	3.25	3.5	3.75	4.25	4.75
$\tilde{\Gamma}_1^{(E,-)}$	0.1129	0.0803	0.0876	0.1378	0.2320	0.5506	1.0323
$\tilde{\Gamma}_1^{(E,+)}$	39.334	39.862	40.542	41.380	42.376	44.845	47.949
$\tilde{q}$	5.25	5.75	6.25	7.5	8.5	10	14
$\tilde{\Gamma}_1^{(E,-)}$	1.6585	2.4089	3.2645	5.7687	8.0726	11.951	24.681
$\tilde{\Gamma}_1^{(E,+)}$	51.680	56.032	60.994	76.034	90.679	116.87	210.39

$\tilde{\Gamma}_1^{(E,+)}$, computed numerically by solving (6.41) and (6.45) for $Ra = 1650$ and $Pr = 8$. For horizontal wave numbers larger than $\tilde{q} \simeq 14$ we encountered numerical convergence problems in the calculation of the decay rates: thus we do not present numerical results for $\tilde{q}$ larger than 14. Values displayed in Table 6.1 will be later used to calculate the nonequilibrium structure factor for the case of two rigid boundaries.

Figure 6.2 shows the dimensionless decay rates $\tilde{\Gamma}_1^{(E,-)}$ and $\tilde{\Gamma}_1^{(E,+)}$ as a function of $\tilde{q}_{\parallel}$ for $Ra = 1650$ and $Pr = 8$ and for $Ra = 150$ and $Pr = 8$. We observe in Fig. 6.2 that the slowest (first) decay rate develops a prominent minimum for $\tilde{q}_{\parallel}$ values close to $\tilde{q}_{\parallel,c}$, when Ra approaches Ra_c. An interesting question is how the wave number $\tilde{q}_m$ corresponding to the the the minimum of the decay rate depends on the Prandtl and Rayleigh numbers. This has been studied by Domínguez Lerma et al. (1984), as will be discussed in Sect. 8.6.

In the the previous subsections we obtained analytic expressions for the decay rates of modes with even vertical parity in the limits $q_{\parallel} \to 0$ and $q_{\parallel} \to \infty$. For arbitrary $q_{\parallel}$, the decay rates of a fluid between two rigid boundaries can only be investigated numerically. We have verified that the numeral results displayed in Fig. 6.2 reproduce the predicted asymptotic behavior in these limits. We recall that the asymptotic limit for large $q_{\parallel}$ depends on whether the decay rate corresponds to even modes of the $(+)$ or the $(-)$ kind, while it is independent of the index N of the eigenvalue, the Ra number or the Pr number. From Eqs. (6.58) and (6.59) we see that in the $q_{\parallel} \to \infty$ limit:

$$\tilde{\Gamma}_1^{(E,-)}(q_{\parallel}) = \frac{1}{Pr}\, \tilde{q}_{\parallel}^2 + \mathcal{O}(1), \qquad \tilde{\Gamma}_1^{(E,+)}(q_{\parallel}) = \tilde{q}_{\parallel}^2 + \mathcal{O}(1). \qquad (6.65)$$

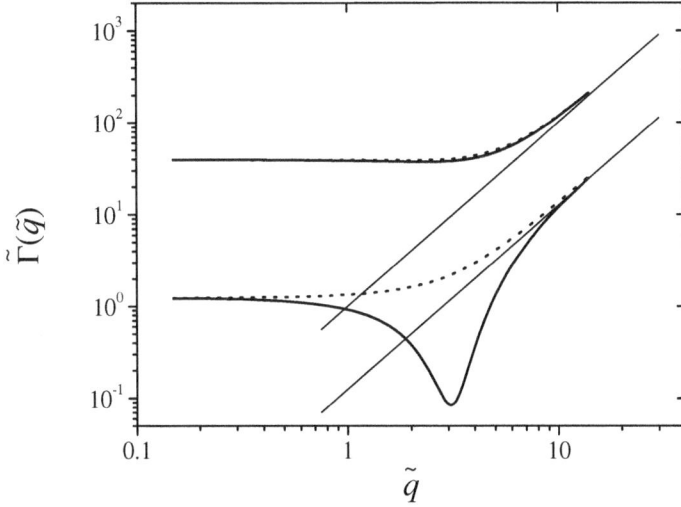

Figure 6.2: Slowest pair of decay rates for rigid boundaries as a function of $\tilde{q}$. The two upper curves represent $\Gamma_1^{(+)}$, and the two lower curves $\Gamma_1^{(-)}$. Solid curves are for $Ra = 1650$, which is close to the instability ($\epsilon = -0.035$). Dotted curves are for $Ra = 150$. $Pr = 8$ in all cases. Thin lines indicate the asymptotic behavior for large $\tilde{q}$ in accordance with Eq. (6.65).

The decay rates computed numerically do indeed agree with the expected asymptotic behavior (6.65) indicated by the two thin lines in Fig. 6.2. It is interesting that the decay rates (6.32) for two free boundaries have the same asymptotic behavior for large $q_\parallel$ as the decay rates for rigid boundaries, *i.e.*, Eq. (6.65) is common for the two sets of BC. Furthermore, this asymptotic behavior for large (horizontal) wave number is identical to the large-q limit of the decay rates (4.17) of the "bulk" fluctuations, that were obtained without considering BC in the vertical direction. This result is to be expected, because fluctuations with large wave number correspond to very small wavelengths at which the presence of boundaries should no longer affect the fluctuations.

In Fig. 6.2 we see that the decay rates reach a constant limit for small-$q_\parallel$, in accordance which Eq. (6.47). Indeed, from Eq. (6.49) it follows that the first decay rate should have the following exact asymptotic limit for $q_\parallel \to 0$:

$$\tilde{\Gamma}_1^{(E,-)}(\tilde{q}_\parallel) = \frac{\pi^2}{Pr} + \mathcal{O}(\tilde{q}_\parallel^2), \tag{6.66}$$

independent of the Ra number. The slowest decay rate, when computed numerically, does indeed agree with Eq. (6.66) for $q_\parallel \to 0$.

As we did for free boundaries, the parameter τ_0 defined by Eq. (6.63) can be evaluated for rigid boundaries. We do not give details, but the procedure is as follows: Upon substitution of Eq. (6.63) into Eq. (6.41) with $Ra = Ra_c(1 + \epsilon)$, and by solving the resulting expression perturbatively, we obtain the roots $\tilde{\lambda}_j$ up to $\mathcal{O}(\epsilon)$ for $\tilde{q}_\parallel = \tilde{q}_{\parallel c}$. Then, substituting those roots into Eq. (6.45) and expanding in powers of ϵ, we see that it is possible to cancel the $\mathcal{O}(\epsilon)$ term if (Hohenberg and Swift, 1992):

$$\tau_0^{-1} = \frac{19.649 \, Pr}{Pr + 0.512}. \tag{6.67}$$

This demonstrates that RB convection for rigid boundaries belongs to the same instability class as for free boundaries, *i.e.*, the growth rate at critical wave number increases linearly with the distance to the instability.

In closing this section we remark that it is possible to extend the calculation sketched above to obtain an expansion of the slowest decay rate in powers of ϵ not only for $\tilde{q}_\parallel = \tilde{q}_{\parallel c}$, but also for horizontal wave numbers close to, but different from, $\tilde{q}_{\parallel c}$. Such an analysis has been presented by Cross (1980) and Cross and Hohenberg (1993). We shall further discuss this topic in Sect. 8.3, in the context of the so-called Swift and Hohenberg (1977) approximation.

Chapter 7

Thermal nonequilibrium fluctuations in one-component-fluid layers

With the procedure developed in the preceding chapter, we are now ready to evaluate and analyze the nonequilibrium structure factor of a one-component fluid confined between horizontal plates and subjected to a stationary temperature gradient. We start this program by considering in Sect. 7.1 the nonequilibrium structure factor for a fluid layer between two free boundaries for which explicit analytic expressions can be obtained. In Sect. 7.2 we consider the more realistic case of a fluid layer between two rigid boundaries. As discussed in Sect. 6.3, the decay rates for a fluid between two rigid boundaries as functions of the wave number can be computed only numerically; as a consequence, only computational results for the nonequilibrium structure factor will be presented in this section. However, it is possible to obtain explicit expressions for the asymptotic behavior of the nonequilibrium structure factor of a fluid between two rigid boundaries in the limits of small and large wave numbers, as will be shown in Sect. 7.3.

For practical purposes, such as fitting experimental data, it is useful to have analytical expressions for the structure factor. Therefore, in Sect. 7.4 we present a Galerkin-approximation method, which will be shown to provide simple and good approximations for the nonequilibrium structure factor of a fluid between two rigid boundaries. We then proceed in Sect. 7.5 to investigate the equal-time density autocorrelation function in real space, thus elucidating the spatially long-ranged nature of the thermal nonequilibrium

141

fluctuations. We conclude this chapter with some comments in Sect. 7.6 concerning the contribution of nonequilibrium fluctuations to heat transfer across a fluid layer subjected to a temperature gradient.

7.1   A fluid confined between two free boundaries

For a fluid layer between two free boundaries, the nonequilibrium dynamic structure factor is obtained by substituting into the general expression (6.26) for $S(\omega, q_\parallel, z, z')$ Eq. (6.37) for B_N and using Eqs. (6.39)-(6.40) for the mode-coupling coefficients C_{NM}. Due to the simplicity of the slip-free BC, one readily arrives at an explicit expression for dynamic structure factor:

$$S(\omega, q_\parallel, z, z') = \alpha_p^2 \rho^2 \sum_{N=0}^{\infty} \frac{H_N(z)\, H_N(z')}{|D_N|^2} \left\{ \frac{C_{NN}^{(++)}}{\left|i\omega + \Gamma_N^{(+)}\right|^2} + \frac{C_{NN}^{(--)}}{\left|i\omega + \Gamma_N^{(-)}\right|^2} \right.$$
$$\left. - 2\mathrm{Re}\left[\frac{C_{NN}^{(+-)}}{\left[-i\omega + \Gamma_N^{(+)*}\right]\left[i\omega + \gamma_N^{(-)}\right]} \right] \right\} \quad (7.1)$$

with coefficients $D_N(q_\parallel)$ defined by Eq. (6.38). Notice that $S(\omega, q_\parallel, z, z')$ in Eq. (7.1) is a real number, independent of the decay rates being real (positive Ra) or forming complex conjugate pairs (propagating modes).

Here we are interested in the *static* structure factor $S(q_\parallel, z, z')$, which is obtained by integrating the dynamic structure factor over the frequency ω in accordance with Eq. (6.29):

$$\frac{S(q_\parallel, z, z')}{\alpha_p^2 \rho^2} = \sum_{N=0}^{\infty} \frac{H_N(z)\, H_N(z')}{|D_N(q_\parallel)|^2} \left\{ \frac{C_{NN}^{(--)}}{2\mathrm{Re}\left(\Gamma_N^{(+)}\right)} + \frac{C_{NN}^{(++)}}{2\mathrm{Re}\left(\Gamma_N^{(-)}\right)} \right.$$
$$\left. - 2\mathrm{Re}\left[\frac{C_{NN}^{(+-)}}{\Gamma_N^{(+)*} + \Gamma_N^{(-)}} \right] \right\}, \quad (7.2)$$

We assume $Ra < Ra_c$, so that the system is below the convective instability and $\Gamma_N^{(\pm)}(q_\parallel) > 0$ for any value of $q_\parallel$. To sum the series (7.2) we use the fact, discussed in Sect. 6.2, that in the most general case the decay rates for free boundaries form K-pairs of complex conjugate numbers (propagating modes) plus an infinite set of real numbers. For Ra numbers larger than the RHS of Eq. (6.33), the number K of complex conjugate pairs is zero, *i.e.*, there are no propagating modes. We can thus make a rearrangement of the terms in the series, so that:

$$
\frac{S(q_\parallel, z, z')}{\alpha_p^2 \rho^2} = \sum_{N=0}^{K-1} \left\{ \frac{C_{NN}^{(++)}}{\mathrm{Re}\left(\Gamma_N^{(+)}\right)} + \mathrm{Re}\left[\frac{C_{NN}^{(+-)}}{\Gamma_N^{(+)}}\right] \right\} \frac{H_N(z)\,H_N(z')}{\left|D_N(q_\parallel)\right|^2} \tag{7.3}
$$
$$
+ \sum_{N=K}^{\infty} \left\{ \frac{C_{NN}^{(++)}}{2\Gamma_N^{(+)}} + \frac{C_{NN}^{(--)}}{2\Gamma_N^{(-)}} - \frac{2C_{NN}^{(+-)}}{\Gamma_N^{(+)} + \Gamma_N^{(-)}} \right\} \frac{H_N(z)\,H_N(z')}{\left[D_N(q_\parallel)\right]^2},
$$

where we have made use of the relation $C_N^{(++)} = C_N^{(--)}$ valid for the modes with index N for which the decay rates form a complex conjugate pair, see Eq. (6.39). In addition we have used that for real decay rates the normalization coefficients $D_N(q_\parallel)$ are real numbers.

Substituting Eq. (6.39) into Eq. (7.3), after some algebra and using the explicit expression (6.32) for the decay rates, we can decompose the static structure factor of a fluid layer in the presence of a temperature gradient into an equilibrium and a nonequilibrium contribution:

$$
S(q_\parallel, z, z') = S_E \frac{\gamma - 1}{\gamma} \left\{ S_E(z, z') + \tilde{S}_{NE}^0\, S_{NE}(q_\parallel, z, z') \right\}, \tag{7.4}
$$

where:

$$
S_E(z, z') = \frac{2}{L} \sum_{N=1}^{\infty} H_N(z)\,H_N(z') = \delta(z - z') \tag{7.5}
$$

and:

$$
S_{NE}(q_\parallel, z, z') = \frac{2}{L} \sum_{N=1}^{\infty} \frac{\tilde{q}_\parallel^2\,H_N(z)\,H_N(z')}{(N^2\pi^2 + \tilde{q}_\parallel^2)^3 - Ra\,\tilde{q}_\parallel^2}. \tag{7.6}
$$

In these equations $S_E(\gamma-1)/\gamma$ represents the dimensionless amplitude of the Rayleigh line for a fluid in equilibrium, given by Eq. (4.24), while $\tilde{S}_{NE}^0$ is the strength of the "bulk" nonequilibrium fluctuations, as given by Eq. (4.30). In Eq. (7.5), the expansion of the delta function in a Fourier series for the interval $[-L/2, L/2]$ has been used. It is worth noting that the result obtained in order N no longer depends on whether the N-mode is propagating or not, i.e., on whether the corresponding decay rates are real numbers or form a complex conjugate pair. Equation (7.5) equals the result previously obtained by Ortiz de Zárate and Sengers (2001) with a different, more direct method.

The equilibrium part of Eq. (7.4) equals the static structure factor of a divergence-free (incompressible) fluid in thermodynamic equilibrium; it is short ranged, proportional to a delta function, and it is not affected by any finite-size effects (Berne and Pecora, 1976). The second part of Eq. (7.4) represents the nonequilibrium enhancement of the structure factor. This nonequilibrium enhancement is proportional to $(\nabla T_0)^2$ through the expression (4.30) for $\tilde{S}^0_{NE}$; it depends on the gravitational acceleration constant g through the appearance of the Rayleigh number in Eqs. (7.6) and (4.30), and it depends on the finite height L of the fluid layer explicitly in Eq. (7.6) and also through the Rayleigh number and $\tilde{q}_{\parallel} = q_{\parallel} L$. In the limit $Ra \rightarrow 0$ we recover from (7.6) the expression for $S(q_{\parallel}, z, z')$ previously obtained by Ortiz de Zárate et al. (2001), where buoyancy was neglected, but finite-size effects included. In that case, $\tilde{S}^0_{NE}$ is to be obtained by taking the limit the $g \rightarrow 0$ limit in Eq. (7.4). Most importantly, the $q_{\parallel}$ dependence of (7.6) makes the equal-time autocorrelation function of density fluctuations to be spatially long-ranged, as discussed in Sect. 7.5.

Equations (7.4)-(7.6) incorporate finite-size effects for the case of free-slip BC. Equation (4.29) for the bulk fluctuations was valid only for negative Ra. However, Eq. (7.6), that includes finite-size effects is valid for both negative and positive Rayleigh numbers, provided that the denominator in the series terms is not zero for any N and/or $q_{\parallel}$. The denominator is zero (for $N = 1$, the closest case) when the instability condition (6.62) holds. As is well known, Eq. (6.62) can be satisfied only for Rayleigh numbers such that (Chandrasekhar, 1961):

$$Ra < Ra_c = \frac{27\pi^4}{4}. \tag{7.7}$$

For $Ra \geq Ra_c$ there always exist values of $q_{\parallel}$ for which the RHS of Eq. (7.6) diverges. However, for $Ra < Ra_c$, including any negative value of Ra, the series (7.6) is convergent, and its sum can be calculated for any value of $q_{\parallel}$. It is worth noticing that we recover from an analysis of the divergence in the expression for the nonequilibrium structure factor, the same value for the critical Rayleigh number Ra_c that results form a linear stability analysis of the Boussinesq equations.

The appearance of divergences in (7.6) means that the present calculation breaks down for $Ra \geq Ra_c$, because the fluctuations around the conductive state are no longer small. Obviously, using the linearized fluctuating Boussinesq equations to represent the spatiotemporal evolution of the fluctuations is no longer justified for $Ra \geq Ra_c$. As is well known, when Ra is larger than the critical value (7.7), convection cells or patterns (sometimes referred to as *dissipative structures*) develop in the system. We note the connection between divergences in the amplitude of fluctuations and the development of (deterministic) patterns.

The nonequilibrium enhancement of fluctuations is expressed in Eq. (7.6) as a series of trigonometric functions. The summation of this series can be performed exactly (Ortiz de Zárate and Sengers, 2001). In this book we are specifically interested in nonequilibrium fluctuations as they can be observed by small-angle light-scattering experiments (Law and Sengers,

1989; Law et al., 1988, 1990; Segrè et al., 1992; Vailati and Giglio, 1996, 1997b) or by shadowgraph experiments (Wu et al., 1995). In accordance with Eq. (6.28), the experimentally relevant quantity is obtained upon integration of Eq. (7.4) over the vertical variables z and z'. Summing the series for the nonequilibrium contribution in Eq. (7.5) and performing the integration over z and z', we obtain (Ortiz de Zárate and Sengers, 2001):

$$S(\mathbf{q}) = S_{\mathrm{E}} \frac{\gamma - 1}{\gamma} \left\{ 1 + \tilde{S}_{\mathrm{NE}}^0 \; \tilde{S}_{\mathrm{NE}}(\tilde{q}_\parallel, \tilde{q}_\perp) \right\}, \tag{7.8}$$

where we have introduced

$$\tilde{S}_{\mathrm{NE}}(\tilde{q}_\parallel, \tilde{q}_\perp) = \frac{\tilde{q}_\parallel^2}{\tilde{q}^4 - \tilde{q}_\parallel^2 \, Ra}$$

$$+ \frac{2\tilde{q}_\parallel}{3} \sum_{j=0}^{2} \frac{\lambda_j \sqrt{\lambda_j - \tilde{q}_\parallel^{4/3}}}{\left[\tilde{q}_\parallel^2 - \lambda_j \tilde{q}_\parallel^{2/3} \right]^2} \frac{\cos(\tilde{q}_\perp) - \cos\left(\tilde{q}_\parallel^{1/3} \sqrt{\lambda_j - \tilde{q}_\parallel^{4/3}} \right)}{\sin\left(\tilde{q}_\parallel^{1/3} \sqrt{\lambda_j - \tilde{q}_\parallel^{4/3}} \right)}. \tag{7.9}$$

In Eq. (7.9) λ_j ($j = 0, 1, 2$) represent the three complex cubic roots of the Rayleigh number:

$$\lambda = |Ra|^{1/3} \begin{pmatrix} \exp\left(i\frac{\phi}{3} \right) \\ \exp\left(i\frac{\phi + 2\pi}{3} \right) \\ \exp\left(i\frac{\phi + 4\pi}{3} \right) \end{pmatrix} \tag{7.10}$$

where ϕ is the phase of Ra when considered as a complex number, *i.e.*, $\phi = 0$ for $Ra > 0$ and $\phi = \pi$ when $Ra < 0$. If we substitute the first term of Eq. (7.9) into Eq. (7.8), we recover Eq. (4.29) for the bulk structure factor of the fluid. Hence, the finite-size effects are accounted for by the second line of Eq. (7.9).

For the interpretation of low-angle light-scattering or shadowgraph experiments, as will be discussed in more detail in Chapter 10, one may consider the small-angle approximation $q_\parallel \to q$ and $q_\perp \to 0$, see Eq. (10.21). If we take the $q_\perp \to 0$ limit in Eq. (7.9), the resulting expression for the nonequilibrium structure factor becomes somewhat simpler and, most importantly, will depend only on the magnitude of the horizontal wave number $\tilde{q}$. For the remainder of this section, we adopt the small-angle approximation. We note that the divergences discussed below Eq. (7.5) remain the same in this approximation.

It is interesting to look at the asymptotic behavior of the nonequilibrium enhancement of the structure factor for $q \to 0$ and for $q \to \infty$. From Eq. (7.9) one readily deduces for small q:

$$\tilde{S}_{\mathrm{NE}}^{\mathrm{F}}(\tilde{q}) \xrightarrow{\tilde{q} \to 0} \frac{17}{20160} \, \tilde{q}^2, \tag{7.11}$$

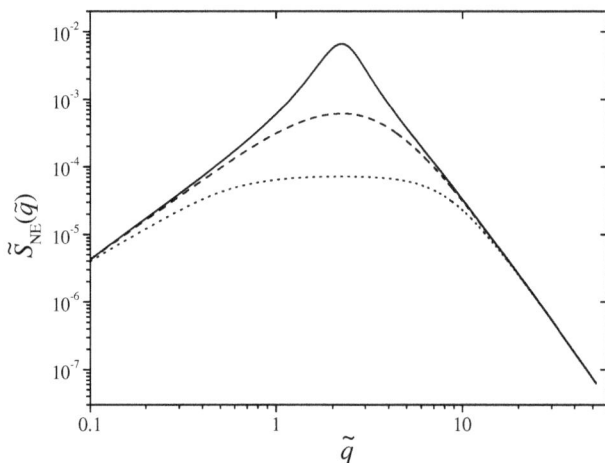

Figure 7.1: Normalized nonequilibrium enhancement $\tilde{S}_{\mathrm{NE}}(\tilde{q})$ of the structure factor for a fluid between two free boundaries, as given by Eq. (7.9). The solid curve is for $Ra = 600$, which is near the convective instability, the dashed curve is for $Ra \simeq 0$ and the dotted curve is for $Ra = -5000$. Reproduced from Ortiz de Zárate and Sengers (2001).

and for large $\tilde{q}$:

$$\tilde{S}_{\mathrm{NE}}(\tilde{q}) \xrightarrow{\tilde{q} \to \infty} \frac{1}{\tilde{q}^4}. \tag{7.12}$$

We note from Eqs. (7.11) and (7.12) that neither of the two limiting behaviors of $\tilde{S}_{\mathrm{NE}}(\tilde{q})$ depends on the value of the Rayleigh number Ra nor on the value of the Prandtl number Pr.

In Fig. 7.1 we show on a double logarithmic scale the nonequilibrium enhancement $\tilde{S}_{\mathrm{NE}}(\tilde{q})$ as a function of $\tilde{q}$ for a fluid between two free boundaries, as calculated from Eq. (7.9) in the small-angle approximation $q_\parallel \to q$ and $q_\perp \to 0$. Results are shown for three different values of the Rayleigh number, namely for a large negative value $Ra = -5000$, for a value $Ra \simeq 0$ corresponding to $g \simeq 0$, and for a value $Ra = 600$ which is close to the critical value $Ra_c \simeq 656$ for free boundaries. At larger values of $\tilde{q}$, $\tilde{S}_{\mathrm{NE}}(\tilde{q})$ varies as $\tilde{q}^{-4}$, independently of the Rayleigh number, in accordance with Eq. (7.12). As the wave number $\tilde{q}$ decreases, $\tilde{S}_{\mathrm{NE}}(\tilde{q})$ goes through a maximum, and for very small values of $\tilde{q}$, $\tilde{S}_{\mathrm{NE}}(\tilde{q})$ decreases as $\tilde{q}^2$ in agreement with Eq. (7.11). Notice that for positive Ra, $\tilde{S}_{\mathrm{NE}}(\tilde{q})$ develops a very prominent peak close to $\tilde{q}_c = \pi/\sqrt{2}$, whose height eventually diverges as $Ra \to Ra_c$, see Chapter 8. We conclude that a major effect of the additive noise terms in the Boussinesq equations is the appearance of long-ranged (fluctuating) patterns in the fluid, even below the convective instability. The presence of fluctuating patterns is a major feature of stochastic differential equations when compared to deterministic equations, as has been discussed by several authors (García Ojalvo and Sancho, 1999; Staliunas, 2001).

7.2 A fluid confined between two rigid boundaries

For a fluid layer confined between two rigid boundaries it is not possible to obtain a simple and compact explicit expression for the nonequilibrium structure factor. Instead, from the general theory of Sect. 6.1 and the expressions for the hydrodynamic modes and decay rates in Sect. 6.3, the nonequilibrium structure factor can be computed numerically. However, analytical expressions can be obtained in the limits $q_\parallel \to 0$ and $q_\parallel \to \infty$ by using the results of Sects. 6.3.1 and 6.3.2, respectively.

For $0 < Ra < Ra_c$ the decay rates (6.8) are known to be real and positive numbers (Schmitz and Cohen, 1985b). Hence, from the discussion of Sect. 6.3, we conclude that the corresponding hydrodynamic modes can be normalized so that they are real-valued functions. In the preceding section we found for a fluid between free boundaries that the presence (for negative Ra values) of complex conjugate pairs of decay rates (propagating modes) add mathematical difficulties to the calculation of the static structure factor. However, the final result for $S(q_\parallel, z, z')$, Eqs. (7.8) and (7.9), was valid for any Ra number, independently of whether there are propagating modes or not. This observation gives us confidence that the results for the static structure factor to be obtained in this section for positive Ra can also be extended to negative Ra. Assuming that the decay rates and the hydrodynamic modes are real numbers and using Eq. (6.8) and the orthogonality condition (6.16), Eq. (6.24) for the mode-coupling coefficients is conveniently split as the sum of two contributions:

$$C_{NM}(q_\parallel) = -\frac{k_B T^2}{\rho c_p} \, 2\alpha_p g q_\parallel^2 \, \Gamma_N(q_\parallel) \, B_N(q_\parallel) \, \delta_{NM} + C_{NM}^{\mathrm{NE}}(q_\parallel), \qquad (7.13)$$

where $C_{NM}^{\mathrm{NE}}(q_\parallel)$ are nonequilibrium mode-coupling coefficients given by:

$$C_{NM}^{\mathrm{NE}}(q_\parallel) = \frac{k_B T^2}{\rho c_p} \, 2q_\parallel^2 \left\{ \tilde{S}_{\mathrm{NE}}^0 \frac{Pr+1}{Pr^2} \frac{\nu^3}{L^4} \int_{-\frac{L}{2}}^{\frac{L}{2}} W_N \left[D^2 \, W_M \right] \, dz \right.$$

$$\left. -\frac{Ra}{Pr} \frac{\nu^2}{L^4} (\Gamma_N + \Gamma_M) \int_{-\frac{L}{2}}^{\frac{L}{2}} W_N \left[D \, W_M \right] \, dz \right\}. \qquad (7.14)$$

Here $\tilde{S}_{\text{NE}}^0$ is again the dimensionless strength of the nonequilibrium enhancement as previously defined by (4.30) for the "bulk" fluctuations. It is
obvious from Eq. (7.14) that $C_{NM}^{\text{NE}}(q_{\parallel})$ vanishes for $\nabla T_0 = 0$. However, it
should also be noted that the first term in the RHS of Eq. (7.13) contains
nonequilibrium contributions through the dependence of the decay rates and
the normalization constants $B_N(q_{\parallel})$ on Ra. Hence, the decomposition (7.13)
does not split the dynamic structure factor $S(\omega, q_{\parallel}, z, z')$ into an equilibrium
and a nonequilibrium contribution[‡]. However, if after substitution of (7.13)
into Eq. (6.27) for $S(\omega, q_{\parallel}, z, z')$ we integrate over the frequency to obtain
the *static* structure factor $S(q_{\parallel}, z, z')$ in accordance with Eq. (6.29), it can
be demonstrated that the decomposition (7.13) splits $S(q_{\parallel}, z, z')$ into the
sum of an equilibrium plus a nonequilibrium contribution. Consequently,
Eq. (7.4) found for two free boundaries also applies for two rigid boundaries.
In the version of Eq. (7.4) for rigid boundaries, the equilibrium contribution
$S_{\text{E}}(z, z')$ arises from the first term in the RHS of Eq. (7.13) and is expressed
as:

$$S_{\text{E}}(z, z') = \sum_{N=1}^{\infty} \frac{-\alpha_p g q_{\parallel}^2}{B_N(q_{\parallel})} \, \Theta_N(q_{\parallel}, z) \, \Theta_N(q_{\parallel}, z'). \tag{7.15}$$

The summation in Eq. (7.15) can be performed exactly and yields the same
result as Eq. (7.5) for free boundaries. Thus, Eq. (7.15) is indeed independent of the Ra number or the horizontal wave number $q_{\parallel}$. To elucidate this,
let us consider the two-dimensional vector function:

$$\mathbf{G}(z') = \begin{pmatrix} 0 \\ \delta(z - z') \end{pmatrix}. \tag{7.16}$$

For $z' \in [-L/2, L/2]$ the vector function (7.16) satisfies the no-slip
BC (6.10), independently of the value of z. Hence, as we did in (6.19)
for $\mathbf{U}(z)$, the vector function (7.16) may be expanded in a series of right
eigenfunctions by projection onto the set of left eigenfunctions. Thus, for
real hydrodynamic modes and provided that z is a point located inside the
interval $[-L/2, L/2]$, we obtain:

$$0 = \sum_{N=1}^{\infty} \frac{1}{B_N(q_{\parallel})} \, \Theta_N(q_{\parallel}, z) \, W_N(q_{\parallel}, z')$$

$$\delta(z - z') = \sum_{N=1}^{\infty} \frac{-\alpha_p g q_{\parallel}^2}{B_N(q_{\parallel})} \, \Theta_N(q_{\parallel}, z) \, \Theta_N(q_{\parallel}, z'). \tag{7.17}$$

Substitution of the second line of Eq. (7.17) into (7.15) we not only recover (7.5) for free boundaries, but also the expression for the Rayleigh

---

[‡]However, when buoyancy is neglected, the decay rates and hydrodynamic modes do
not depend on Ra and the dynamic structure factor does split additively into equilibrium
plus nonequilibrium contributions

structure factor of a fluid in equilibrium, discussed in Chapter 3. We confirm that the structure factor in equilibrium is not affected by BC, which is to be expected because the equilibrium structure factor is spatially short ranged (proportional to delta functions), and cannot be affected by what happens at the boundaries.

In the case of two rigid boundaries it is not possible to obtain a compact expression for the normalized nonequilibrium contribution $\tilde{S}_{\mathrm{NE}}(q_\parallel, z, z')$ like Eq. (7.6) for two free boundaries. Substituting the nonequilibrium part of Eq. (7.13) into Eq. (6.27), and performing the integration over ω, as specified by Eq. (6.29), we obtain:

$$S_{\mathrm{NE}}(q_\parallel, z, z') = \sum_{N,M=1}^{\infty} \frac{\tilde{C}_{NM}^{\mathrm{NE}}(q_\parallel)\,\Theta_N(q_\parallel, z)\,\Theta_M(q_\parallel, z')}{B_N(q_\parallel)B_M(q_\parallel)[\Gamma_N(q_\parallel) + \Gamma_M(q_\parallel)]}, \tag{7.18}$$

where the second line of Eq. (6.27), being an odd function of ω, does not contribute to the static structure factor when the decay rates are real and positive numbers. From the first line of Eq. (7.17) it follows that the term proportional to Ra in the expression for $C_{NM}^{\mathrm{NE}}(q_\parallel)$, $i.e.$, the second line of Eq. (7.14), does not contribute to the static structure factor. Therefore, the mode-coupling coefficients $\tilde{C}_{NM}^{\mathrm{NE}}(q_\parallel)$ to be substituted in (7.18) can be simply expressed as:

$$\tilde{C}_{NM}^{\mathrm{NE}}(q_\parallel) = 2\frac{\nu^3}{L^6}\,\tilde{q}_\parallel^2\,\frac{Pr+1}{Pr^2}\int_{-\frac{L}{2}}^{\frac{L}{2}} W_N(q_\parallel, z)\left[\mathrm{D}^2\,W_M(q_\parallel, z)\right]\,dz. \tag{7.19}$$

We thus conclude that the nonequilibrium static structure factor (7.14) for a fluid between two rigid boundaries, (7.6) for a fluid between for two free boundaries, and (4.29) for the bulk fluid, are all directly proportional to the dimensionless parameter $\tilde{S}_{\mathrm{NE}}^0$ defined by Eq. (4.30).

Furthermore, Eq. (7.19) implies that $\tilde{C}_{NM}^{\mathrm{NE}}(q_\parallel) = 0$, when the indexes N and M correspond to hydrodynamic modes of different vertical parity. Hence, Eq. (7.18) may be rewritten as the sum of two contributions with a similar structure, one containing only even hydrodynamic modes and the other containing only odd modes.

In deriving Eq. (7.18) we assumed, just as for free boundaries, that the decay rates $\Gamma_N(q_\parallel)$ are positive for any values of N and $q_\parallel$. Again, when $\Gamma_N(q_\parallel) = 0$, the conductive solution becomes unstable and the linear theory for the fluctuations will no longer be valid. While the validity of Eq. (4.29) for the "bulk" structure factor is restricted to negative values of the Rayleigh number Ra, the expression (7.18) will be also valid for positive values of Ra up to $Ra_c \approx 1707.76$, the critical value of Ra that corresponds to the first convective instability for a fluid layer between two rigid boundaries (Chandrasekhar, 1961). As was the case for free boundaries, inclusion of finite-size effects again extends the range of applicability of linearized fluctuating hydrodynamics for the fluctuations from negative Ra to a range of positive Ra. We note that for two rigid boundaries, the interval of positive Rayleigh numbers for which the structure factor can be calculated with a linear theory, is larger than for two free boundaries. This is a consequence of the fact that "no-slip" boundary conditions are more stabilizing than "free-slip" boundary conditions.

Next, we recall that the structure factor $S(\mathbf{q})$, measured in light-scattering or shadowgraph experiments is obtained by integrating $S_{\mathrm{NE}}(q_\parallel, z, z')$ in accordance with Eq. (6.28). Just as in Eq. (7.8) for the case of two free boundaries, the experimental static structure factor for two rigid boundaries may be expressed as:

$$S(\mathbf{q}) = S_{\mathrm{E}} \frac{\gamma - 1}{\gamma} \left\{ 1 + \tilde{S}_{\mathrm{NE}}^0 \, \tilde{S}_{\mathrm{NE}}(\tilde{q}_\parallel, \tilde{q}_\perp) \right\}. \tag{7.20}$$

For the dimensionless nonequilibrium enhancement $\tilde{S}_{\mathrm{NE}}(\tilde{q}_\parallel, \tilde{q}_\perp)$ of the fluctuations in the case of rigid boundaries, a more compact result is obtained in the approximation of small scattering angles so that $q_\parallel \simeq q$, $q_\perp \simeq 0$. Then the contribution from the hydrodynamic modes of odd vertical parity vanishes after integrating Eq. (7.18) over $[-L/2, L/2]$. We thus obtain:

$$\tilde{S}_{\mathrm{NE}}(\tilde{q}) = \frac{1}{L} \sum_{N,M=1}^{\infty} \frac{\tilde{C}_{NM}^{\mathrm{NE,E}}(q)}{\Gamma_N^{\mathrm{E}}(q) + \Gamma_M^{\mathrm{E}}(q)} \, \Xi_N^{\mathrm{E}}(q) \, \Xi_M^{\mathrm{E}}(q), \tag{7.21}$$

where the coefficients $\Xi_N^{\mathrm{E}}(q)$ were defined in Eq. (6.46) as the ratio of the vertical integral of the temperature component of the even eigenfunctions to the normalization constants. The superscript (E) in Eq. (7.21) indicates that both indices N and M correspond to even eigenfunctions.

It is not possible to obtain a more explicit exact expression for the normalized nonequilibrium enhancement of fluctuations than Eqs. (7.19)-(7.21), so we proceed with a numerical analysis of the solution. Similarly to Fig. 7.1 for free boundaries, we show in Fig. 7.2 a plot of the normalized nonequilibrium enhancement of the fluctuations as a function of the wave number $\tilde{q}$. The curves in Fig. 7.2 have been calculated by substituting into Eq. (7.21) the numerical values for the first couple of decay rates displayed numerically in Tab. 6.1 and graphically in Fig. 6.2. As will be discussed in more detail later in Sect. 8.2, in spite of having used only two decay rates (namely: $\tilde{\Gamma}_1^{(+)}$ and $\tilde{\Gamma}_1^{(-)}$), the data displayed in Fig. 7.2 give a very good numerical representation of the nonequilibrium enhancement. For reference we also indicate in Fig. 7.2 by two thin lines the exact asymptotic behaviors for large and small wave numbers, to be derived in Sect. 7.3.

The data displayed in Fig. 7.2 resembles qualitatively the nonequilibrium enhancement of the fluctuations shown in Fig. 7.1 for the case of free boundaries. First, we observe again crossover from a $\tilde{q}^2$ behavior at small

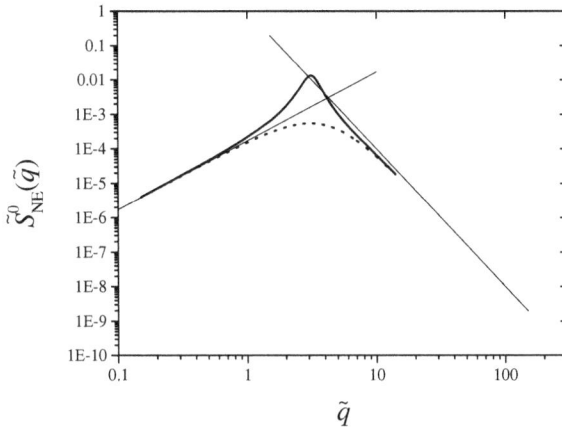

Figure 7.2: Normalized nonequilibrium enhancement $\tilde{S}_{NE}(\tilde{q})$ of the structure factor of a fluid between two rigid boundaries with $Pr = 8$. The solid curve is for $Ra = 1650$, which is close to the instability($\epsilon = -0.035$) and dotted curve is for $Ra = 150$. The thin lines indicate the common asymptotic behaviors for small and large $\tilde{q}$.

$\tilde{q}$ to a $\tilde{q}^{-4}$ behavior at large $\tilde{q}$. Separating both asymptotic behaviors, the nonequilibrium enhancement shows a single maximum for $\tilde{q}$ values close to $\tilde{q}_c$. The height of the maximum grows as $Ra \to Ra_c$, and eventually diverges at $Ra = Ra_c$ as further discussed in Chapter 8. However, in addition to different values of $\tilde{q}_c$ and Ra_c, Fig. 7.2 for rigid boundaries and Fig. 7.1 for free boundaries also differ in the role of the Pr number. In the case of free boundaries all dependence in Pr is collected in the prefactor $\tilde{S}_{NE}^0$, and the $\tilde{S}_{NE}(q)$ values in Fig. 7.1 are independent of Pr. However, for rigid boundaries, the $\tilde{S}_{NE}(q)$ values do still depend on Pr. Data in Fig. 7.2 are for $Pr = 8$.

For negative Ra the numerical computation of the decay rates for rigid boundaries is complicated by the appearance of propagating modes and associated cross and branch points (Schmitz and Cohen, 1985b), a phenomenon shown in Fig. 6.1 for free boundaries. One should also remember that the derivation of Eq. (7.21) is strictly valid only for positive Ra, when all the decay rates are real. For these two reasons we have not displayed numerical data for negative Ra in Fig. 7.2. We believe that the simple expression (7.21) for the nonequilibrium enhancement may be extended to negative Ra, but such an extension lies outside the scope of the present chapter. Since for negative Ra gravity has a damping effect (opposite to a resonant effect for positive Ra), it is physically expected that the height of the maximum in $\tilde{S}_{NE}$ will diminish for $Ra < 0$ and eventually will become very flat, as was the case in Fig. 7.1 for free boundaries. This will be further discussed in Sect. 7.4, where a single-mode analytical Galerkin approximation to the nonequilibrium enhancement will be developed for a fluid confined between two rigid boundaries.

7.3 Limiting behavior of the structure factor for rigid boundaries at small and large wave numbers

In this section we employ the results of Sects. 6.3.1 and 6.3.2 to obtain exact analytical expressions for the limiting behavior of the nonequilibrium enhancement of the structure factor for rigid boundaries, $\tilde{S}_{NE}(\tilde{q})$. We start with the small-q limit for which the results of Sect. 6.3.1 are relevant.

We recall from Sect. 6.3.1 that there are two kinds of hydrodynamic modes and corresponding decay rates for rigid boundaries in the small-q limit. For the $(-)$ modes of even vertical parity, the coefficients of the small-q expansion of the decay rates are given by (6.50), while for the $(-)$ even decay rates the coefficients are given by (6.49). The normalization coefficients and the results of the vertical integration of the temperature fluctuations are given by Eqs. (6.53) and (6.54), respectively. The only additional information we need in order to to derive the structure factor from Eq. (7.21) are the mode-coupling coefficients $\tilde{C}_{NM}^{NE,E}(q)$. They can be readily obtained by substituting into their definition (7.19) the expression (6.42) for the modes, performing the resulting integrals, and then using the small-q expansions (6.47) for the decay rates and (6.48) for the $\tilde{\lambda}_j$ roots of the secular equation (6.41). We start by evaluating the coefficients for two (even) modes of the first kind $(-)$ with the same index N, for which we obtain:

$$\tilde{C}_{NN}^{(--)}(q) = \frac{\alpha_p^2 g^2 \nu^3}{L^9} \frac{(Pr+1)}{Pr^2} \left[1 + (Pr+1)\tan^2\left(\frac{N\pi}{2\sqrt{Pr}}\right) \right. \tag{7.22a}$$

$$\left. - \frac{2\sqrt{Pr}(3Pr+1)}{N\pi(Pr-1)}\tan\left(\frac{N\pi}{2\sqrt{Pr}}\right) \right] \tilde{q}^6 + \mathcal{O}(\tilde{q}^8)$$

for odd $N = 1, 3, 5 \ldots$ Since in this section we only deal with even modes, we shall drop the superscript (NE,E) in the mode-coupling coefficients. For $N \neq M$, the first term in the small-q expansion of the $(--)$ mode-coupling coefficients is:

$$\tilde{C}_{NM}^{(--)}(q) = \frac{\alpha_p^2 g^2 \nu^3}{L^9} \frac{(Pr^2-1)}{Pr^2} \frac{4MN\sqrt{Pr}}{\pi(N^2-M^2)} \tag{7.22b}$$

$$\times \left[\frac{N}{N^2 - PrM^2} \tan\left(\frac{M\pi}{2\sqrt{Pr}}\right) + \frac{M}{PrN^2 - M^2} \tan\left(\frac{N\pi}{2\sqrt{Pr}}\right) \right] \tilde{q}^6 + \mathcal{O}(\tilde{q}^8)$$

for odd $N \neq M = 1, 3, 5 \ldots$. For the modes of the second kind, the mode-coupling coefficients may be calculated by a similar procedure, but now the coefficients (6.50) of the small-q expansion of the corresponding decay rates should be used. We obtain:

$$\tilde{C}_{NM}^{(++)}(q) = \frac{\alpha_p^2 g^2 \nu^3}{L^9 \, Pr^2} (Pr + 1) \, \tilde{q}^6 \, \delta_{NM} + \mathcal{O}(\tilde{q}^8). \tag{7.22c}$$

Finally, we have to consider the mode-coupling coefficients among modes of different kind. Substituting (6.49) and (6.50) into (7.19), performing the integrations and expanding in powers of $\tilde{q}_{\parallel}^2$, we obtain:

$$\tilde{C}_{NM}^{(+-)}(q) = \frac{\alpha_p^2 g^2 \nu^3}{L^9 \, Pr^2} \frac{4(Pr^2 - 1)\sqrt{Pr} \, MN^2}{\pi(N^2 Pr - M^2)(N^2 - M^2)} \tan\left(\frac{\pi M}{2\sqrt{Pr}}\right) \tilde{q}^6 \tag{7.22d}$$

$$+ \mathcal{O}(\tilde{q}^8),$$

where $N = 2, 4, 6 \ldots$ is an even integer corresponding to a $(+)$ mode and $M = 1, 3, 5, \ldots$ is an odd integer corresponding to a $(-)$ mode. The other cross mode-coupling coefficients $\tilde{C}_{NM}^{(+-)}(q)$ are obtained from (7.22d) by exchanging the indices N and M. As we noticed in Sect. 6.3.1 for the decay rates and hydrodynamic modes, Eqs. (7.22) for the mode-coupling coefficients exhibit divergences for an infinite set of Pr numbers. However, again it turns out that upon calculating the lower-order approximation to the structure factor, all such divergences cancel, yielding a physically meaningful result for any positive value of Pr.

Next, substituting Eqs. (7.22), (6.49), (6.50) and (6.54) into Eq. (7.21) for $\tilde{S}_{NE}(q)$, we can obtain the small-q limit of the static structure factor for rigid boundaries. Just as in Eq. (7.11) for the case of free boundaries, $\tilde{S}_{NE}(\tilde{q})$ becomes again proportional to $\tilde{q}^2$. Also, the proportionality coefficient does not depend on the Ra number, but it does depend on Pr:

$$\tilde{S}_{NE}^{R}(\tilde{q}) \xrightarrow{\tilde{q} \to 0} \frac{17}{20160} F(Pr) \, \tilde{q}^2, \tag{7.23}$$

where the function $F(Pr)$ has been normalized by a factor $17/20160$ to facilitate a comparison with the asymptotic limit for free boundaries previously reported in Eq. (7.11). After some algebra and by using formulas for the sum of several trigonometric series, one finds that the function $F(Pr)$ can be expressed as:

$$F(Pr) = \frac{(2Pr - 1)(Pr + 1)}{(Pr - 1)} \tag{7.24}$$

$$- \frac{161280}{17\pi^7} \sqrt{Pr} \sum_{\substack{N=1 \\ N, \, odd}}^{\infty} \frac{1}{N^7} \left[1 + \frac{2\tanh\left(N\frac{\pi}{2}\right)}{N\pi(Pr - 1)} \right] \coth\left(\frac{N\pi}{2\sqrt{Pr}}\right).$$

It is interesting to note that in the limit $Pr \to 0$, $F(Pr) \to 1$, so that the small-q expansion of the nonequilibrium enhancement of the structure factor reduces to that obtained in (7.11) for free boundaries. From this value of unity, $F(Pr)$ decreases monotonically with increasing Pr, reaching a constant nonzero limit at $Pr \to \infty$. The function $F(Pr)$ is analytic and continuous for any positive Pr, including $Pr = 1$, although the latter case may not be completely evident from (7.24). We have found that for $Pr > 2$, the proportionality coefficient is very well represented (within 1%) by the two first terms of the asymptotic expansion for large Pr of (7.24), namely:

$$F(Pr) \simeq 0.185 + \frac{0.146}{Pr} + \mathcal{O}\left(\frac{1}{Pr^2}\right). \tag{7.25}$$

To evaluate the static nonequilibrium structure factor in the large-q limit, we start by substituting the series expansion (6.55) of the decay rates and the expansion (6.56) of the dimensionless exponents $\tilde{\lambda}_i$ into the representation (6.42) of the hydrodynamic modes of even vertical parity. We then substitute the resulting expression into Eq. (7.19) for the mode-coupling coefficients, we perform the corresponding integral and finally expand the result into power series in terms of $q_\parallel^{-1}$. This procedure yields:

$$\tilde{C}_{NM}^{(++)}(q) = \frac{\alpha_p^2 g^2 \nu^3}{L^9} \frac{1}{N^2 \pi^2} \delta_{NM} \, \tilde{q}^8 + \mathcal{O}(\tilde{q}^7), \tag{7.26}$$

$$\tilde{C}_{NM}^{(+-)}(q) = \frac{\alpha_p^2 g^2 \nu^3}{L^9} \frac{\sqrt{Pr}\left[\sqrt{Pr} - \sqrt{Pr-1}\right]}{\pi^2 N^2} \delta_{NM} \, \tilde{q}^8 + \mathcal{O}(\tilde{q}^7),$$

$$\tilde{C}_{NM}^{(--)}(q) = \frac{\alpha_p^2 g^2 \nu^3}{L^9} \frac{Pr\left[\sqrt{Pr} - \sqrt{Pr-1}\right]^2}{\pi^2 N^2} \delta_{NM} \, \tilde{q}^8 + \mathcal{O}(\tilde{q}^6),$$

for odd integers $N = 1, 3, 5\dots$. Next, using expansions (7.26) for the mode-coupling coefficients, the expansions (6.61) for the ratios $\tilde{\Xi}_N(q)$, and (6.55) for the even decay rates, we obtain for the limiting behavior of the nonequilibrium structure factor in the limit of large q :

$$\tilde{S}_{NE}^{R}(\tilde{q}) \xrightarrow{\tilde{q} \to \infty} \left\{ \sum_{N=0}^{\infty} \frac{8}{\pi^2 (2N+1)^2} \right\} \frac{1}{\tilde{q}^4} = \frac{1}{\tilde{q}^4}, \tag{7.27}$$

where we have used the fact that the large-q expansions of the decay rates in Sect. 6.3.2 are for odd integers N. Again, we find that the large-q behavior of the nonequilibrium enhancement is independent of the BC and reduces again to that of a bulk fluid as it should.

7.4 A Galerkin approximation for two rigid boundaries

In Sect. 7.2 we showed how a numerical solution for the nonequilibrium structure factor of a fluid between two rigid boundaries can be obtained. However, for practical applications, like the analysis of experimental data, approximate analytical expressions are highly desirable. In this section we show how one can obtain an analytical solution in a Galerkin approximation (Ortiz de Zárate and Muñoz Redondo, 2001; Ortiz de Zárate and Sengers, 2002; Ortiz de Zárate et al., 2004) and discuss its validity. Another advantage of the Galerkin approximation method is its simple extension for nonlinear problems. For instance, the well-known Lorenz (1964) nonlinear convection model, where chaos was first identified, is simply obtained by projecting the (nonlinear) Boussinesq equations onto a conveniently chosen set of Galerkin test functions (Manneville, 1990), by a procedure similar to the one described below.

Our goal is to solve the eigenvalue problem (6.8) with the BC (6.6). The idea of the Galerkin method is to expand the hydrodynamic modes $\mathbf{U}_N^R$ in a suitable basis of integrable functions that satisfy the appropriate BC. Specifically, we assume that a given hydrodynamic mode $\mathbf{U}^R(z)$ can be represented by a polynomial expansion of the form:

$$\mathbf{U}^R(z) = \sum_{N=0}^{\infty} \alpha_p g \; w_N(\mathbf{q}_{\parallel}) \; \mathbf{W}_N(z) + \frac{\nu}{L^2} \; \theta_N(\mathbf{q}_{\parallel}) \; \mathbf{\Theta}_N(z) \qquad (7.28)$$

with

$$\mathbf{W}_N(z) = \left(\frac{1}{4} - \frac{z^2}{L^2} \right)^2 \binom{z^N}{0}, \qquad \mathbf{\Theta}_N(z) = \left(\frac{1}{4} - \frac{z^2}{L^2} \right) \binom{0}{z^N}. \quad (7.29)$$

The polynomials, given by Eq. (7.29), satisfy the required boundary conditions (6.6). Furthermore, if we assume that the hydrodynamic modes are analytic functions of the vertical variable z, they can indeed be expanded into a series expansion like (7.28). A Galerkin approximation is obtained by truncating the series (7.28) at some order, K. The $2K$ coefficients $w_N(\mathbf{q}_{\parallel})$ and $\theta_N(\mathbf{q}_{\parallel})$ are then computed by requiring that the expansion (7.28) is an exact solution of the eigenvalue problem (6.8) in the $2K$-dimensional subspace generated by the finite (truncated) basis (7.29). That is, the coefficients are obtained by solving the $2K$ set of algebraic equations obtained upon projection of the eigenvalue problem (6.8) onto the $2K$ elements of the basis (7.29):

$$\sum_{N=0}^{K} \alpha_p g \ w_N(q_\parallel) \int_{-\frac{1}{2}L}^{\frac{1}{2}L} \mathbf{W}_M \left\{ (\mathcal{H} - \Gamma \mathcal{D}) \mathbf{W}_N \right\} dz \qquad (7.30a)$$

$$+ \frac{\nu}{L^2} \ \theta_N(q_\parallel) \int_{-\frac{1}{2}L}^{\frac{1}{2}L} \mathbf{W}_M \left\{ (\mathcal{H} - \Gamma \mathcal{D}) \boldsymbol{\Theta}_N \right\} dz = 0,$$

$$\sum_{N=0}^{K} \alpha_p g \ w_N(q_\parallel) \int_{-\frac{1}{2}L}^{\frac{1}{2}L} \boldsymbol{\Theta}_M \left\{ (\mathcal{H} - \Gamma \mathcal{D}) \mathbf{W}_N \right\} dz \qquad (7.30b)$$

$$+ \frac{\nu}{L^2} \ \theta_N(q_\parallel) \int_{-\frac{1}{2}L}^{\frac{1}{2}L} \boldsymbol{\Theta}_M \left\{ (\mathcal{H} - \Gamma \mathcal{D}) \boldsymbol{\Theta}_N \right\} dz = 0.$$

For having a solution of Eq. (7.30) different from zero, the corresponding secular equation has to be satisfied. Solving the $2K$-order secular equation for Γ produces a set of $2K$ approximate decay rates, $\Gamma_N(q_\parallel)$. Then, for each one of these decay rates, the corresponding set of approximate hydrodynamic modes can be calculated. Once we have approximate decay rates and hydrodynamic modes, we obtain an approximate nonequilibrium structure factor by following the general procedures described in Sect. 7.2. We remark that the Galerkin method is equivalent to a finite-mode approach with approximate decay rates and corresponding modes. The advantage of this scheme is that, at lowest order, it produces analytical expressions for the decay rates and the nonequilibrium structure factor. This Galerkin approach is the same "variational" method described by Chandrasekhar (1961) for the study of the linear instability of the solution of the hydrodynamic equations.

The quality of the approximate solution obtained from truncating the expansion (7.28) will depend on the choice of the test functions (7.29) used to force the solution to match the BC. The polynomials, given by Eq. (7.29), are the simplest possible test functions. These polynomials have been previously used by Manneville (1990) for a linear stability analysis. Furthermore, the Galerkin polynomials (7.28) are considered to be optimal for studying the convective instability, because of the variational structure of the underlying problem (Manneville, 1990). However, other choices of test functions are also possible; for example Ortiz de Zárate and Muñoz Redondo (2001) have used trigonometric functions for the temperature fluctuations and our same polynomials for the vertical velocity. Chandrasekhar (1961) and more recently Niederländer et al. (1991) also used trigonometric functions for the temperature, but so-called Chandrasekhar functions for the velocity. Ultimately, the quality of the Galerkin approximation adopted has to be evaluated *a posteriori*, by comparing the approximate result obtained from truncating the series (7.28) with cases for which exact results are known.

Here we shall consider only a zeroth-order approximation. That is, we retain only the terms of order $N = 0$ in the expansion (7.28), to be substituted into Eq. (7.30). After evaluating the various integrals, we obtain the corresponding secular equation as:

$$\det \left[\mathsf{H}(q_{\|}) - \tilde{\Gamma} \mathsf{D}(q_{\|}) \right] = 0 \tag{7.31}$$

where the 2×2 dimensionless matrices $\mathsf{H}(q_{\|})$ and $\mathsf{D}(q_{\|})$ are given by:

$$\mathsf{H}(q_{\|}) = \begin{bmatrix} \dfrac{(12 + \tilde{q}_{\|}^2)^2}{630} + \dfrac{4}{7} & -\dfrac{\tilde{q}_{\|}^2}{140} \\[3mm] -\dfrac{Ra}{140 Pr} & \dfrac{10 + \tilde{q}_{\|}^2}{30 Pr} \end{bmatrix} \tag{7.32a}$$

and

$$\mathsf{D}(q_{\|}) = \begin{bmatrix} \dfrac{(12 + \tilde{q}_{\|}^2)}{630} & 0 \\[3mm] 0 & \dfrac{1}{30} \end{bmatrix}. \tag{7.32b}$$

Solving the secular equation we obtain the zeroth-order Galerkin approximation for the slowest decay rates as:

$$\tilde{\Gamma}_1^{(\pm)} = \frac{1}{2} \frac{(\tilde{q}_{\|}^2 + 10)[Pr + \tilde{A}(\tilde{q}_{\|})]}{Pr \, \tilde{A}(\tilde{q}_{\|})} \tag{7.33}$$

$$\times \left\{ 1 \pm \sqrt{1 - \frac{Pr \tilde{A}(\tilde{q}_{\|})}{7[Pr + \tilde{A}(\tilde{q}_{\|})]^2} \left[28 - \frac{27 \tilde{q}_{\|}^2 Ra}{(\tilde{q}_{\|}^2 + 10)(\tilde{q}_{\|}^4 + 24\tilde{q}_{\|}^2 + 504)} \right]} \right\},$$

where the function $\tilde{A}(\tilde{q}_{\|})$ is given by:

$$\tilde{A}(\tilde{q}_{\|}) = \frac{(\tilde{q}_{\|}^2 + 10)(\tilde{q}_{\|}^2 + 12)}{(\tilde{q}_{\|}^4 + 24\tilde{q}_{\|}^2 + 504)}. \tag{7.34}$$

We note that both zeroth-order test functions (7.29) have a single maximum in the interval $[-L/2, L/2]$ and have even parity in the vertical variable z, so that they indeed correspond to the slowest even decay rate specifically discussed in Sect 6.4. The corresponding hydrodynamic modes may be readily derived from Eqs. (7.32). From Eq. (7.33) with $Ra < 0$, approximate predictions can be made for the threshold of propagating modes, but we shall not discuss it here and we move on to the calculation of the nonequilibrium structure factor.

We now have all the information needed to obtain the static structure factor $S(q_\parallel, z, z')$. We first substitute the approximate hydrodynamic modes into expressions (6.17) and (7.19), so as to obtain approximate values for the normalization constant $B_N(q_\parallel)$ and the mode-coupling coefficients C $(q_\parallel)$. Next, from expressions (7.15) for the equilibrium part and (7.18) for the nonequilibrium part, we find that the static structure factor for two rigid boundaries, in the zeroth-order Galerkin approximation, can be written as (Ortiz de Zárate and Sengers, 2002):

$$\frac{\gamma \; S(q_\parallel, z, z')}{S_{\mathrm{E}}(\gamma - 1)} = \left[\frac{30}{L} + \tilde{S}^0_{\mathrm{NE}} \; C^{\mathrm{G}}_0(\tilde{q}_\parallel) \right] \; \left(\frac{z}{L} - \frac{z^2}{L^2} \right) \; \left(\frac{z'}{L} - \frac{z'^2}{L^2} \right). \quad (7.35)$$

In Eq. (7.35) we have introduced the quantity $C^{\mathrm{G}}_0(\tilde{q}_\parallel)$ which represents the normalized nonequilibrium enhancement for two rigid boundaries in the zeroth-order Galerkin approximation, given by:

$$C^{\mathrm{G}}_0(\tilde{q}_\parallel) = \frac{30}{L} \frac{Pr + 1}{Pr + \tilde{A}(\tilde{q}_\parallel)} \frac{27\tilde{q}^2_\parallel}{28(\tilde{q}^2_\parallel + 10)[(\tilde{q}^2_\parallel + 12)^2 + 360] - 27\tilde{q}^2_\parallel \; Ra}. \quad (7.36)$$

As an alternative procedure, a more direct calculation of the nonequilibrium structure can be performed by assuming that the solution of the original problem (6.4) can be represented similarly to the slowest decay rate in Eq. (7.30). Then, substituting Eqs. (6.4) into the linearized random Boussinesq and projecting the results onto the set of the two Galerkin test functions (7.29), a set of algebraic equations is obtained from which the nonequilibrium structure factor may be calculated. This more direct approach was actually adopted in the original derivation of Ortiz de Zárate and Sengers (2002). Next, we shall comment briefly on the result (7.35) obtained in our zeroth-order Galerkin approximation.

The first term in the RHS of Eq. (7.35 represents the equilibrium static structure factor when $\nabla T_0 = 0$. As we have seen, the static structure factor is short ranged and should become proportional to a delta function $\delta(z - z')$, when a complete and exact set of hydrodynamic modes is used. However, our approximate solution for two rigid boundaries contains in (7.35) a constant multiplying the Galerkin polynomials. Actually, what we obtained in (7.35) for the equilibrium structure factor is the first term of the series expansion of the delta function in terms of the Galerkin polynomials. Therefore, this shortcoming is a consequence of having performed the calculation in zeroth order only.

Regarding the nonequilibrium contribution, it should be noticed that the coefficient $C^{\mathrm{G}}_0(\tilde{q}_\parallel)$ has a structure very similar to the general term appearing under the sum in Eq. (7.6) for the case of free boundaries. With regards to the dependence on the Prandtl number, for free boundaries the normalized nonequilibrium enhancement is independent of Pr, while in Eq. (7.36) for rigid boundaries a rational function of Pr appears, that rapidly approaches unity as $\tilde{q}_\parallel$ increases. The dependence on the Rayleigh number is also similar in the two cases: the factor $(\tilde{q}_\parallel^2 + N^2\pi^2)^3$ in the series (7.6) for free boundaries is replaced with a polynomial of sixth order in $\tilde{q}_\parallel$.

Our approximate solution for rigid boundaries breaks down when there is a zero in the denominator of (7.35), so that:

$$Ra = \frac{28}{27} \frac{(\tilde{q}_\parallel^2 + 10)[(\tilde{q}_\parallel^2 + 12)^2 + 360]}{\tilde{q}_\parallel^2}. \tag{7.37}$$

Equation (7.37) represents the threshold condition, $\Gamma_1^{(-)} = 0$, for the convective instability, as obtained from the approximate Eq. (7.33), and discussed more extensively by Manneville (1990). The minimum of Eq. (7.37) as a function of $\tilde{q}_\parallel^2$ occurs at:

$$\tilde{q}_c^2 = \left(\frac{29107}{27} + 2\sqrt{\frac{846790}{3}}\right)^{\frac{1}{3}} + \left(\frac{29107}{27} - 2\sqrt{\frac{846790}{3}}\right)^{\frac{1}{3}} - \frac{17}{3}, \tag{7.38}$$

corresponding to $\tilde{q}_c \simeq 3.1165$. Consequently, for Eq. (7.37) to hold, the Rayleigh number Ra has to be larger than a critical Rayleigh number Ra_c obtained by substitution of (7.38) into (7.37), or:

$$Ra_c = \frac{28}{135}(\tilde{q}_c^2 + 10)^2(\tilde{q}_c^2 + 12) \simeq 1750. \tag{7.39}$$

In the literature (Chandrasekhar, 1961; Manneville, 1990) these approximate critical values are considered to be in good agreement with the exact threshold values for the case of two rigid boundaries, which as already quoted, are $R_c \simeq 1708$ and $\tilde{q}_c \simeq 3.1163$ (Manneville, 1990). These good results for the prediction of the instability give us confidence in the quality of our Galerkin approximation.

From the divergence of the structure factor we recover the well-known results for the appearance of convection from a linear-stability analysis. In addition we note that the introduction of boundary conditions in the calculation of the structure factor, again results in an extension of the validity of Eq. (4.29) for the structure factor at negative Rayleigh numbers to a finite range of positive Rayleigh numbers. The derivation of the nonequilibrium structure factor in Sect. 7.2 assumed $0 < Ra < Ra_c$. However, Eq. (7.35) gives meaningful results for any negative Ra number; even for Rayleigh numbers for which the approximate expression for the decay rates (7.33) predicts the existence of propagating modes. This fact gives us confidence that the method developed in Sect. 7.2 can be extended to negative Ra.

To obtain the experimental nonequilibrium structure factor at small scattering angles, we substitute Eq. (7.35) into Eq. (6.28), perform the double integration in Eq. (6.28) and take the limit $q_\perp \to 0$ and $q_\parallel \to q$. Instead of the exact Eq. (7.20), we then obtain for $S(\tilde{q})$:

$$S(\tilde{q}) = S_E \frac{\gamma - 1}{\gamma} \left[\frac{5}{6} + \tilde{S}_{NE}^0 \, \tilde{S}_{NE}^G(\tilde{q}) \right]. \tag{7.40}$$

Here $\tilde{S}_{NE}^G(\tilde{q})$ is a (zeroth-order) Galerkin approximation to the normalized nonequilibrium enhancement for rigid boundaries:

$$\tilde{S}_{NE}^G(\tilde{q}) = \frac{L}{36} \, C_0^G(\tilde{q}) \tag{7.41}$$

where $C_0^G(\tilde{q})$ is given by Eq. (7.36) with $\tilde{q} = \tilde{q}_\parallel$. The exact nonequilibrium enhancement is given by Eq. (7.21) as an expansion in the hydrodynamic modes. The advantage of Eqs. (7.40)-(7.41) is that the (approximate) nonequilibrium structure factor is expressed as a function of q, Ra and Pr, with no numerical computations to be carried out. A first evidence of the approximate character of Eq. (7.40) is that, as a consequence of having performed a single mode approximation, the result obtained for equilibrium is $1 - (5/6) \simeq 20\%$ lower than the expected result.

To compare the Galerkin approximation to the nonequilibrium contribution in Eq. (7.41) with the exact numerical results shown in Fig. 7.1, we have plotted in Fig. 7.3 the normalized nonequilibrium enhancement $\tilde{S}_{NE}(\tilde{q})$ as a function of $\tilde{q}$ for $Pr = 8$, a Prandtl-number value that approximately corresponds to that of pure toluene at 20°C (Li et al., 1994a); the two lower curves (undistinguishable at the scale of the figure) correspond to $Ra = 650$; the two upper curves correspond to $Ra = 1650$, which is close to the instability $\epsilon = -0.035$. The solid curves represent the exact numerical results from the equations in Sect. 7.2, while the dotted curves represent the Galerkin approximation, given by Eq. (7.41). The results in Fig. 7.3 indicate that the Galerkin approximation is excellent for moderate values of the Ra number, while at first sight the agreement deteriorates at values of Ra close to $Ra_c = 1708$. However, if we scale the Ra number in Eq. (7.41) by $Ra_c = 1750$ corresponding to the Galerkin approximation, *i.e.*, if we substitute Ra by $1750(1 + \epsilon)$, the resulting scaled Galerkin approximation matches almost perfectly the exact numerical result obtained in Sect. 7.2 for the same value of ϵ. For instance, for the value $\epsilon = -0.035$ displayed in the upper curves of Fig. 7.3 the difference between the exact numerical result and the Galerkin approximation is less than 1% for the entire range of $\tilde{q}$, and the corresponding curves become practically undistinguishable. We conclude that the Galerkin approximation, Eq. (7.41) with the Ra number conveniently scaled, yields an excellent representation of the nonequilibrium structure factor in the Rayleigh-Bénard problem.

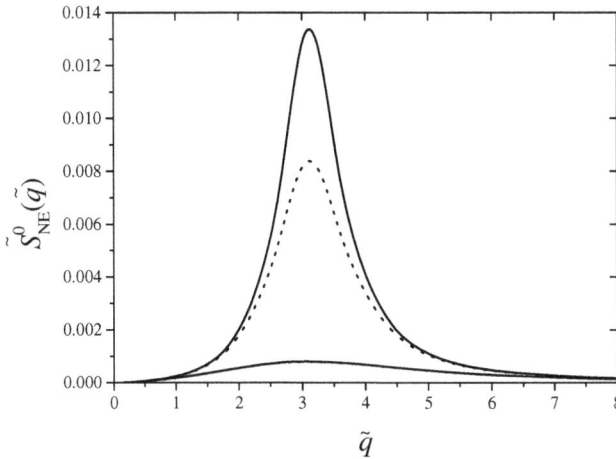

Figure 7.3: Normalized nonequilibrium enhancement of fluctuations as a function of $\tilde{q}$. The two lower curves (undistinguishable at the scale of the figure) are for $Ra = 650$, while the two upper curves are for $Ra = 1650$. $Pr = 8$ in all cases. Solid curves represent the exact numerical result of Sect. 7.2, while the dotted curves represent the Galerkin approximation Eq. (7.41). See text for further comment.

We finalize this section by studying the large and small-q limits of the dimensionless static structure factor in the zeroth-order Galerkin approximation. Taking the appropriate limits in (7.36), we obtain for small $\tilde{q}$:

$$\tilde{S}_{\mathrm{NE}}^{\mathrm{G}}(\tilde{q}) \xrightarrow{\tilde{q}\to 0} \frac{3(Pr+1)}{896(21Pr+5)}\, \tilde{q}^2, \tag{7.42}$$

and for large $\tilde{q}$:

$$\tilde{S}_{\mathrm{NE}}^{\mathrm{G}}(\tilde{q}) \xrightarrow{\tilde{q}\to\infty} \frac{45}{56}\frac{1}{\tilde{q}^4}. \tag{7.43}$$

These results should be compared with the exact results given by Eq. (7.23) for small-q and Eq. (7.27) for large-q. We note that the Galerkin approximation reproduces for both limits the correct asymptotic behavior qualitatively. Quantitatively, the prefactor multiplying $\tilde{q}^2$ in the small-q limit differs from the exact prefactor (7.23) by less than 2% for any value of Pr. In the large-q limit the Galerkin approximation, as the exact Eq. (7.27), has a prefactor multiplying q^{-4} that does not depend on Pr. However, the numerical value of the prefactor is 45/56, which is a 20% less than the value unity expected from the exact result (7.27). The Galerkin approximation adopted here yields a slightly better result in the small-q limit than in the large-q limit, as is to be expected from a single-mode approximation. Overall, our simple Galerkin approximation (with Ra scaled where appropriate) gives a good representation of the actual structure factor for a fluids between two rigid boundaries.

As mentioned earlier, the current Galerkin approximation gives meaningful results for any negative Ra. As expected, due to the damping effect of gravity when heating from above, the structure factor as calculated from Eqs. (7.40)-(7.41) presents a very flat maximum for negative Ra. We shall not further discuss it here and we refer the interested reader to Ortiz de Zárate and Sengers (2002).

7.5   Correlations in real space

That the nonequilibrium equal-time correlation functions are spatially long ranged is more evident when we consider these functions in real space. Thus far we have studied correlation functions in Fourier space; to revert to real space we need to apply an inverse Fourier transformation. From the relationship (6.30) between the static structure factor $S(q_\parallel, z, z')$ and the temperature autocorrelation function, the equation of state (4.3) in the Boussinesq approximation, and Eq. (7.4) for $S(q_\parallel, z, z')$, applying inverse Fourier transformations in the horizontal wave vectors $\mathbf{q}_\parallel$ and $\mathbf{q}'_\parallel$, we obtain the equal-time density autocorrelation function for the RB problem as:

$$\frac{\langle \delta\rho(\mathbf{r}, t) \cdot \delta\rho(\mathbf{r}', t)\rangle}{\rho m_0} = S_{\mathrm{E}} \frac{\gamma - 1}{\gamma} \left[\delta(\mathbf{r} - \mathbf{r}') + \tilde{S}^0_{\mathrm{NE}}\ \tilde{G}_{\mathrm{NE}}(r_\parallel, z, z') \right], \quad (7.44)$$

with

$$\tilde{G}_{\mathrm{NE}}(r_\parallel, z, z') = \int_0^\infty q_\parallel\ J_0(q_\parallel r_\parallel)\ \tilde{S}_{\mathrm{NE}}(q_\parallel, z, z')\ dq_\parallel. \quad (7.45)$$

Here $r_\parallel^2 = (x - x')^2 + (y - y')^2$ is the distance in the horizontal plane between $\mathbf{r}$ and $\mathbf{r}'$, and $J_0(x)$ the Bessel function of the first kind and of order zero (Gradstein and Ryzhik, 1994). As discussed in previous sections, Eqs. (7.4) and (7.5) apply for two free boundaries and for two rigid boundaries. Hence, Eqs. (7.44) and (7.45) will be valid for both sets of BC. However, the normalized nonequilibrium enhancement $\tilde{S}_{\mathrm{NE}}(q_\parallel, z, z')$ to be substituted in (7.45) depends on the BC. It will be given by (7.6) for two free boundaries or (7.18) for two rigid boundaries (and $0 < Ra < Ra_c$).

When $\nabla T_0 = 0$ (and, thus, $\tilde{S}^0_{\mathrm{NE}} = 0$), Eq. (7.44) reproduces the result (3.32) for a fluid in thermodynamic equilibrium. The equilibrium structure factor is spatially short ranged, *i.e.*, at the level of a hydrodynamic description it is proportional to a delta function. On the other hand, the nonequilibrium contribution $\tilde{G}_{\mathrm{NE}}(r_\parallel, z, z')$ is spatially long ranged. To

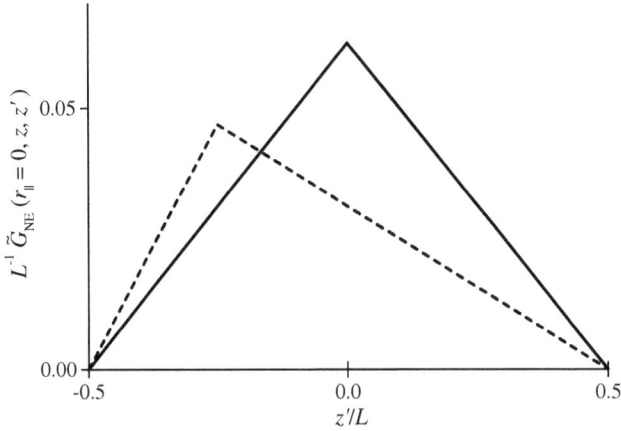

Figure 7.4: Normalized nonequilibrium component of the real space density autocorrelation in the vertical direction, $\tilde{G}_{NE}(r_\parallel = 0, z, z')$ from (7.46), as a function of $\tilde{z}'$. Solid line is for $\tilde{z} = 0$ and dashed line for $\tilde{z} = -1/4$. The long-ranged nature of such autocorrelation function seems evident.

best visualize the spatial long-ranged behavior, we need to perform the integration in Eq. (7.45).

The integral (7.45) cannot be evaluated analytically in general. Ortiz de Zárate et al. (2001) have considered the simplest case of a fluid layer between two free boundaries and neglecting buoyancy. When buoyancy is neglected, $\tilde{S}_{NE}(q_\parallel, z, z')$ for two free boundaries has a very simple expression obtained by substituting $Ra = 0$ in Eq. (7.6). However, even in that simplest case, the function $\tilde{G}_{NE}(r_\parallel, z, z')$ can still not be evaluated analytically in general. Fortunately, there are two particular cases in which the integral (7.45) can be obtained analytically.

(i) The first situation corresponds to $r_\parallel = 0$, which yields the nonequilibrium equal-time density autocorrelation function along the vertical direction of the stationary temperature gradient. After performing the integration (7.45), one obtains in this case (Ortiz de Zárate et al., 2001):

$$\tilde{G}_{NE}(r_\parallel = 0, z, z') = \frac{L}{4}\left[\frac{1}{4} - \frac{z\,z'}{L^2} - \frac{|z - z'|}{2L}\right]. \qquad (7.46)$$

As an example, we show in Fig. 7.4 the normalized correlation function as a function of z'/L, for two different values of z/L. This nonequilibrium real-space equal-time correlation function along the z-axis is always positive, has a maximum at $z = z'$ and decreases monotonically from this maximum, reaching zero at both ends of the interval. The real-space correlation function in the vertical (parallel to the gradient) direction is nonlocal, long ranged and does not involve any intrinsic length scale, *i.e.*, the correlation encompasses the entire system only to be cut off by the finite size of the system itself. The algebraic exponent characterizing the long-ranged nature of the correlation in the direction coincident with the temperature gradient equals $+1$. We note that for points near the center of the liquid layer, where $z \simeq z' \simeq 0$, expression (7.46) reduces to:

$$\tilde{G}_{\mathrm{NE}}(r_\| = 0, z, z') = \frac{L}{16}\left[1 - 2\,\frac{|z - z'|}{L}\right]. \tag{7.47}$$

(ii) The integral (7.45) for free boundaries and neglecting buoyancy can also be evaluated exactly for the situation $z = z' = 0$. In that case $\tilde{G}_{\mathrm{NE}}$ depends only on the radial variable $r_\|$, and represents the nonequilibrium equal-time density autocorrelation function in the plane parallel to the boundaries at mid-height in the cell, *i.e.*, the plane at maximum distance from the boundaries. Integration of (7.45) in this case yields:

$$\tilde{G}_{\mathrm{NE}}(r_\|, 0, 0) = \frac{r_\|}{2\pi}\left[\sum_{\substack{N=1 \\ N,\ \mathrm{odd}}}^{\infty} \frac{1}{N}\,K_1(N\pi\tilde{r}_\|) - \frac{\pi}{2}\tilde{r}_\| \sum_{\substack{N=1 \\ N,\ \mathrm{odd}}}^{\infty} K_0(N\pi\tilde{r}_\|)\right], \tag{7.48}$$

where $\tilde{r}_\| = r_\|/L$ and K_0 and K_1 are modified Bessel functions of the second kind. In Fig. 7.5 we have plotted the normalized correlation function $\tilde{G}_{\mathrm{NE}}(r_\|, 0, 0)$ as a function of $\tilde{r}_\|$. From (7.48) we conclude that for $\tilde{r}_\| \ll 1$ the real-space correlation function will vary with $r_\|$ as:

$$\tilde{G}_{\mathrm{NE}}(r_\|, 0, 0) = \frac{L}{16}\left(1 - \frac{3\,r_\|}{L}\right), \tag{7.49}$$

which confirms the existence of long-ranged correlations in the direction perpendicular to the temperature gradient. On the other hand, for $\tilde{r}_\| \gg 1$, using the asymptotic expansion of the Bessel functions, we obtain:

$$\tilde{G}_{\mathrm{NE}}(r_\|, 0, 0) = \frac{L(2 - \pi\tilde{r}_\|)}{8\pi}\,\sqrt{2\tilde{r}_\|}\,\exp(-\pi\tilde{r}_\|). \tag{7.50}$$

Hence, the finite size of the liquid layer not only restricts the long-ranged nature of the correlations in the direction coincident with the temperature gradient, but also in the direction perpendicular to the temperature gradient. Figure 7.5 shows how the correlations in a plane perpendicular to the temperature gradient rapidly vanish as the radial distance $r_\|$ approaches

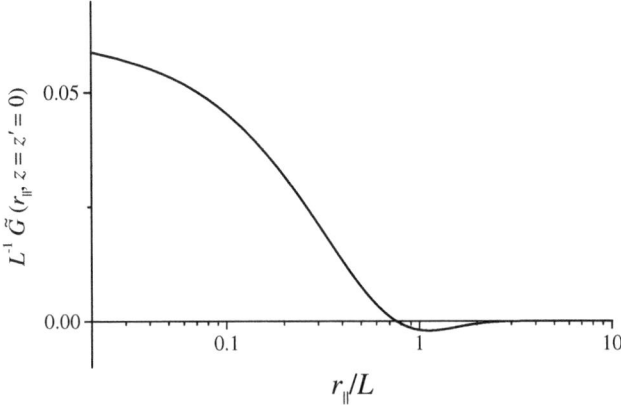

Figure 7.5: Semi-logarithmic plot of the normalized nonequilibrium component of the real space density autocorrelation in the central plane of the liquid layer, $\tilde{G}_{\mathrm{NE}}(r_\parallel, z = 0, z' = 0)$ from (7.48), as a function of $\tilde{r}_\parallel$.

values of the order of L. This is also evident from the $\tilde{r}_\parallel$ dependence of the correlation function, which switches from a linear dependence in Eq. (7.49) for $\tilde{r}_\parallel \ll 1$ to an exponential dependence in Eq. (7.50) for $\tilde{r}_\parallel \gg 1$.

We mention that the results presented here are consistent with the work of Schmitz and Cohen (1985b) regarding the spatial nature of the correlations in the absence of boundary conditions. If one substitutes $r_\parallel = 0$ into Eq. (3.13a) of Schmitz and Cohen (1985b), one obtains the result that the nonequilibrium part of the correlation function is directly proportional to $-2|z - z'|$ in agreement with Eq. (7.47) for the correlation function in interior points of the liquid layer. However, Eq. (7.47) differs from the result of Schmitz and Cohen (1985b) by an additional term proportional to L which diverges in the limit $L \to \infty$ and, therefore, was not considered by Schmitz and Cohen (1985b). Furthermore, if one substitutes $z_1 = z_2$ in into Eq. (3.13a) of Schmitz and Cohen (1985b), one obtains the result that the nonequilibrium part of the correlation function is proportional to $-3r_\parallel$, to be compared with our Eq. (7.49) for the correlations in the direction perpendicular to the temperature gradient and far from the boundaries, when $\tilde{r}_\parallel \ll 1$. As can be seen from Fig. 7.5, the correlation function in the direction parallel to the boundaries has a minimum close to $r_\parallel = L$, (actually for our boundary conditions at $\tilde{r}_\parallel \simeq 1.1542$). This value corresponds approximately to the size of the rolls that appear at the first convective instability. It seems that the first convective instability somehow pre-exists in the nonequilibrium structure factor of the fluid, even when buoyancy is not taken into account.

The results of this section apply to the simplest case of a fluid layer subjected to a temperature gradient, between two free boundaries and without including any buoyancy effects. One may wonder how the present results are modified for realistic BC and when buoyancy is included. It is not possible to obtain nice and compact analytical expressions in that cases. However, the qualitative behavior of $G_{\mathrm{NE}}(r_{\parallel}, z, z')$ will only slightly differ from the one displayed in Figs. 7.4 and 7.5. For instance, inclusion of buoyancy in Fig. 7.4 causes the vertex appearing at $z = z'$ to become rounded, the corresponding function being continuous and with continuous derivative over the entire $[-L/2, L/2]$ interval. Numerical values will change, of course, but the fact that the correlation function is long ranged and encompasses the entire interval remains valid.

7.6   Contribution of nonequilibrium fluctuations to heat transfer

It is interesting to note that, in a fluid layer subjected to a temperature gradient, the cross correlation between temperature fluctuations and vertical velocity fluctuations is generically nonzero: $\langle \delta T(\mathbf{r}, t) \cdot \delta v_z(\mathbf{r}', t) \rangle \neq 0$. As a result, a small contribution of fluctuations to heat transfer across the layer is expected (van Beijeren and Cohen, 1988b). We shall not discuss this effect in any detail here, but mention that it becomes more pronounced close to the convective instability, where enhancement of the nonequilibrium fluctuations is large. Close but below the convective instability, heat transfer due to nonequilibrium fluctuations has been theoretically studied by Ahlers et al. (1981) in the framework of the Swift and Hohenberg (1977) approximation (see Sect. 8.3). However, experimental verification of the theoretical predictions have remained somewhat inconclusive (Ahlers et al., 1981; Meyer et al., 1991). More detailed theoretical approaches, including modifications to the amplitude equation due to periodic driving (Osenda et al., 1998; Schmitt and Lücke, 1991), or inclusion of nonlinear modes (Osenda et al., 1997), seem to give a closer agreement with experiments. However the question is far from settled and further theoretical and experimental work will be required on this topic.

Cross-correlation between temperature and velocity fluctuations for a fluid subjected to a uniform temperature gradient has also been considered by Malek Mansour et al. (1987), who verified the long-ranged nature of such correlations. A related problem is the possible contribution of nonequilibrium concentration fluctuations to diffusive flux, which has been analyzed by Brogioli and Vailati (2001).

Chapter 8

Thermal fluctuations close to the Rayleigh-Bénard instability

Probably one of the most intensively studied topics in the theory of nonequilibrium fluctuations is thermal noise near the convective instability (Zaitsev and Shliomis, 1971; Swift and Hohenberg, 1977; Kirkpatrick and Cohen, 1983; Schmitz and Cohen, 1985b; van Beijeren and Cohen, 1988b; Hohenberg and Swift, 1992). As was elucidated in the previous chapter, the enhancement of the intensity of nonequilibrium fluctuations exhibits a maximum as a function of the horizontal wave number $q_\parallel$. When the convective instability is approached, i.e., when $Ra \to Ra_c$, the peak grows larger and larger. Evidently this mechanism is responsible for the spontaneous appearance of patterns once the threshold Ra_c is surpassed. In a sense, patterns are just nonequilibrium fluctuations that have become macroscopically large. In addition, near the instability fluctuations with wave number close to $q_{\parallel c}$ take a long time to decay, a phenomenon known as critical slowing down. Thus, close to but below the threshold for convection, fluctuations take longer to decay and the intensity of the fluctuations diverges; above the threshold spatial translational symmetry is broken and the fluctuations do not decay in time and give rise to stationary convective patterns. There is an obvious parallel with the physics of second-order equilibrium phase transitions, so that instabilities like the RB convection are sometimes referred to as nonequilibrium phase transitions. However, such an analogy should not be taken too seriously. Equilibrium phase transitions appear because the system minimizes a certain free energy (or maximizes entropy if isolated, see Sect. 3.4), while for nonequilibrium states the existence of such a functional (or nonequilibrium entropy) is dubious. Nevertheless, close analogies do indeed exist regarding the mathematical treatment of thermal noise close to nonequilibrium instabilities and the theory of second-order equilibrium phase transitions.

In this chapter we review what can be obtained from the linear theory of nonequilibrium fluctuations developed in Chapter 6 when the first convective instability is approached from below. Extremely close to the convection threshold, thermal fluctuations are no longer small and a linear theory is expected to fail ultimately. Nevertheless, the linear theory does yield interesting predictions up to values of the Ra number rather close to the critical Ra_c for the appearance of convection. In this respect it is worth noticing that Normand et al. (1977) estimated that, for the RB problem, the values of Ra close to critical where nonlinear effects are important is given by $|\epsilon| = |Ra - Ra_c|/Ra_c \lesssim 10^{-5}$. Therefore, we shall continue to study the nature of the thermal fluctuations on the basis of a linear theory. For nonlinear theories of fluctuations the reader is referred to to the relevant literature elsewhere (Swift and Hohenberg, 1977; Hohenberg and Swift, 1992).

We shall first discuss in Sect. 8.1 the critical slowing down of nonequilibrium fluctuations, which will allow us to introduce in Sect. 8.2 the so-called most-unstable mode approximation for the intensity of fluctuations close to, but below, the convective instability. In subsequent sections of this chapter, we shall discuss some other approximation schemes for characterizing the intensity of nonequilibrium fluctuations near the onset of convection. We shall find a difference between the wave number of maximum enhancement of fluctuations and the wave number of maximum growth rate. This effect is a consequence of the intrinsic stochastic nature of the fluctuations.

8.1   Critical slowing down of nonequilibrium fluctuations

For $0 < Ra < Ra_c$ we know from Chapter 6 that all decay rates are real numbers. Furthermore, the hydrodynamic modes can be normalized so that normalization constants and mode-coupling coefficients are also real numbers. The consequences of this fact for the static structure factor were already discussed in Sect. 7.2. However, in this section we shall consider the consequences for the experimentally accessible dynamic structure factor, $S(\omega, q)$ as defined by Eq. (6.28). Then, integration over the vertical variables of Eq. (6.27) for $S(\omega, q_\parallel, z, z')$, when all decay rates are real numbers and in the small-angle approximation, yields:

$$S(\omega, q) = \sum_{\substack{N,M=1 \\ \text{even}}}^{\infty} \frac{C_{NM}\, \Xi_N\, \Xi_M}{\Gamma_N + \Gamma_M} \left\{ \frac{2\Gamma_N}{\omega^2 + \Gamma_N^2} + \frac{2\Gamma_M}{\omega^2 + \Gamma_M^2} \right\}, \tag{8.1}$$

where the summation is over the even decay rates, and where the ratios $\Xi_N(q)$ were defined in Eq. (6.46). To shorten the expression we do not denote in the RHS of Eq. (8.1) the dependence on the wave number q explicitly. An important feature concerning Eq. (8.1) is that the second line of Eq. (6.27) for the mode-coupling coefficients does not contribute. We mentioned in Sect. 6.1.1 that $S(\omega, q)$ is always real. Since, for $0 < Ra < Ra_c$, $\Gamma_N(q)$, $\Xi_N(q)$ and $C_{NM}(q)$ are real numbers, it follows that the contribution from the second line of Eq. (6.27) is purely imaginary and, hence, must vanish. Obviously, upon integrating Eq. (8.1) over ω, and using Eq. (7.13) for $C_{NM}(q)$, we recover Eq. (7.21) for the nonequilibrium *static* structure factor. A first conclusion from (8.1) is that for $0 < Ra < Ra_c$ the experimental nonequilibrium structure factor can be expressed as a sum of simple lorentzians, which confirms the absence of propagating modes in this case.

Rather than the dynamic structure factor, we prefer to consider the time-dependent correlation function $\tilde{S}_{NE}(q, \tau)$, since that is the quantity actually measured in photon-counting light scattering or in shadowgraph experiments (see Chapter 10). Applying an inverse Fourier transformation to Eq. (8.1), we obtain $\tilde{S}_{NE}(q, \tau)$ as a sum of exponentials:

$$S(q, \tau) = \sum_{\substack{N=1 \\ \text{even}}}^{\infty} A_N(q) \, e^{-\Gamma_N(q) \, |\tau|} \tag{8.2}$$

with amplitudes

$$A_N(q) = \Xi_N(q) \sum_{\substack{M=1 \\ \text{even}}}^{\infty} \frac{2 \, C_{NM}(q) \, \Xi_M(q)}{\Gamma_N(q) + \Gamma_M(q)}. \tag{8.3}$$

As discussed in Sect. 6.4, when $Ra \to Ra_c$ the first decay rate $\Gamma_1^{(-)}(q)$ approaches zero for wave numbers of the order of q_c. This means that the fluctuations with horizontal wave numbers around q_c will decay very slowly. Critical slowing down of nonequilibrium fluctuations in the Rayleigh-Bénard problem has been observed experimentally by Oh et al. (2004), as further discussed in Chapter 10.

8.2 The most-unstable-mode approximation

From Eq. (8.2) we observe that, for $0 < Ra < Ra_c$, the time correlation function of the nonequilibrium system consists of a series of exponentials with decay rates $\Gamma_N(q)$. When the convective instability is approached, the slowest decay rate goes to zero for values of q around q_c. As a consequence, the series (8.2) will be completely dominated by the first term. Since the decay rates form a discrete well-separated set of positive numbers, the other exponential terms in (8.2) will have completely decayed at values of τ for which the first term is still important. The most unstable-mode approximation consists in retaining only the first term in the series Eq. (8.2). The single-mode approximation is only correct at Rayleigh numbers below Ra_c. The assumption fails in situations where also Rayleigh numbers just above Ra_c play a role as discussed by van Beijeren and Cohen (1988a,b).

Alternatively, the most-unstable mode approximation may be discussed in terms of the static structure factor. In Chapter 7 it was shown that the intensity of nonequilibrium fluctuations can be expressed as the sum of a double series, Eq. (7.21), over the hydrodynamic modes with even vertical parity. This series is equivalent to taking $\tau = 0$ in Eq. (8.3). The decay rates appear in the denominator of the terms of the double series (7.21). When $Ra \lesssim Ra_c$, the value of the slowest decay rate will be close to zero for $q_\parallel$ values close to the critical wave number. Consequently, the sum of the series (7.21) will be dominated by the term $N = M = 1$, which contains $\Gamma_1^{(-)}$ in the denominator. Again, the most-unstable-mode approximation for the nonequilibrium enhancement of the intensity of the fluctuations (Zaitsev and Shliomis, 1971) is obtained by retaining only the term $[N = 1, M = 1]$ in the double series (7.21):

$$\tilde{S}_{\mathrm{NE}}(\tilde{q}) \simeq \frac{1}{2L} \frac{\tilde{C}_{11}^{\mathrm{NE,E}}(q)}{\Gamma_1^{\mathrm{E}}(q)} \,\Xi_1^{\mathrm{E}}(q)^2. \tag{8.4}$$

For the case of two free boundaries, we have earlier derived analytic expressions for the decay rates and corresponding hydrodynamic modes yielding a very compact expression (7.6) for the nonequilibrium enhancement of the fluctuations. The most-unstable-mode approximation for free boundaries corresponds to truncating Eq. (7.6) at $N = 1$, so that:

$$\tilde{S}_{\mathrm{NE}}^{\mathrm{F}}(\tilde{q}_\parallel, z, z') \simeq \frac{\tilde{q}_\parallel^2}{(\tilde{q}_\parallel^2 + \pi^2)^3 - Ra\,\tilde{q}_\parallel^2} \, U(z) \, U(z'), \tag{8.5}$$

where

$$U(z) = \sqrt{\frac{2}{L}} \, \sin\left(\frac{\pi z}{L}\right), \tag{8.6}$$

are the most unstable hydrodynamic modes at critical wave number, normalized in such a way that:

$$\frac{1}{L} \int_{-\frac{1}{2}L}^{\frac{1}{2}L} U^2(z) \, dz = 1. \tag{8.7}$$

In practice, the equilibrium fluctuations can be neglected compared with the nonequilibrium fluctuations, and the most-unstable-mode approximation for the nonequilibrium enhancement, when multiplied by the normalization $S_E \tilde{S}_{NE}^0$ can be identified with the structure factor of the nonequilibrium fluid.

The most-unstable-mode approximation to the actual (experimental) static structure factor for free boundaries will be obtained by substitution of Eq. (8.5) into Eq. (6.31) and performing the vertical integrations. In the small-angle approximation ($q_\parallel \simeq q$, $q_\perp \simeq 0$) we obtain:

$$\tilde{S}_{NE}^F(\tilde{q}) \simeq \frac{8}{\pi^2} \frac{\tilde{q}^2}{(\tilde{q}^2 + \pi^2)^3 - Ra\tilde{q}^2} = \frac{8}{\pi^2} \frac{\tilde{q}^2}{(\tilde{q}^2 + 4\pi^2)(\tilde{q}^2 - \tilde{q}_c^2)^2 - Ra_c\tilde{q}^2\epsilon}, \quad (8.8)$$

where in the RHS we introduced the critical wave number $\tilde{q}_c = \pi/\sqrt{2}$ and critical Rayleigh number $Ra_c = 27\pi^4/4$ corresponding to free boundaries; the quantity $\epsilon = (Ra - Ra_c)/Ra_c$ is a measure of the distance to the instability. We note in Eq. (8.8) that the most-unstable-mode approximation reproduces qualitatively the correct asymptotic behavior of the nonequilibrium structure factor, namely $\propto q^2$ for small q and $\propto q^{-4}$ for large q. This is evident from Eq. (8.8) for free boundaries, while for rigid boundaries it may be deduced from the asymptotic decay rates and vertical ratios $\Xi_N(q)$, discussed in Sects. 6.3.1 and 6.3.2.

Furthermore, because the decay rates form a discrete set of well-separated numbers, the most-unstable-mode approximation (8.8), yields an excellent approximation not only for $Ra \lesssim Ra_c$ and $q \simeq q_c$, but also for arbitrary Rayleigh numbers, especially for relatively small $\tilde{q}$. For instance, in the limit of small q for two free boundaries, the difference between the proportionality coefficient of the exact result and of the most-unstable-mode approximation is $(17/20160) - (8/\pi^8) \simeq 0.01\%$. For large q the result is worse, the difference being $1 - 8/\pi^2 \simeq 20\%$.

As an example, we show in Fig. 8.1 a plot of the normalized nonequilibrium structure factor for a fluid between two free boundaries at $\epsilon = -10^{-2}$. The solid curve in Fig. 8.1 represents the exact solution, as given by Eq. (7.9). The dotted curve is the most-unstable-mode approximation, as given by Eq. (8.8). The dashed curve is the Swift-Hohenberg approximation, to be discussed in the next section. The most-unstable-mode is indeed an excellent approximation; in Fig. 8.1 it is indistinguishable from the exact solution, except for the asymptotic behavior at large q.

For the case of two rigid boundaries, it is not possible to obtain analytical expressions for the decay rates and even the most-unstable-mode approximation can only be evaluated numerically. In this connection it should be mentioned that the curves displayed in Fig. 7.2 actually correspond to a most-unstable-mode approximation, since only the two first decay rates have been considered for the numerical evaluation. Having verified the good

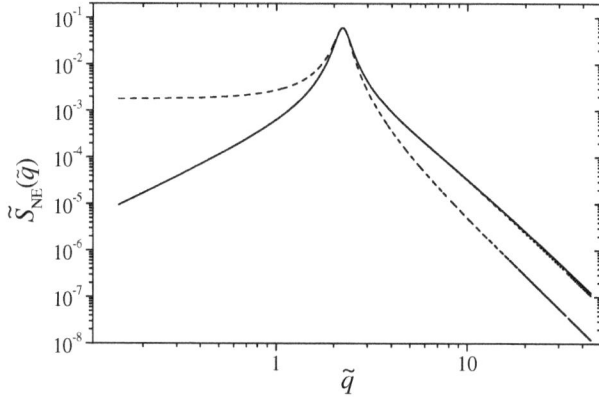

Figure 8.1: Normalized nonequilibrium enhancement of fluctuations as a function of the horizontal wave number for a fluid between two free boundaries at $\epsilon = -10^{-2}$. The solid curve is the exact result, given by Eq. (7.9), the dotted curve is the most-unstable-mode approximation, Eq. (8.8), and the dashed curve is the Swift-Hohenberg approximation.

quality of the most-unstable-mode approximation for the case of free boundaries, we expect that the curves shown in Fig. 7.2 give a fair representation of the actual nonequilibrium structure factor for the case of rigid boundaries as well. Moreover, inclusion of additional decay rates in the numerical calculation indicates that the difference is only a few percent.

We conclude this section by noting that the Galerkin approximation presented in Sect. 7.4 can be also be interpreted as a kind of most-unstable-mode approximation. To elucidate the analogy we note that Eq. (7.41) may be rewritten as:

$$\tilde{S}_{\mathrm{NE}}^{\mathrm{G}}(\tilde{q}) = \frac{30}{36} \frac{Pr + 1}{Pr + \tilde{A}(\tilde{q})} \frac{27\tilde{q}^2}{28(\tilde{q}^2 + 2\tilde{q}_c^2 + 34)(\tilde{q}^2 - \tilde{q}_c^2)^2 - 27\tilde{q}^2 \, Ra_c\epsilon}, \quad (8.9)$$

to be compared with the most-unstable-mode approximation (8.8) for free boundaries. Of course, in Eq. (8.9) the critical wave number and Rayleigh number correspond to (7.38) for the Galerkin approximation. The good results obtained for the nonequilibrium structure factor in the zeroth-order Galerkin approximation, as discussed in Fig. 7.3, confirm the general adequacy of a most-unstable-mode approximation.

8.3 The Swift-Hohenberg approximation

The denominator in the generic most-unstable-mode approximation (8.4) becomes zero when $Ra = Ra_c$ and $\tilde{q} = \tilde{q}_c$. A popular approximation scheme, valid for $q_\parallel$ values close to q_c is obtained by expanding the denominator of the most-unstable-mode approximation in powers of $\tilde{q}_\parallel^2$ around $\tilde{q}_c^2$. For instance, in the case of free boundaries we have:

$$\frac{(\pi^2 + \tilde{q}_\parallel^2)^3}{\tilde{q}_\parallel^2} - Ra \simeq Ra_c \left[-\epsilon + \frac{4}{3\pi^4}(\tilde{q}_\parallel^2 - \tilde{q}_c^2)^2 \right] + \mathcal{O}\left[(\tilde{q}_\parallel^2 - \tilde{q}_c^2)^3\right], \quad (8.10)$$

where $\epsilon = (Ra - Ra_c)/Ra_c$ is again a measure of the distance from the convective instability. Upon substituting the expansion (8.10) into the denominator of the most-unstable-mode approximation (8.5), one obtains the linear Swift-Hohenberg (SH) approximation (Swift and Hohenberg, 1977; Cross, 1980; Cross and Hohenberg, 1993). The linear SH approximation to the enhancement of nonequilibrium fluctuations is usually expressed in the generic form:

$$\frac{Pr\,\tilde{S}_{\mathrm{NE}}(\tilde{q}_\parallel, z, z')}{Pr + 1} = \frac{\tilde{c}^2}{Ra_c^3}\,\frac{\tau_0^{-1}}{\tilde{\xi}_0^4\,(\tilde{q}_\parallel^2 - \tilde{q}_c^2)^2 - \epsilon}\,U(z)\,U(z'). \quad (8.11)$$

In Eq. (8.11), τ_0 is the same linear coefficient in the expansion (6.63) of the slowest decay rate at critical wave number; for the RB problem it is a function of the Prandtl number. The hydrodynamic modes $U(z)$ containing the vertical dependence are subjected to the same normalization condition (8.7) as in the most-unstable mode approximation. In the LHS of Eq. (8.11) we have introduced the dimensionless prefactor $Pr/(Pr + 1)$, because the normalizing factor for the total intensity of fluctuations in the standard presentation of the SH approximation (Swift and Hohenberg, 1977) differs slightly from our definition of $\tilde{S}_{\mathrm{NE}}^0$. Substitution of Eq. (8.10) into the denominator of Eq. (8.5) and comparison with the generic SH (8.11), allows for the determination of the various parameters: $\tilde{\xi}_0^2$, $\tilde{c}$, of the linear SH approximation for nonequilibrium fluctuations in the RB problem for free boundaries. The parameter τ_0 has to be obtained independently from an analysis of the slowest decay rate, for free boundaries it was previously discussed in Eq. (6.64).

The relevance of the SH approximation is that the nonequilibrium structure factor close to the convective instability can be cast in the general form (8.11) independently of the choice for the BC. That is, for the case of two rigid boundaries it is possible to obtain an approximation of the form (8.11) with appropriate values of the parameters Ra_c, q_c, $\tilde{\xi}_0^2$ and τ_0 (Swift and Hohenberg, 1977; Hohenberg and Swift, 1992). Furthermore, it can be shown that any spatially extended system whose evolution is described by (linear) stochastic differential equations and that exhibits an I_s instability, admit a SH-like approximation for the enhancement of the fluctuations close to (but below) the instability (Cross and Hohenberg, 1993). This can be demonstrated through a general approximation scheme,

Table 8.1: Values of the parameters in the SH approximation, Eq. (8.11), for different BC, (F: free; R: rigid).

	Ra_c	$\tilde{q}_c$	$\tilde{\xi}_0^2$	$\tau_0^{-1}(Pr)$	$\tilde{c}$
F[a]	$\dfrac{27\pi^4}{4}$	$\dfrac{\pi}{\sqrt{2}}$	$\dfrac{2}{\sqrt{3}\,\pi^2}$	$\dfrac{3\pi^2}{2}\dfrac{Pr}{Pr+1}$	$\dfrac{9}{4}\sqrt{6}\pi^3$
	657	2.221	0.017		171
R[b]	Eq(7.39)	Eq(7.38)	$\sqrt{\dfrac{28(3\tilde{q}_c^2+34)}{27\,\tilde{q}_c^2 Ra_c}}$	$\dfrac{(\tilde{q}_c^2+10)Pr}{Pr+(5/\tilde{q}_c^2)}$	$\dfrac{Ra_c}{\sqrt{\tilde{q}_c^2+10}}$
	1750	3.1165	0.062	$\dfrac{19.713Pr}{Pr+0.515}$	394
R[c]	1708	3.117	0.062	$\dfrac{19.65Pr}{Pr+0.512}$	385

[a] The values for two free boundaries are exact.

[b] Values based on the zeroth-order Galerkin approximation of Sect. 7.4.

[c] Exact values reproduced from Hohenberg and Swift (1992)

called the amplitude-equation approach (Cross, 1980; Cross and Hohenberg, 1993), which projects the stochastic equations over the critical hydrodynamic modes, and uses the distance ϵ to the instability as a small parameter to obtain approximate expressions for the decay rates and the intensity of the fluctuations. A discussion of the amplitude-equation approach is outside the scope of the present book and the reader is referred to Cross and Hohenberg (1993).

For two free boundaries, the parameters $\tilde{c}$ and $\tilde{\xi}_0^2$ can be readily obtained from the most-unstable-mode approximation (8.5), as elucidated above. For two rigid boundaries, the parameter τ_0 is given by Eq. (6.67), obtained when we studied the slowest decay rate. The other parameters of the SH approximation have been investigated elsewhere (Cross, 1980; Cross and Hohenberg, 1993), and we refer the reader to these articles for the details of their calculation. However, as pointed out in the previous section, the Galerkin approach of Sect. 7.4 is a most-unstable-mode approximation. Hence, it is also possible to deduce an SH-approximation from the Galerkin expression of the nonequilibrium structure factor, by expanding the denominator of the second quotient in the LHS of Eq. (8.9) in powers of $q_\parallel^2$ around $q_{\parallel c}^2$, and substituting $\tilde{A}(\tilde{q})$ by $\tilde{A}(\tilde{q}_c)$ in the first quotient. This procedure yields a zeroth-order Galerkin approximation to the SH-parameters $\tilde{c}$ and $\tilde{\xi}_0^2$ for the case of two rigid boundaries. The corresponding zeroth-order Galerkin approximation to the SH-parameter τ_0 is to be obtained from the expression (7.33) for the slowest decay rate, by a procedure identical to the one discussed in Sect. 6.4. As expected, it turns out that the normalization constant $\tilde{c}$ does not depend on the Pr number. The Galerkin approximations to the SH-parameters for rigid boundaries are to be compared with the exact values reported in the literature (Cross, 1980; Cross and Hohenberg, 1993).

Such a comparison is presented in Table 8.1. The first two rows give the exact values obtained for the SH parameters in the case of two free boundaries. The third and fourth rows of Table 8.1 give the approximate values obtained for two rigid boundaries from the Galerkin approximation. Finally, the last row of Table 8.1 gives the exact values of the SH parameters reproduced from Cross and Hohenberg (1993), see also Eq. (6.67). A comparison between the last two rows of Table 8.1 yields further confirmation of the adequacy of the Galerkin approximation developed in Sect. 7.4, when the Ra number is expressed in terms of the distance ϵ to the instability.

The linear SH approximation for the enhancement of nonequilibrium fluctuations has to be compared with the exact results of Chapter 7. For the case of two free boundaries, substitution of Eq. (8.11) into Eq. (6.31) and use of the small-angle limit ($q_{\parallel} \simeq q$, $q_{\perp} \simeq 0$), gives the SH approximation for the experimental enhancement of nonequilibrium fluctuations. We recall that the SH model uses the hydrodynamic modes with the normalization (8.7), so that the SH modes for free boundaries are the same as those of the most-unstable mode approximation, given by (8.6). To test the quality of the SH approximation, we have added to Fig. 8.1 as a dotted curve the SH-results for the value $\epsilon = -0.01$, using the SH-parameters for a fluid layer with two free boundaries, displayed in the first row of Tab. 8.1. The dotted curve in Fig. 8.1 is to be compared with the solid curve representing the exact solution (7.9). We see that the SH approximation reproduces the structure factor in the vicinity of the peak near the critical waver number for the onset of convection. It has the correct asymptotic q^{-4} behavior proportional to q^{-4} for large q, but the proportionality factor is off by more of 50%. In the limit of small q, the SH approximation yields a qualitatively incorrect asymptotic behavior. We conclude that the most-unstable-mode approximation yields a considerable improvement over the SH approximation.

8.4 Power of thermal fluctuations

A feature of nonequilibrium fluctuations close to the convective instability that has received attention in the literature, particularly in the context of the SH model, is the so-called power of thermal fluctuations, $\langle \delta T^2 \rangle$. It is defined as the average of the equal-time temperature fluctuations:

$$\langle \delta T^2 \rangle = \frac{1}{(\Delta V)^2} \int_{\Delta V} d\mathbf{r} \int_{\Delta V} d\mathbf{r}' \, \langle \delta T^*(\mathbf{r}, t) \cdot \delta T(\mathbf{r}', t) \rangle, \tag{8.12}$$

where ΔV is a small volume element inside the fluid. Equation (8.12) is similar to the average mass-density fluctuation defined in Eq. (3.34). Using the results of Sect. 3.3.2, we obtain for a fluid in equilibrium:

$$\langle \delta T^2 \rangle_{\mathrm{E}} = \frac{S_{\mathrm{E}} (\gamma - 1)}{\alpha_p^2 \rho^2 \gamma} \frac{1}{\Delta V}, \tag{8.13}$$

which is independent of the position $\mathbf{r}$ where the volume element ΔV is located, as expected from the spatial translational symmetry of an equilibrium state (in the absence of external forces). Physically, $\langle \delta T^2 \rangle$ is a measure of the intensity of the thermal fluctuations and has units of squared temperature. For a fluid subjected to a temperature gradient, the presence of boundaries breaks the translational symmetry in the vertical direction, so that the power of thermal fluctuations is expected to depend on the vertical position z where the volume element ΔV is located, $\langle \delta T^2 \rangle(z)$. One thus considers a vertical average of the power of thermal fluctuations (Wu et al., 1995):

$$\overline{\langle \delta T^2 \rangle} = \frac{1}{L} \int_{-\frac{1}{2}L}^{\frac{1}{2}L} \langle \delta T^2 \rangle(z) \, dz. \tag{8.14}$$

Since we shall be concerned with the average power of thermal fluctuations close to the convective instability, we can neglect the equilibrium contribution given by Eq. (8.13). Then, by using Eqs. (6.25) and (6.28), the average power of thermal fluctuations can be related to the static structure factor discussed in the previous chapters by:

$$\overline{\langle \delta T^2 \rangle} = \frac{m_0}{\rho \alpha_p^2} \frac{1}{L} \int_{-\frac{1}{2}L}^{\frac{1}{2}L} dz \int \frac{d^2 q_{\parallel}}{(2\pi)^2} S_{\mathrm{NE}}(q_{\parallel}, z, z), \tag{8.15}$$

where we have assumed that the horizontal surface ΔS of the volume element ΔV is so small that:

$$\frac{1}{(\Delta S)^2} \int_{\Delta S \times \Delta S} e^{i\mathbf{q}_{\parallel}(\mathbf{x}_{\parallel} - \mathbf{x}'_{\parallel})} \, d^2 \mathbf{x}_{\parallel} \, d^2 \mathbf{x}'_{\parallel} \simeq 1.$$

Because of translational symmetry in the horizontal xy-plane, $\overline{\langle \delta T^2 \rangle}$ does not depend on the horizontal position $\mathbf{x}_{\parallel}$, where the volume element ΔV is located, nor does it depend on the time t, since we are dealing with a nonequilibrium steady state. For the case of two free boundaries we can substitute the exact analytic expression (7.6) for the structure factor into Eq. (8.15), so as to obtain (Ortiz de Zárate and Sengers, 2001):

$$\overline{\langle \delta T^2 \rangle} = \frac{m_0 S_{\mathrm{E}}(\gamma - 1)}{\alpha^2 \rho \gamma L} \frac{\tilde{S}_{\mathrm{NE}}^0}{2\pi} \sum_{N=1}^{\infty} \int_0^{\infty} \frac{\tilde{q}^2 \, q \, dq}{(\tilde{q}^2 + N^2 \pi^2)^3 - R \, \tilde{q}^2}. \tag{8.16}$$

The integral in Eq. (8.16) can be performed analytically, but the result is long and not particularly interesting. Expanding the results in powers of ϵ we find:

$$\overline{\langle \delta T^2 \rangle} = \frac{m_0 S_{\mathrm{E}}(\gamma - 1)}{\alpha^2 \rho^2 \gamma L^3} \left(\tilde{S}_{\mathrm{NE}}^0 \right)_{\mathrm{c}} \frac{\sqrt{3}}{54\pi^2 \sqrt{-\epsilon}} + \mathcal{O}(1). \tag{8.17}$$

In Eq. (8.17), the symbol $(\tilde{S}_{\mathrm{NE}}^0)_{\mathrm{c}}$ means that the normalized amplitude of the nonequilibrium enhancement has to be evaluated at the critical temperature gradient, i.e., the temperature gradient for which $Ra = Ra_{\mathrm{c}}$. Using Eqs. (3.28) and (4.30), neglecting the contribution of the adiabatic temperature gradient, we rewrite Eq. (8.17) as:

$$\overline{\langle \delta T^2 \rangle} = k_{\mathrm{B}} T \frac{(\Delta T_{\mathrm{c}})^2}{\rho L \nu^2} \frac{Pr^2}{Pr + 1} \frac{\sqrt{3}}{54\pi^2} \frac{1}{\sqrt{-\epsilon}} + \mathcal{O}(1), \tag{8.18}$$

where ΔT_{c} is the temperature difference which corresponds to the critical Rayleigh number Ra_{c}. Notice that Eq. (8.18) is exact, since it has been obtained from the exact structure factor (7.6) for free boundaries. However, the same asymptotic result (8.18) is obtained if the most-unstable-mode approximation (8.5) is used, because only the $N = 1$ mode contributes to the $\epsilon^{-1/2}$ divergent part of $\overline{\langle \delta T^2 \rangle}$.

The power of the thermal fluctuations, $\overline{\langle \delta T^2 \rangle}$, can be studied within the framework of the standard SH approximation. For this purpose Eq. (8.11) should be substituted into Eq. (8.15). Then, by using the normalization (8.7) and by expanding in powers of ϵ the result of the integration over wave vectors, one obtains (Wu et al., 1995; Bodenschatz et al., 2000):

$$\overline{\langle \delta T^2 \rangle} \simeq k_{\mathrm{B}} T \frac{(\Delta T_{\mathrm{c}})^2}{\rho L \nu^2} \frac{Pr \tilde{c}^2}{Ra_{\mathrm{c}}^3} \frac{\tau_0^{-1}}{4\tilde{\xi}_0^2 \sqrt{-\epsilon}} + \mathcal{O}(1). \tag{8.19}$$

We note that for any problem with a linear $\mathrm{I_s}$ instability for which an SH approximation can be developed, Eq. (8.19) is valid, and one just needs to adopt the parameter values corresponding to the physical problem under consideration, including the BC. For the RB problem, both for free and rigid boundaries, the numerical values for these parameters have been reproduced in Table 8.1.

As already mentioned, upon substitution of the parameters in Table 8.1 for free boundaries into Eq. (8.19), exact numerical agreement with

Eq. (8.18) is obtained. The SH approximation reproduces exactly the leading (divergent) term of the power of thermal fluctuations as $\epsilon \to 0$ from below. This feature is a consequence of the construction of the SH approximation based on the amplitude equation, *i.e.*, a systematic perturbative calculation of the decay rates and the hydrodynamic modes using the distance ϵ to the instability as small parameter. We refer the interested reader to Cross (1980).

For the case of rigid boundaries, we developed in Sect. 7.4 a Galerkin approximation for the nonequilibrium structure factor, given by (7.35). Substituting Eq. (7.35) into Eq. (8.15), we can obtain a Galerkin approximation for the power of thermal fluctuations. The integration of the coupling coefficient $C_0^{\mathrm{G}}(\tilde{q})$, given by Eq. (7.36), over the wave number q is complicated. However, since we are really interested only in the asymptotic behavior close to the instability, $C_0^{\mathrm{G}}(\tilde{q})$ can be expressed in terms of ϵ, and an asymptotic expansion of the integral can be readily performed, with the result (Ortiz de Zárate and Sengers, 2002):

$$\overline{\langle \delta T^2 \rangle} = k_{\mathrm{B}}T \; \frac{(\Delta T_{\mathrm{c}})^2}{\rho L \nu^2} \; \frac{Pr^2 \, \tilde{q}_{\mathrm{c}}}{Pr + \dfrac{5}{\tilde{q}_{\mathrm{c}}^2}} \; \sqrt{\frac{21}{3\tilde{q}_{\mathrm{c}}^2 + 34}} \; \frac{3}{56\sqrt{-\epsilon}} + \mathcal{O}(1). \qquad (8.20)$$

From a comparison of Eq. (8.20) with the result obtained by substituting into Eq. (8.19) the parameters corresponding to rigid boundaries in the standard SH approximation (*c.f.* last row of Table 8.1), we make the following observations:

- As in the case of free boundaries, we recover the divergence of $\overline{\langle \delta T^2 \rangle}$ proportional to $1/\sqrt{-\epsilon}$, predicted by the standard SH model and confirmed experimentally by Wu et al. (1995), see Chapter 10.

- We also recover, in good approximation, the dependence of $\overline{\langle \delta T^2 \rangle}$ on the Prandtl number: a factor $Pr^2/(Pr + 0.515)$, to be compared with $Pr^2/(Pr + 0.512)$ predicted by the standard SH model.

- The prefactor for the total power of the thermal noise calculated with the Galerkin approximation, Eq. (8.20), is 3% smaller than the same quantity calculated with the standard SH model (which, as mentioned, is asymptotically exact). This small difference again shows that the Galerkin approximation introduced in Sect. 7.4 is a very good approximation indeed.

Several authors (Kirkpatrick and Cohen, 1983; Schmitz and Cohen, 1985b; van Beijeren and Cohen, 1988b; Hohenberg and Swift, 1992) have represented the q dependence of the structure factor near the instability in terms of a lorentzian profile centered at $\tilde{q}_{\mathrm{c}}$ with a width proportional to ϵ;

Wu et al. (1995) have analyzed their experimental data in terms of such a lorentzian profile. We remark that both the exact solution for free boundaries and the SH asymptotic expansion for rigid boundaries do not yield a lorentzian profile close to the instability. Moreover, a lorentzian does not recover the proper asymptotic behavior for either small q or large q. It even leads to an apparent divergence when one tries to calculate the power of the fluctuations by integrating $S(q)$ over all two-dimensional wave vectors (Wu et al., 1995; Hohenberg and Swift, 1992).

From the information in Fig. 8.1 in Sect. 8.3 we concluded that the most-unstable-mode Galerkin approximation presented in Sect. 7.4 represents the q dependence of the nonequilibrium structure factor better than a lorentzian or the form implied by the standard SH model. Nevertheless, although the SH approximation underestimates the asymptotic behavior of $\tilde{S}_{NE}^{R}(\tilde{q})$ for large q, it gives, by construction, the exact power of the thermal noise asymptotically close to the instability.

8.5 Wave number of maximum enhancement of fluctuations

One of the interesting features of nonequilibrium fluctuations in the Rayleigh-Bénard problem close to the convective instability is the divergence of the structure factor proportional to $(Ra_c - Ra)^{-1}$, as originally predicted by Zaitsev and Shliomis (1971) from linear perturbation theory. The same divergence follows from the more complete solutions presented in Chapter 6, both for the case of free boundaries and for the case of rigid boundaries.

To analyze this divergence we first determine the wave numbers, $\tilde{q}_{max}^{F}$ and $\tilde{q}_{max}^{R}$, at which the nonequilibrium structure factor exhibits a maximum for the case of free boundaries and for the case of rigid boundaries, respectively. The wave number $\tilde{q}_{max}^{F}$ has been evaluated by Ortiz de Zárate and Sengers (2001). Asymptotically close to the instability it has an expansion of the form:

$$\tilde{q}_{max}^{F} = \tilde{q}_{c}^{F} \left\{ 1 + \frac{81}{4} \sum_{N=2}^{\infty} \frac{(N^2-1)(1+2N^2)^2(1-\cos N\pi)}{N^2\left[(1+2N^2)^3 - 27\right]^2} \epsilon^2 + \mathcal{O}(\epsilon^3) \right\}$$

$$\simeq \frac{\pi}{\sqrt{2}} \left\{ 1 + 2.847 \times 10^{-4} \epsilon^2 \right\}. \tag{8.21}$$

For rigid boundaries, the position of the maximum in the nonequilibrium structure factor, $\tilde{q}_{max}^{R}$, may be calculated from the numerical solution developed in Chapter 6. However, since we are really interested only in the maximum asymptotically close to the instability, it will be a very good approximation to calculate it from the most-unstable zeroth-order Galerkin approximation of Sect. 7.4 with the Ra number expressed in terms of ϵ. When close to the instability, the $\tilde{q}^{R}_{max}$ corresponding to the maximum of the amplitude (7.36), may be expanded as (Ortiz de Zárate and Sengers, 2002):

$$\tilde{q}^{R}_{max} = \tilde{q}^{R}_{c}\left\{1 + \frac{14}{27}\frac{(\tilde{q}^{2}_{c}+10)^{2}}{Ra^{R}_{c}\tilde{q}^{2}_{c}(3\tilde{q}^{2}_{c}+34)}\frac{\tilde{q}^{4}_{c}+384\tilde{q}^{2}_{c}+4104}{\tilde{q}^{2}_{c}Pr+5}\epsilon + \mathcal{O}(\epsilon^{2})\right\}$$

$$\simeq 3.1165\left\{1 + \frac{0.154}{Pr+0.515}\epsilon\right\}, \tag{8.22}$$

where the critical wave number $\tilde{q}^{R}_{c} \simeq 3.1165$ and Rayleigh number $Ra^{R}_{c} \simeq 1750$ corresponds to the zeroth-order Galerkin approximation for two rigid boundaries. It is interesting to note that the difference between $\tilde{q}_{max}$ and $\tilde{q}_{c}$ has a quadratic dependence on the distance ϵ from the instability for free boundaries, while this difference depends linearly on ϵ for rigid boundaries. Furthermore, for free boundaries, the maximum in the structure factor moves to *larger* wave numbers as ϵ goes to lower negative; while, in the case of rigid boundaries, the maximum in the structure factor moves to *smaller* wave numbers as ϵ goes to lower negative values. In Fig. 8.2, we have plotted on a double-log scale the absolute value of the difference $|\tilde{q}_{max} - \tilde{q}_{c}|$ as a function of $-\epsilon$ for two rigid boundaries (solid line) and for two free boundaries (dashed line). The curves displayed in Fig. 8.2 have been obtained numerically from Eqs. (7.9) and (7.41), confirming the linear dependence on ϵ for rigid boundaries as opposed to a quadratic dependence on ϵ for free boundaries. Even more significantly, the effect is orders of magnitude larger for rigid boundaries than that for free boundaries.

Having determined the position of the maximum, we can next study the divergence of the height of the maximum as the convective instability is approached. For the case of two free boundaries, we substitute Eq. (8.21) into Eq. (7.9) and conclude that the structure factor, which is proportional to the intensity of the scattered light, diverges when the convective instability is approached as:

$$\tilde{S}^{F}_{NE}(\tilde{q}_{max}) = \frac{-1\,864}{\epsilon}\frac{1}{\pi^{4}}\sum_{N=2}^{\infty}\frac{1-\cos N\pi}{N^{2}\left[(1+2N^{2})^{3}-27\right]} + \mathcal{O}(\epsilon). \tag{8.23}$$

For the case of two rigid boundaries, close to the instability the height of the maximum will continue to be well represented by the most-unstable Galerkin approximation expressed in terms of ϵ. Thus, substituting Eq. (8.22) into Eq. (7.41) we find that the structure factor diverges when the convective instability is approached as:

$$\tilde{S}^{R}_{NE}(\tilde{q}_{max}) = \frac{Pr+1}{Pr+\dfrac{5}{\tilde{q}^{2}_{c}}}\left\{\frac{5}{6Ra^{R}_{c}}\frac{-1}{\epsilon} + \mathcal{O}(1)\right\}, \tag{8.24}$$

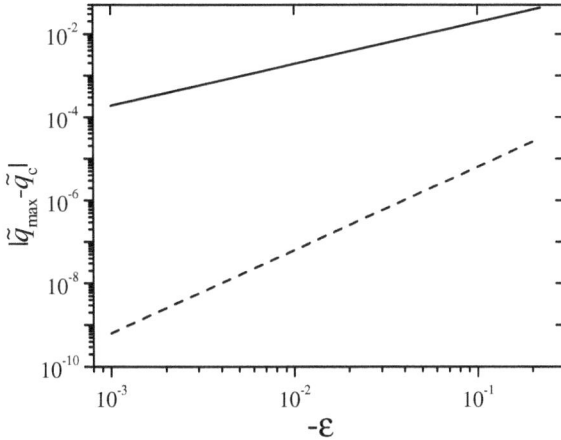

Figure 8.2: Double-logarithmic plot of the absolute value of the difference between the critical wave number $\tilde{q}_c$ and the position of the maximum in the nonequilibrium structure factor $\tilde{q}_{max}$, as a function of $-\epsilon$. The solid line is for the case of two rigid boundaries. The dashed line is for the case of two free boundaries.

with $Ra_c^R \simeq 1750$ corresponding to Eq. (7.39). In both cases, we recover the linear divergence of $\tilde{S}_{NE}$ as a function of $(Ra - Ra_c)^{-1}$ originally obtained by Zaitsev and Shliomis (1971) and confirmed by Swift and Hohenberg (1977); Hohenberg and Swift (1992).

We conclude this section by noting that our results for the divergence of the structure factor at the convective instability have been obtained within a linear fluctuation theory. Extremely close to the instability nonlinear effects are expected to cause a smearing out of the divergence (Swift and Hohenberg, 1977; Hohenberg and Swift, 1992; Normand et al., 1977; Graham, 1974; Graham and Pleiner, 1975), but this effect will only be noticeable for very small ϵ values, estimated by Normand et al. (1977) as $|\epsilon| \lesssim 2.9 \times 10^{-5}$. Hence, observation of the linear divergence of the intensity of the fluctuations is possible in experiments (Wu et al., 1995). Deviations from linear fluctuation theory have been observed by Scherer et al. (2000) in the case of electroconvection.

8.6 Wave number of fluctuations with maximum growth rate

In the previous section, we derived an expression for the wave number $\tilde{q}_{max}$ corresponding to the maximum enhancement of the intensity of the nonequilibrium fluctuations as a function of $\epsilon = (Ra - Ra_c)/Ra_c$ and of the Prandtl

number Pr. It is worth noticing that the wave number $\tilde{q}_{max}$ corresponding to the maximum intensity of the fluctuations cannot be identified with
the wave number $\tilde{q}_m$ corresponding to the minimum decay rate (maximum
growth rate) considered by other investigators (Chandrasekhar, 1961; Manneville, 1990; Westried et al., 1978; Cross, 1980; Domínguez Lerma et al.,
1984). While both $\tilde{q}_{max}$ and $\tilde{q}_m$ approach $\tilde{q}_c$ as $\epsilon \to 0^{(-)}$, the two wave
numbers differ for finite negative values of ϵ and this difference depends on
the Prandtl number. As elucidated below, this fact has consequences when
studying the wave-number selection below the instability.

To illustrate the difference between $\tilde{q}_{max}$, the wave number of maximum
enhancement of fluctuations, and $\tilde{q}_m$, the wave number corresponding to
the maximum growth rate, we first consider the case of two free boundaries,
for which there exists an analytic expression (6.32) for the slowest decay
rate $\Gamma_1^{(-)}$. From Eq. (6.32) we already know that at $Ra = Ra_c$ the decay
rate reaches the value zero at a single finite value of the wave number
$\tilde{q}_\| = \tilde{q}_c = \pi/\sqrt{2}$. For $Ra < Ra_c$, the slowest decay rate given by Eq. (6.32)
is always positive, independent of Pr or $\tilde{q}_\|$; this means that the conductive
solution is stable. For $Ra > Ra_c$ there are ranges of $q_\|$ for which some of
the decay rates are negative, indicating that an instability develops in the
system. A plot of Eq. (6.32) shows that for $Ra \lesssim Ra_c$ there is a minimum in
the wave-number dependence of the slowest decay rate for a particular value
$\tilde{q}_\| = \tilde{q}_m$, similar to the curve in Fig. 6.2 for rigid boundaries. At $Ra = Ra_c$,
the maximum of the nonequilibrium structure factor is located at $\tilde{q}_c$ and the
corresponding value of the decay rate is zero. From Eq. (6.32) one readily
deduces an analytical expression for the position of the minimum decay rate
at $R \lesssim R_c$:

$$\tilde{q}_m = \tilde{q}_c \left[1 + \frac{1}{4}\epsilon + \frac{3Pr^2 + 2Pr + 3}{16(Pr + 1)^2} \, \epsilon^2 + \mathcal{O}(\epsilon^3) \right]. \tag{8.25}$$

A comparison of Eq. (8.25) with Eq. (8.21), reveals a significant difference
between the wave number $\tilde{q}_m$ of maximum growth rate (minimum decay
rate) and the wave number $\tilde{q}_{max}$ of maximum enhancement of fluctuations.
From Eq. (8.21) we see that $\tilde{q}_{max}$, in the case of free boundaries, does not
contain the term linear in ϵ.

For the case of rigid boundaries, the slowest decay rate cannot be calculated analytically, but has been discussed numerically in Sect. 6.4. Plots
of the slowest decay rate for rigid boundaries were presented in Fig. 6.2 for
two values of the Ra number, and they are qualitatively very similar to a
plot of Eq. (6.32) for the case of free boundaries. For $Ra \lesssim Ra_c$, there is
a minimum of the decay rate $\tilde{\Gamma}_1^{(-)}(\tilde{q}_\|)$ at a particular value $\tilde{q}_\| = \tilde{q}_m$. At
$Ra = Ra_c$, the minimum is located at $\tilde{q}_c$ and the value of the decay rate at
the minimum is zero.

In addition to the numerical results of Sect. 6.4, we have an analyti-

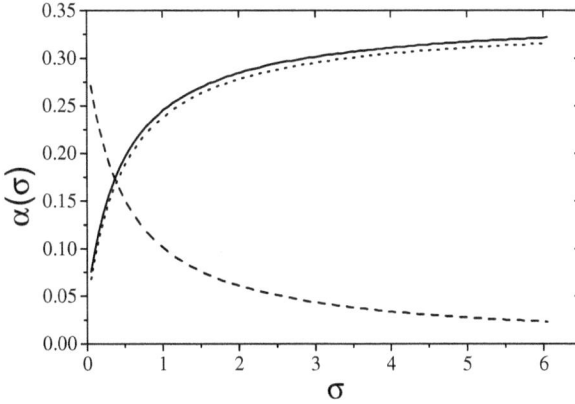

Figure 8.3: Values of the linear coefficient $\alpha(Pr)$ in the expansion, Eq. (8.26), for the wave number $\tilde{q}_m$ of minimum decay rate. Solid curve: obtained analytically from a Galerkin approximation [Eq. (8.27)]. Dotted curve: calculated numerically by Domínguez Lerma et al. (1984). Dashed curve: linear coefficient in the expansion, Eq. (8.22), for the wave number $\tilde{q}_{max}$ of maximum enhancement of fluctuations.

cal zeroth-order Galerkin approximation (7.33) for the slowest decay rate $\tilde{\Gamma}_1^{(-)}(\tilde{q}_\parallel)$. Taking the derivative of $\tilde{\Gamma}_1^{(-)}(\tilde{q}_\parallel)$ with respect to $\tilde{q}_\parallel$, we readily deduce from Eq. (7.33) an analytical expression for the position of the minimum at $Ra \lesssim Ra_c$:

$$\tilde{q}_m = \tilde{q}_c \left[1 + \alpha(Pr)\,\epsilon + \mathcal{O}(\epsilon^2)\right] \tag{8.26}$$

with

$$\alpha(Pr) = \frac{35\,(\tilde{q}_c^2 + 12)^4(\tilde{q}_c^2 + 10)^2}{12\,\tilde{q}_c^6(61987\tilde{q}_c^4 + 165730\tilde{q}_c^2 + 16132820)} \cdot \frac{21(\tilde{q}_c^2 + 12)Pr + 5(\tilde{q}_c^2 - 4)}{\tilde{q}_c^2 Pr + 5}$$

$$\simeq \frac{0.339\,(Pr + 0.063)}{Pr + 0.515}. \tag{8.27}$$

As already mentioned in Sect. 6.4, the wave number corresponding to the maximum of the linear growth rate (minimum decay rate) has been investigated numerically by Domínguez Lerma et al. (1984) for the case of rigid boundaries, who have proposed for the coefficient $\alpha(Pr)$ in Eq. (8.26) the empirical equation:

$$\alpha(Pr) = 0.0494 + \frac{0.295 Pr}{Pr + 0.509}. \tag{8.28}$$

In Fig. 8.3 we have plotted $\alpha(Pr)$ as a function of the Prandtl number: the solid curve represents Eq. (8.27), obtained here analytically using a

Galerkin approximation, while the dotted curve represents the empirical relationship, Eq. (8.28), proposed by Domínguez Lerma et al. (1984). A simple inspection of Fig. 8.3 shows that the Galerkin method provides a very good approximation for the wave number of the maximum growth rate for the range of Prandtl numbers displayed in the figure, the difference between the value of $\alpha(Pr)$ from the Galerkin approximation (8.27) and from the numerical results of Domínguez Lerma et al. (1984) is less than 2%. This gives further evidence for the adequacy of the most-unstable-mode Galerkin approximation developed in Sect. 7.4, when Ra is expressed in terms of the distance ϵ to the instability.

We can next compare Eqs. (8.26)-(8.28) for the wave number $\tilde{q}_{\mathrm{m}}$ of maximum linear growth rate of with Eq. (8.22) in the previous section for the wave number $\tilde{q}_{\mathrm{max}}$ of the maximum enhancement of the fluctuations. We observe that, although at $Ra = Ra_{\mathrm{c}}$ both maxima are located at $\tilde{q}_{\mathrm{c}}$, for $Ra \lesssim Ra_{\mathrm{c}}$ significant differences do appear between their positions. To show these difference we have added in Fig. 8.3, as a dashed curve, the function $\alpha(Pr)$ deduced from Eq. (8.22) for the wave number q_{max} of the maximum enhancement of the fluctuations. We see from Fig. 8.3 that the difference is significant and it increases with the Prandtl number. This difference is due to the fact that the enhancement of the fluctuations is not simply proportional to the inverse of the decay rates, but depends also on the mode-coupling coefficients which, in turn, are expressed in terms of the random dissipative fluxes, see for instance Eq. (7.21). Such a result can be obtained only on the basis of fluctuating hydrodynamics. The quantity to be observed experimentally by shadowgraphy or by light scattering will be the wave number $\tilde{q}_{\mathrm{max}}$ of the maximum enhancement of the fluctuations and not $\tilde{q}_{\mathrm{m}}$. We conclude that for the interpretation of some experimental results a stochastic fluid-dynamics treatment is required, while an analysis of such experiments in terms of deterministic fluid dynamics will be incorrect.

Chapter 9

Thermal nonequilibrium fluctuations in binary-fluid layers

In Chapters 4 and 5 we considered fluctuations in fluids and fluid mixtures subjected to a temperature gradient at wave numbers that are sufficiently large so that finite-size effects due to the presence of boundaries could be neglected. For the case of one-component fluid layers we then extended the theory in Chapters 6 and 7 to fluctuations with smaller wave numbers for which the effects of boundaries need to be incorporated. In the present chapter we consider the boundary effects on nonequilibrium fluctuations in binary-fluid layers.

There are many studies in the literature that have dealt with the two-component Rayleigh-Bénard problem incorporating effects due to the presence of boundaries (Schechter et al., 1972, 1974; Platten and Chavepeyer, 1975; Platten and Legros, 1984; Cross and Hohenberg, 1993; Hollinger and Lücke, 1998). Most of these studies are based on the deterministic Boussinesq equations, *i.e.*, without considering random dissipative fluxes. The primary aim of those investigators was to study the stability of the "conductive" (linear) solution or other nonlinear solutions. However, as we discussed in Chapter 5, from the random Boussinesq equations interesting long-ranged nonequilibrium concentration fluctuations can be deduced even when the liquid mixture is in a hydrodynamically quiescent stable state, far away from any convective instability. How the results of Chapter 5 are modified when BC are incorporated is the primary goal of the present chapter.

An analysis of such finite-size effects on nonequilibrium concentration fluctuations is not only interesting theoretically, but also from an experimental perspective. As will be discussed in Chapter 10, experimental detection of nonequilibrium fluctuations at the extremely small wave numbers where finite-size effects arise, requires in practice to use shadowgraph techniques. Shadowgraphy is only feasible when the thermal noise is intrinsically strong; for this reason actual shadowgraph experiments have been performed in one-component fluids only in supercritical conditions and not far from the critical point (Wu et al., 1995; Oh et al., 2004). The interpretation of those experiments is obviously complicated by the proximity to the critical region, which introduces non-Boussinesq effects that are difficult to analyze (Oh et al., 2004). Most of these problems can be avoided by using binary liquids, where the hydrodynamic fluctuations are usually dominated by strong Soret-driven concentration fluctuations (Lekkerkerker and Laidlaw, 1977; Giglio and Vendramini, 1977). Consequently, shadowgraph experiments can be performed in liquid mixtures far away from any critical point (Brogioli et al., 2000b), where a simple Boussinesq approximation is adequate.

An initial study of finite-size effects on nonequilibrium concentration fluctuations was made by Sengers and Ortiz de Zárate (2001). They adopted a large-Lewis-number approximation and assumed slip-free boundary conditions for the velocity and fully permeable walls for the concentration. The simplicity of these BC enabled the authors to obtain an exact analytical expression for the nonequilibrium structure factor, in a way similar to free slip for a one-component fluid in Sect. 7.1. However, these simplistic BC lead to an asymptotic behavior of the structure factor when $q_\parallel \to 0$ that is unrealistic under actual experimental conditions (Ortiz de Zárate et al., 2004). For this reason, we shall consider here only nonequilibrium fluctuations for realistic BC.

This chapter is organized as follows. First, in Sect. 9.1 we present the general theory for solving the linearized random binary Boussinesq equations incorporating BC for the fluctuating fields. As was the case for a one-component fluid, in principle the nonequilibrium structure factor of a fluid mixture for realistic BC can only be obtained numerically. However, instead of developing a numerical investigation like in previous chapters, we shall present in Sects. 9.2 and 9.3 a Galerkin-approximation scheme for realistic BC, which will turn out to yield reasonable analytical results for positive separation ratios. In the remainder of the chapter we shall discuss consequences that may be obtained from such a Galerkin approximation method. First in Sect. 9.4 we consider the structure factor of a fluid mixture in thermal equilibrium. Since the exact expressions for the structure factor of a binary fluid in thermal equilibrium are well known, see Chapter 5, the thermal-equilibrium case will provide a test for checking the quality of the Galerkin-approximation procedure adopted. We then proceed in Sect. 9.5 with obtaining the expressions for the nonequilibrium temperature and concentration fluctuations on the basis of the previously presented Galerkin approximation. Finally, in Sect. 9.6 we develop a simple and useful expression for the nonequilibrium concentration fluctuations, taking advantage of the fact that, for common mixtures, the Lewis number is large. Since the physical insights of this chapter can be easily hidden by long and cumbersome algebra, we summarize the main results of this Chapter in Sect. 9.7.

9.1 The hydrodynamic operator for binary Boussinesq

The spatiotemporal evolution of fluctuations around the stationary conductive solution of the Rayleigh-Bénard problem for fluid mixtures is governed by the stochastic binary Boussinesq equations earlier presented in Eq. (5.5), to be supplemented now with appropriate BC. The physically realistic boundary conditions for our problem are rigid and impermeable walls (Lhost et al., 1991; Schechter et al., 1974; Hollinger and Lücke, 1998):

$$\delta v_z = \partial_z \, \delta v_z = \delta T = \partial_z (\beta \, \delta c + \psi \alpha_p \, \delta T) = 0, \qquad \text{at } z = \pm \tfrac{1}{2} L. \qquad (9.1)$$

The last condition, in accordance with Eq. (2.91), ensures that the solute flux vanishes at the walls, including the solute transport due to the Soret effect. Notice that in Eq. (9.1), consistent with the Boussinesq approximation, the thermophysical properties of the mixture are assumed to be constants independent of the temperature. When comparing with the work of some previous investigators (Lhost et al., 1991; Hollinger and Lücke, 1998) one should remember that we use here, as in Chapter 5, the definition (5.4) of the separation ratio ψ with the sign convention recommended in the book edited by Köhler and Wiegand (2002).

In view of the last of the boundary conditions in Eq. (9.1), it is convenient to use, instead of δc, the dimensionless fluctuating variable $\zeta = \beta \, \delta c + \psi \alpha_p \, \delta T$. In terms of this new variable, the stochastic binary Boussinesq equations read (Lhost et al., 1991):

$$\frac{\partial}{\partial t} \left(\nabla^2 \delta v_z \right) = \nu \, \nabla^2 \left(\nabla^2 \delta v_z \right) + g \, \left(\partial_x^2 + \partial_y^2 \right) [(1 + \psi) \alpha_p \delta T - \zeta]$$

$$+ \frac{1}{\rho} \left\{ \nabla \times [\nabla \times (\nabla \cdot \delta \Pi)] \right\}_z, \qquad (9.2a)$$

$$\frac{\partial \, \delta T}{\partial t} = a_T \, \nabla^2 \delta T - \delta v_z \, \nabla T_0 - \frac{a_T}{\lambda} \, \nabla \delta \mathbf{Q}', \qquad (9.2b)$$

$$\frac{\partial \zeta}{\partial t} = D \nabla^2 \zeta + \alpha_p \psi a_T \nabla^2 \delta T - \frac{\beta}{\rho} \, \nabla \delta \mathbf{J} - \frac{\alpha_p \psi a_T}{\lambda} \, \nabla \delta \mathbf{Q}', \qquad (9.2c)$$

where we have taken the Dufour effect ratio $\epsilon_D \simeq 0$, as usual in the context of the Boussinesq approximation. With this change of variables, the BC can now be specified in a simpler way:

$$\delta v_z = \partial_z \, \delta v_z = \delta T = \partial_z \, \zeta = 0, \qquad \text{at } z = \pm \tfrac{1}{2} L. \tag{9.3}$$

To solve the set of Eqs. (9.2) for the fluctuating fields, while accommodating the BC (9.3) in the z-direction, we apply as in Chapter 6 a Fourier transformation in time and in the horizontal xy-plane. With this procedure, the set of stochastic linear differential equations (9.2) can be expressed in a form completely similar to Eq. (6.4) for a one-component fluid:

$$\{i\omega \, \mathcal{D} + \mathcal{H}\} \cdot \mathbf{U} = \mathbf{F}, \tag{9.4}$$

where the fluctuating-fields vector $\mathbf{U}$ has now three components, namely:

$$\mathbf{U} = \begin{pmatrix} \delta v_z(\omega, \mathbf{q}_\|, z) \\ \delta T(\omega, \mathbf{q}_\|, z) \\ \zeta(\omega, \mathbf{q}_\|, z) \end{pmatrix}. \tag{9.5}$$

The linear differential operators $\mathcal{H}$ and $\mathcal{D}$ appearing in Eq. (9.4), for the present case of a binary mixture, are:

$$\mathcal{H}(q_\|) = \begin{bmatrix} \nu(q_\|^2 - \partial_z^2)^2 & -\alpha g q_\|^2 (1 + \psi) & g q_\|^2 \\ \nabla T_0 & a(q_\|^2 - \partial_z^2) & 0 \\ 0 & \alpha \psi a(q_\|^2 - \partial_z^2) & D(q_\|^2 - \partial_z^2) \end{bmatrix}, \tag{9.6a}$$

$$\mathcal{D}(q_\|) = \begin{bmatrix} (q_\|^2 - \partial_z^2) & 0 & 0 \\ 0 & 1 & 0 \\ 0 & 0 & 1 \end{bmatrix}. \tag{9.6b}$$

The vector $\mathbf{F}$ of the spatiotemporal Langevin random-noise terms is now to be expressed in terms of the Fourier-transformed random dissipative fluxes as:

$$\mathbf{F} = \frac{1}{\rho} \begin{bmatrix} iq_x(q_\|^2 + \partial_z^2)\delta\Pi_{xz} + iq_y(q_\|^2 + \partial_z^2)\delta\Pi_{yz} + q_\|^2\partial_z\delta\Pi_{zz} \\ \qquad -\partial_z\left(2q_x q_y\delta\Pi_{xy} + q_x^2\delta\Pi_{xx} + q_y^2\delta\Pi_{yy}\right) \\[4pt] -\dfrac{1}{c_p}\left[iq_x\delta Q'_x + iq_y\delta Q'_y + \partial_z\delta Q'_z\right] \\[4pt] -\beta\left[iq_x(\delta J_x + \frac{\alpha_p\psi}{\beta c_p}\delta Q'_x) + iq_y(\delta J_y + \frac{\alpha_p\psi}{\beta c_p}\delta Q'_y)\right. \\ \left. \qquad +\partial_z(\delta J_z + \frac{\alpha_p\psi}{\beta c_p}\delta Q'_z)\right] \end{bmatrix}, \tag{9.7}$$

where the dissipative fluxes in the RHS are Fourier transformed in time and the horizontal plane, so that $\mathbf{F}(\omega, \mathbf{q}_\|, z)$ is a function of the frequency, the horizontal wave vector and the vertical coordinate. The correlation functions between the different components of the random noise $\mathbf{F}(\omega, \mathbf{q}_\|, z)$ can be deduced from the FDT (3.7c) for the random stress of a divergence-free (incompressible) binary mixture; and from Eqs. (3.9)-(3.10) for the random diffusive and heat flows. These correlation functions will be discussed later in more detail, but we can anticipate that, in view of the expressions for the FDT, $\mathbf{F}$ plays in Eq. (9.4) the role of additive random white noise. In Eqs. (9.4)-(9.7), ω and $\mathbf{q}_\parallel$ are, as in Chapter 6, the Fourier variables representing the frequency and the (horizontal) wave vector of the fluctuations, respectively.

It is evident that we can solve Eqs. (9.4) for the fluctuating fields incorporating the BC (9.3) by adopting a procedure similar to the one developed in Chapter 6 for a one-component fluid. Thus, one has to consider the eigenvalue problem defined by

$$\{\mathcal{H}(q_\parallel) \cdot \mathbf{U}^R(q_\parallel, z)\} = \Gamma(q_\parallel) \, \{\mathcal{D}(q_\parallel) \cdot \mathbf{U}^R(q_\parallel, z)\}, \tag{9.8}$$

but with the differential operators $\mathcal{H}(q_\parallel)$ and $\mathcal{D}(q_\parallel)$ corresponding to Eqs. (9.6) for a binary fluid. As in Eq. (6.19), the solution for the fluctuating fields is expressed in a series expansion in terms of the right eigenfunctions or hydrodynamic modes. The coefficients of this series are determined by projection onto the set of left eigenfunctions. Finally, the nonequilibrium structure factor is calculated by computing the mode-coupling coefficients which, as in Eq. (6.22), will be the autocorrelation functions of the projection of the vector of random forces $\mathbf{F}(\omega, \mathbf{q}_\parallel, z)$ onto the set of left eigenfunctions. As discussed for a one-component fluid in Chapter 6, for this procedure to work the real part of the decay rates must be different from zero for any value of $q_\parallel$. Again, the conditions under which it is possible to calculate a nonequilibrium structure factor in a linear approximation follow from classical linear stability analysis.

Such a numerical calculation has not yet been reported in the literature for the realistic BC (9.3). Instead, we adopt here a simple single-mode Galerkin approximation (Ortiz de Zárate et al., 2004), much in the spirit of the approximation presented in Sect. 7.4 for a one-component fluid. As we shall see, such a Galerkin approximation represents the nonequilibrium structure factor for positive separation ratio, while being at the same time simple enough to be treated analytically quite well.

9.2 A Galerkin approximation for rigid and impermeable boundary conditions

The idea behind using a single-mode Galerkin approximation method is similar to the one discussed in Sect. 7.4 for a one-component fluid. In zeroth-order approximation, the first eigenfunction of the problem (9.8) is expressed as a linear combination of Galerkin test functions satisfying the BC of the problem. Such a procedure was implemented by Ortiz de Zárate et al. (2004), who adopted the same Galerkin test functions employed previously by Lhost et al. (1991) with excellent results for performing a linear stability analysis. Hence, we shall look for approximate solutions to the eigenvalue problem (9.8) whose z-dependence in a zeroth-order Galerkin approximation may be expressed in the form:

$$\mathbf{U}^R(q_\|, z) = \alpha_p g \, w_0(q_\|) \, \mathbf{G}_{0,0}(z)$$
$$+ \frac{\nu}{L^2} \theta_0(q_\|) \, \mathbf{G}_{0,1}(z) + \frac{\alpha_p \nu}{L^2} \zeta_0(q_\|) \, \mathbf{G}_{0,2}(z), \quad (9.9)$$

with Galerkin test functions

$$\mathbf{G}_{0,0}(z) = \begin{pmatrix} \mathcal{C}_1(z) \\ 0 \\ 0 \end{pmatrix}, \quad \mathbf{G}_{0,1}(z) = \begin{pmatrix} 0 \\ \sqrt{2} \, \cos \dfrac{\pi z}{L} \\ 0 \end{pmatrix}, \quad \mathbf{G}_{0,2}(z) = \begin{pmatrix} 0 \\ 0 \\ 1 \end{pmatrix}. \quad (9.10)$$

In Eq. (9.10), $\mathcal{C}_1(z)$ is the function:

$$\mathcal{C}_1(z) = \frac{\cosh(\Lambda z/L)}{\cosh(\Lambda/2)} - \frac{\cos(\Lambda z/L)}{\cos(\Lambda/2)}, \quad (9.11)$$

with $\Lambda \approx 4.73$, corresponding to the first Chandrasekhar function. This value of Λ assures that the derivative of $\mathcal{C}_1(z)$ vanishes at $z = \pm \frac{1}{2} L$ (Chandrasekhar, 1961). Hence, the three test functions (9.10) satisfy the required BC (9.3). They are linearly independent, so that a basis of the space of functions satisfying the BC (9.3) can be obtained by choosing the three elements (9.10) and an infinite number of linearly independent functions which we don't need to discuss for our purpose here. It is worth noticing that the Galerkin test functions for the vertical velocity and temperature fluctuations in Eq. (9.10) differ from the Galerkin test functions (7.29) used for studying the fluctuations in a one-component fluid. We shall further comment on this point later.

To evaluate the coefficients $w_0(q_\|)$, $\theta_0(q_\|)$ and $\zeta_0(q_\|)$ of the first eigenfunction, we require Eq. (9.8) to be satisfied in the subspace generated by the three functions (9.10). We thus substitute Eq. (9.9) into Eq. (9.8) and project the result onto the three functions (9.10) to obtain a set of three algebraic equations. To deduce a solution different from zero, we consider the secular equation:

$$\det \left[\mathsf{H}(q_\|) - \tilde{\Gamma}(q_\|) \, \mathsf{D}(q_\|) \right] = 0, \quad (9.12)$$

with the dimensionless decay rate $\tilde{\Gamma} = L^2/\nu \, \Gamma$, so that Eq. (9.12) is the same as Eq. (7.31), but where now the two 3×3 matrices $\mathsf{H}(q_\|)$ and $\mathsf{D}(q_\|)$

are to be obtained from the hydrodynamic operator for a binary mixture, resulting in (Ortiz de Zárate et al., 2004):

$$\mathsf{H}(q_\parallel) = \begin{bmatrix} B(\tilde{q}_\parallel) & -(1+\psi)P_1\tilde{q}_\parallel^2 & P_0\tilde{q}_\parallel^2 \\ -P_1\dfrac{Ra}{Pr} & \tilde{q}_\parallel^2\dfrac{C_1(q_\parallel)}{Pr} & 0 \\ 0 & \dfrac{\psi}{Pr}\dfrac{2\sqrt{2}}{\pi}\tilde{q}_\parallel^2 C_1(\tilde{q}_\parallel) & \dfrac{\tilde{q}_\parallel^2}{PrLe} \end{bmatrix}, \quad \mathsf{D}(q_\parallel) = \begin{bmatrix} A(\tilde{q}_\parallel) & 0 & 0 \\ 0 & 1 & 0 \\ 0 & 0 & 1 \end{bmatrix},$$

$$(9.13)$$

with the Prandtl number Pr, the Rayleigh number Ra, and the Lewis number Le, again given by Eqs. (4.7), (4.8) and (5.6), respectively. In Eqs. (9.13) for $\mathsf{H}(q_\parallel)$ and $\mathsf{D}(q_\parallel)$, the various dimensionless projections of the Galerkin test functions are:

$$P_0 = \frac{1}{L}\int_{-L/2}^{L/2} dz\, \mathcal{C}_1(z) = \frac{4}{\Lambda}\tanh\left(\tfrac{\Lambda}{2}\right), \tag{9.14}$$

$$P_1 = \frac{1}{L}\int_{-L/2}^{L/2} dz\, \sqrt{2}\cos\left(\frac{\pi z}{L}\right)\mathcal{C}_1(z) = \frac{4\sqrt{2}\pi\Lambda^2}{\Lambda^4 - \pi^4}, \tag{9.15}$$

$$A(q_\parallel) = L\int_{-L/2}^{L/2} dz\, \mathcal{C}_1(z)\left\{[q_\parallel^2 - \partial_z^2]\cdot\mathcal{C}_1(z)\right\} \tag{9.16}$$

$$= q_\parallel^2 L^2 + \frac{P_0\Lambda^2}{16}\left(P_0\Lambda^2 - 8\right) = \tilde{q}_\parallel^2 + P_2\Lambda^2,$$

and, finally,

$$B(q_\parallel) = L^3\int_{-L/2}^{L/2} dz\, \mathcal{C}_1(z)\left\{[q_\parallel^2 - \partial_z^2]^2\cdot\mathcal{C}_1(z)\right\} \tag{9.17}$$

$$= \tilde{q}_\parallel^4 + \frac{P_0\Lambda^2}{8}\left(P_0\Lambda^2 - 8\right)\tilde{q}_\parallel^2 + \Lambda^4$$

$$= A^2(\tilde{q}_\parallel) + \Lambda^4(1 - P_2^2),$$

with

$$P_2 = \frac{P_0}{16}(P_0\Lambda^2 - 8). \tag{9.18}$$

In Eqs. (9.14) to (9.17), $\tilde{q}_\parallel = q_\parallel L$ is again the dimensionless wave number of the fluctuations. The additional dimensionless function $C_1(q_\parallel)$, appearing in Eq. (9.13), will be defined in Eq. (9.21) below. It is worth noting for future use, that both $A(q_\parallel)$ and $B(q_\parallel)$ are real and positive valued functions that depend only on the magnitude $\tilde{q}_\parallel$ of the dimensionless vector $\tilde{\mathbf{q}}_\parallel$.

Solving the secular equation (9.12) yields a cubic equation in the decay rate $\tilde{\Gamma}$:

$$0 = -A(\tilde{q}_\parallel)\,\tilde{\Gamma}^3 + B(\tilde{q}_\parallel)\,D_1(\tilde{q}_\parallel)\,\tilde{\Gamma}^2$$

$$-\frac{\tilde{q}_\parallel^2}{Pr}\,B(\tilde{q}_\parallel)\,C_3(\tilde{q}_\parallel, Ra)\,\tilde{\Gamma} + \frac{\tilde{q}_\parallel^4}{Pr^2}\,B(\tilde{q}_\parallel)\,C_4(\tilde{q}_\parallel, Ra), \quad (9.19)$$

where we have defined several dimensionless functions:

$$D_1(\tilde{q}_\parallel) = \frac{C_1(\tilde{q}_\parallel) + Pr\,C_2(\tilde{q}_\parallel)}{Pr\,C_2(\tilde{q}_\parallel)} + \frac{1}{PrLe\,C_2(\tilde{q}_\parallel)}, \quad (9.20a)$$

$$C_3(\tilde{q}_\parallel, Ra) = C_1(\tilde{q}_\parallel) + \frac{C_1(\tilde{q}_\parallel) + Pr\,C_2(\tilde{q}_\parallel)}{LePr\,C_2(\tilde{q}_\parallel)} - \frac{P_1^2(1+\psi)\,Ra}{B(\tilde{q}_\parallel)}, \quad (9.20b)$$

$$C_4(\tilde{q}_\parallel, Ra) = \frac{C_1(\tilde{q}_\parallel)}{Le} - P_1^2\left\{\frac{1+\psi}{Le} - \frac{2P_0\sqrt{2}}{\pi P_1}\,C_1(\tilde{q}_\parallel)\,\psi\right\}\frac{Ra}{B(\tilde{q}_\parallel)}. \quad (9.20c)$$

Furthermore, we have introduced in Eqs. (9.20) some functions that will be useful in the remainder of this chapter. They are:

$$C_1(\tilde{q}_\parallel) = \frac{\tilde{q}_\parallel^2 + \pi^2}{\tilde{q}_\parallel^2}, \qquad C_2(\tilde{q}_\parallel) = \frac{B(\tilde{q}_\parallel)}{\tilde{q}_\parallel^2 A(\tilde{q}_\parallel)}. \quad (9.21)$$

Note that all C-functions have a finite limit when $\tilde{q}_\parallel \to \infty$, while diverging as $\tilde{q}_\parallel^{-2}$ when $\tilde{q}_\parallel \to 0$. The D-functions have finite limits at both $\tilde{q}_\parallel \to \infty$ and $\tilde{q}_\parallel \to 0$.

The roots $\tilde{\Gamma}_1(\tilde{q}_\parallel)$, $\tilde{\Gamma}_2(\tilde{q}_\parallel)$ and $\tilde{\Gamma}_3(\tilde{q}_\parallel)$ of Eq. (9.19) give the zeroth-order Galerkin approximation to the first eigenvalues of (9.8), corresponding to the three different decay rates of the thermodynamic fluctuations in a binary mixture (Berne and Pecora, 1976). Explicit expressions may be obtained from the formulas for the roots of a cubic equation, but the resulting expressions are complicated and will not be displayed here.

A nonequilibrium structure factor can be calculated only if the real part of all eigenvalues $\tilde{\Gamma}_i(\tilde{q}_\parallel)$ is nonzero for any value of $\tilde{q}_\parallel$. For the linearized binary Boussinesq equations, a detailed linear instability analysis has already been performed by Lhost et al. (1991) who used the same set of Galerkin test functions (9.10). Indeed, starting from Eq. (9.12), Lhost et al. (1991) deduced a linear-stability diagram for the stick-impermeable BC that resulted qualitatively similar to the diagram for slip-free permeable BC discussed in Fig. 5.1. For instance, it is obvious that, when $C_4(\tilde{q}_\parallel, Ra) = 0$, $\Gamma = 0$ (both real and imaginary parts) is a root of Eq. (9.19). This fact means that the physical system exhibits a stationary Turing-like instability I_s, where a single real eigenvalue becomes zero. The condition $C_4(\tilde{q}_\parallel, Ra) = 0$ for the appearance of such an I_s-instability is equivalent to $Ra = R_s(\tilde{q}_\parallel)$, with the function $R_s(\tilde{q}_\parallel)$ given by (Lhost et al., 1991; Ortiz de Zárate et al., 2004):

$$R_s(\tilde{q}_\parallel) = \frac{(\tilde{q}_\parallel^2 + \pi^2)}{P_1^2 \tilde{q}_\parallel^2} \frac{(\tilde{q}_\parallel^2 + P_2 \Lambda^2)^2 + \Lambda^4 (1 - P_2^2)}{1 + \left[1 + \dfrac{2\sqrt{2} P_0 (\tilde{q}_\parallel^2 + \pi^2)}{P_1 \pi \tilde{q}_\parallel^2} Le\right] \psi}. \tag{9.22}$$

Looking for the minimum as a function of $\tilde{q}_\parallel$ of Eq. (9.22), one can determine the critical wave number for the stationary instability, $\tilde{q}_{\parallel c,s}$. The corresponding critical Rayleigh number is then obtained by evaluating Eq. (9.22) at the minimum: $Ra_{c,s} = R_s(\tilde{q}_{\parallel c,s})$. Depending on the values of ψ and Le, the critical wave number may be either zero or nonzero, although for usual binary mixtures $\tilde{q}_{\parallel c,s} \neq 0$. However, for $\psi > 0$, $Ra_{c,s}$ is always positive (Lhost et al., 1991). Explicit analytical expressions of the critical wave number and Rayleigh number become rather complicated, so that they are usually evaluated numerically, as done by Lhost et al. (1991). It is worth noticing that, by setting $\psi = 0$ in Eq. (9.22), the results obtained by Niederländer et al. (1991) for the first convective instability in a one-component fluid are reproduced.

As discussed in detail by Lhost et al. (1991), there is another possibility for the real part of the roots $\tilde{\Gamma}_i(q_\parallel)$ of Eq. (9.12) to be zero, namely: $Pr D_1 C_2 C_3 = C_4$. In this second possibility, there exists a pair of complex conjugate roots of Eq. (9.12) with zero real parts but nonzero imaginary parts, leading to oscillatory or Hopf-like instability I_o. The condition for the oscillatory I_o-instability, similarly to the one for the I_s instability, can be expressed as $Ra = R_o(\tilde{q})$ where (Lhost et al., 1991; Ortiz de Zárate et al., 2004):

$$R_o(\tilde{q}_\parallel) = \frac{\dfrac{B(q_\parallel)}{Le^2 Pr} [LePr\, C_2(q_\parallel) + 1][Le\, C_1(q_\parallel) + 1]}{P_1^2\, C_2(q_\parallel) \left\{1 + \psi - \dfrac{2\sqrt{2}\, P_0\, C_1(q_\parallel)\, \psi}{\pi P_1 [Pr\, C_2(q_\parallel) + C_1(q_\parallel)]}\right\}}. \tag{9.23}$$

The condition $Ra = R_o(\tilde{q}_\parallel)$, with $R_o(\tilde{q}_\parallel)$ given by Eq. (9.23) with the appropriate change of notation, reproduces the result reported by Lhost et al. (1991) who used the same set of Galerkin test functions. Again by looking for the minimum value of $R_o(\tilde{q}_\parallel)$ (as a function of $\tilde{q}_\parallel$) in Eq. (9.23), one can determine the critical Rayleigh number for the Hopf bifurcation, $Ra_{c,o}$. It turns out that, for $\psi > 0$, $Ra_{c,o} > Ra_{c,s}$, while for $\psi < 0$, $Ra_{c,o} < Ra_{c,s}$ (Lhost et al., 1991). Thus, the Rayleigh instability is critical for mixtures with positive separation ratios, while for negative separation ratios the Rayleigh instability is subcritical, and the Hopf instability critical. The presence of these two competing instability mechanisms is important for the physical interpretation of the nonequilibrium fluctuations, as will be discussed in more detail in Sect. 9.5. For future use it should be noticed that the two functions $R_s(q_{\parallel})$ and $R_o(q_{\parallel})$ become proportional to $q_{\parallel}^4$ in the limit of large $q_{\parallel}$. For $q_{\parallel} \to 0$, $R_o(q_{\parallel}) \propto q_{\parallel}^{-2}$, while $R_s(q_{\parallel})$ reaches a finite limit.

We conclude this section by pointing out that Lhost et al. (1991) present a complete instability diagram, numerically deduced from Eqs. (9.22) and (9.23), which qualitatively resembles the simpler case of free and permeable walls discussed in Fig. 5.1. These Galerkin approximations (9.22) and (9.23), give fairly good approximations for the exact critical Rayleigh and wave numbers (Lhost et al., 1991). For instance, the fact that the stationary instability happens first for positive ψ while the oscillatory instability happens first for negative ψ, is known to be exactly correct (Schechter et al., 1974). These facts give us confidence in the adequacy of the Galerkin-approximation procedure (9.9). Further evidence for the quality of the use Galerkin of test functions (9.9) will be presented later.

9.3 Evaluation of nonequilibrium structure factors

When presenting in Sect. 7.4 the Galerkin approximation for the calculation of the nonequilibrium structure factor of a one-component fluid, we used the exact expression (6.42) of the hydrodynamic modes in terms of the decay rates. For the case examined in this chapter of a binary mixture, we do not have an equivalent expression. However, the zeroth-order Galerkin approximation to the nonequilibrium structure factor can also be obtained by using directly the Galerkin test function in the hydrodynamic equations (Ortiz de Zárate and Sengers, 2002). In the case of a binary fluid, this procedure is implemented by substituting Eq. (9.9) into the linearized random Boussinesq Eqs. (9.4) and by projecting the resulting expressions onto the set of three Galerkin test functions (9.10). This process yields the set of algebraic equations (Ortiz de Zárate et al., 2004):

$$\left[i\tilde{\omega}\, \mathsf{D}(\tilde{q}_{\parallel}) + \mathsf{H}(\tilde{q}_{\parallel}) \right] \begin{pmatrix} w_0(\omega, \mathbf{q}_{\parallel}) \\ \theta_0(\omega, \mathbf{q}_{\parallel}) \\ \zeta_0(\omega, \mathbf{q}_{\parallel}) \end{pmatrix} = \hat{\mathbf{F}}(\omega, \mathbf{q}_{\parallel}), \qquad (9.24)$$

where we have introduced $\tilde{\omega} = \omega\, L^2/\nu$ as the dimensionless frequency of the fluctuations. The 3×3 matrices H and D are the same as those defined by Eq. (9.13), while the amplitudes vector with components w_0, θ_0 and ζ_0 depend now on the frequency ω. These amplitudes specify, in zeroth-order approximation, the solution of the random Boussinesq equations with the appropriate BC. Finally, the random force $\hat{\mathbf{F}}(\omega, \mathbf{q})$ in Eq. (9.24) is the

projection onto the Galerkin test functions of the corresponding component
of the Langevin random noise, so that:

$$\hat{\mathbf{F}}(\omega, \mathbf{q}_{\|}) = \int_{-L/2}^{L/2} \begin{bmatrix} \dfrac{L^3}{\alpha_p g \nu} F_0(\omega, \mathbf{q}_{\|}, z)\, C_1(z) \\[2ex] \dfrac{L^3}{\nu^2} F_1(\omega, \mathbf{q}_{\|}, z)\, \sqrt{2}\cos\dfrac{\pi z}{L} \\[2ex] \dfrac{L^3}{\alpha_p \nu^2} F_2(\omega, \mathbf{q}_{\|}, z) \end{bmatrix} dz, \tag{9.25}$$

where the RHS contains the three components of the random force
$\mathbf{F}(\omega, \mathbf{q}_{\|}, z)$ defined in Eq. (9.7). Inverting the matrix $\left[i\tilde{\omega}\, \mathsf{D}(\tilde{q}_{\|}) + \mathsf{H}(\tilde{q}_{\|})\right]$
in the LHS of Eq. (9.24), we immediately solve for the amplitudes of the
fluctuations: $w_0(\omega, q_{\|})$, etc.

To calculate the correlation functions among the various fluctuating
fields, we also need the correlations among the three components of the
Langevin noise (9.25). They can be computed from the FDT for a binary
mixture, Eqs. (3.9) and (3.10) and the definition of the random noise in
terms of fluctuating dissipative fluxes. As in Eq. (5.28), the resulting correlation functions are conveniently expressed in terms of a correlation matrix,
$\mathsf{C}(q_{\|})$, which is now defined as (Ortiz de Zárate et al., 2004):

$$\langle \hat{F}_\alpha^*(\omega, \mathbf{q}_{\|}) \cdot \hat{F}_\beta(\omega', \mathbf{q}_{\|}') \rangle = (2\pi)^3\, 2k_{\mathrm{B}} \frac{L^5 T^2}{\rho c_p \nu^3}\, \tilde{C}_{\alpha\beta}(q_{\|})\, \delta(\mathbf{q}_{\|} - \mathbf{q}_{\|}')\, \delta(\omega - \omega'), \tag{9.26}$$

where, similar to Eq. (4.54), we have used the strength of the random heat
flux autocorrelation to introduce a dimensionless noise correlation matrix,
which in our current case is given by:

$$\tilde{\mathsf{C}}(q_{\|}) = \tilde{q}_{\|}^2 \begin{bmatrix} \mathcal{R}_T B(\tilde{q}_{\|}) & 0 & 0 \\[2ex] 0 & \dfrac{C_1(\tilde{q}_{\|})}{Pr}\left[1 + \dfrac{\epsilon_{\mathrm{D}}}{Le}\right] & \dfrac{2\sqrt{2}\psi}{\pi Pr}\left[1 + \dfrac{1 + \epsilon_{\mathrm{D}}}{Le}\right] \\[2ex] 0 & \dfrac{2\sqrt{2}\psi}{\pi Pr}\left[1 + \dfrac{1 + \epsilon_{\mathrm{D}}}{Le}\right] & \dfrac{\psi^2}{Pr}\left[1 + \dfrac{(1 + \epsilon_{\mathrm{D}})^2}{\epsilon_{\mathrm{D}} Le}\right] \end{bmatrix}, \tag{9.27}$$

with $B(q_{\|})$ being the function defined by Eq. (9.17), and where the dimen-
sionless parameter $\mathcal{R}_T$ is a measure of the relative strength of random stress
to random heat autocorrelations, namely:

$$\mathcal{R}_T = \frac{\nu^2 c_p}{L^4 T \alpha_p^2 g^2}. \tag{9.28}$$

In deducing Eq. (9.26)-(9.27), the Dufour effect in the FDT (3.9) has been
included, and all thermophysical properties have been evaluated at their

average values in the layer (Ortiz de Zárate et al., 2004). We note that, in order to express Eq. (9.27) in dimensionless form, both the thermal diffusion coefficient k_T and the concentration derivative of the chemical potential have been expressed in terms of the parameters ϵ_D and ψ. Initially, since we have neglected Dufour effect in Eq. (9.13) for $H(\tilde{q}_\parallel)$ and $D(\tilde{q}_\parallel)$, we should also for consistency neglect Dufour effect in the correlation matrix. We have retained it here because, as discussed in more detail later, it shall be needed to obtain consistent approximations[‡]. In this connection, it should be noticed that the dimensionless parameter $\mathcal{R}_T$ may be quite large, so that, for a practical calculation (especially for the case of concentration fluctuations, see below) only the terms of order $\mathcal{O}(\epsilon_D^{-1})$ and $\mathcal{O}(\mathcal{R}_T)$ of the correlation matrix (9.27) will be considered. Finally, we note that in a previous publication Ortiz de Zárate et al. (2004) the cross-term $C_{12}(q_\parallel)$ of the correlation matrix was neglected, in accordance with other investigations (Hollinger and Lücke, 1998).

We now have all the ingredients required to calculate the dynamic structure factor $S(\omega, \mathbf{q})$ of the nonequilibrium fluid mixture accounting for finite-size effects. As in Sect. 5.3, since both temperature and concentration independently cause refractive-index fluctuations, the structure factor $S(\omega, \mathbf{q})$ shall be expressed in terms of partial structure factors, just as for the "bulk" case examined in Eq. (5.30). Furthermore, since we have applied here a spatial Fourier transformation to the fluctuating hydrodynamics equations in the horizontal XY-plane only, the partial structure factors have to be obtained upon integration over the vertical variable, just as in Eq. (6.28) for a one-component fluid. For instance, for the TT-partial structure factor, $S_{TT}(\omega, \mathbf{q}) = S_{TT}(\omega, q_\parallel, q_\perp)$, one has:

$$S_{TT}(\omega, q_\parallel, q_\perp) = \frac{\nu^2}{L^4} \left[A^{-1}(\omega, q_\parallel) \right]_{2i}^* \left[A^{-1}(\omega, q_\parallel) \right]_{2j} 2k_B \frac{L^5 T^2}{\rho c_p \nu^3} \tilde{C}_{ij}(q_\parallel)$$

$$\times \frac{1}{L} \int_{-L/2}^{L/2} dz \int_{-L/2}^{L/2} dz' \, e^{-iq_\perp(z - z')} \{ \mathbf{G}_{0,2}(z) \cdot \mathbf{G}_{0,2}(z') \}, \quad (9.29)$$

where $A(\omega, q) = i\tilde{\omega}D(q_\parallel) + H(q_\parallel)$ is the matrix in the LHS of Eq. (9.24). Just as Eq. (5.31b) for the "bulk" partial structure factor, Eq. (9.29) includes a summation over the (coordinate) indices i and j. The index 2 appears in the components of $A(\omega, q)$ and in the Galerkin test functions $\mathbf{G}_{0,2}$, because the temperature is the second component in the vector of fluctuating fields (9.5). The double integration in the vertical coordinate arises because, in small-angle light-scattering experiments, the scattering volume extends over the entire vertical dimension of the layer (as for the one-component fluid discussed in Sect. 6.1.1). The prefactor ν^2/L^4 arises from the prefactor multiplying the amplitude $\theta_0(q_\parallel)$ of the temperature fluctuations in

---

[‡]This is similar to Chapter 5, where we used a small ϵ_D expansion in the decay rates, but not in the amplitudes, see Eq. (5.41)

Eq. (9.9). The $\zeta\zeta$-partial structure factor as well as the $T\zeta$-partial structure factor have a form similar to Eq. (9.29), with the index 2 replaced by the appropriate index in the definition of the vector of fluctuating fields. Also, instead of ν^2/L^4, the corresponding prefactor in the definition of the dimensionless amplitudes of fluctuating fields in Eq. (9.9) should appear.

As in Chapter 7 for a one-component fluid, in practice we adopt the small-scattering-angle approximation: $q_\| \simeq q$ and $q_\perp \simeq 0$. Then the structure factor will depend only on the magnitude q of the wave vector $\mathbf{q}$. To obtain the structure factor measured in shadowgraph experiments the same approximation applies, as discussed in Sect. 10.1.2.

Before proceeding with an explicit calculation of the various nonequilibrium partial structure factors, we shall first evaluate the equilibrium partial structure factors in the same Galerkin-polynomial approximation by simply taking $Ra = 0$ in Eq. (9.13) for $\mathsf{H}(q_\|)$. By comparing the results with the known structure factor for a fluid mixture, as presented in Sect. 5.3, we can test the quality of using the Galerkin test functions adopted in Eq. (9.10).

9.4 Structure factor in thermal equilibrium

In equilibrium the correlation functions among fluctuating thermodynamic fields are spatially short ranged, so that BC do not play any role. Hence, the "bulk" results of Chapter 5, obtained by performing a full spatiotemporal Fourier transformation (including the direction of the gravity, perpendicular to the bounding plates) of the random binary Boussinesq equations, represent the complete solution for the structure factor in thermal equilibrium ($\nabla T_0 = 0$) for any BC that might be considered at the bounding plates (Berne and Pecora, 1976; Boon and Yip, 1980). For the dynamic structure factor $S(\omega, q)$, the relevant expressions are Eqs. (5.33) if we include the Dufour effect or Eq. (5.37) if we neglect it. For the static structure factor, *i.e.*, upon integration over ω of $S(\omega, q)$, the exact equilibrium result is given by Eq. (5.34), independent of whether the Dufour effect is neglected or not (see discussion in Sect. 5.3).

To obtain the corresponding results in our Galerkin approximation we substitute $Ra = 0$ into Eq. (9.13) for $\mathsf{A} = i\omega\mathsf{D} + \mathsf{H}$, invert the resulting matrix and evaluate $S_{TT}^{(\mathrm{E})}(\omega, q)$, as specified by Eq. (9.29). In the small-scattering-angle approximation we get:

$$S_{TT}^{(\mathrm{E})}(\omega, q) = \frac{\nu^2}{L^4} \, 2k_{\mathrm{B}} \frac{L^5 T^2}{\rho c_p \nu^3} \frac{Pr \, \tilde{q}^2 \, C_1(\tilde{q})}{Pr^2\tilde{\omega}^2 + \tilde{q}^4 \, C_1^2(\tilde{q})} \frac{8}{\pi^2} L. \tag{9.30}$$

In the derivation we have neglected the Dufour effect. Since the terms $\mathcal{O}(\epsilon_{\mathrm{D}}^{-1})$ and $\mathcal{O}(\mathcal{R}_T)$ of the correlation matrix $\tilde{\mathsf{C}}(\tilde{q})$ in Eq. (9.27) do not contribute to the TT-partial structure factor in equilibrium, to arrive at (9.30) we had to consider the $\mathcal{O}(1)$ terms in $\tilde{\mathsf{C}}(\tilde{q})$. Equation (9.30) is to be compared to the $\epsilon_D \to 0$ limit of the "bulk" Eq. (5.33b). We find that Eq. (9.30) is indeed in good agreement with Eq. (5.33b) in that limit. The only difference is the appearance of the function $C_1(q)$, that rapidly approaches unity with increasing q, and an overall factor $8/\pi^2 \simeq 0.811$ that is close to unity.

We also want to make a comparison with the partial structure factor $S_{cc}^{(\mathrm{E})}(\omega, q)$ as given by Eq. (5.33a). Since in this chapter, to accommodate for BC, we are using $\zeta = \beta\,\delta c + \psi\alpha_p\,\delta T$ as variable, we have to express $S_{cc}^{(\mathrm{E})}(\omega, q)$ in terms of the TT-partial structure factor, the $\zeta\zeta$-partial structure factor, as well as the cross $T\zeta$-partial structure factor. Moreover, since we are only interested in comparing with (5.33a) in the $\epsilon_D \to 0$ limit, it is sufficient to retain only the $\mathcal{O}(\epsilon_D^{-1})$ terms in the dimensionless correlation matrix (9.27). Furthermore, it should be noticed that, if we had retained higher-order terms in the Dufour effect ratio, the resulting expressions for the cc-partial structure factor will be unphysical in the small-q limit. Taking into account these facts, we obtain:

$$S_{cc}^{(\mathrm{E})}(\omega, q) = \frac{\alpha_p^2 \nu^2}{L^4}\, 2k_B \frac{L^5 T^2}{\rho c_p \nu^3}\, \frac{PrLe\psi^2 \tilde{q}^2}{\epsilon_D \left(Pr^2 Le^2\, \tilde{\omega}^2 + \tilde{q}^4\right)}\, L, \qquad (9.31)$$

which equals Eq. (5.33a) in the $\epsilon_D \to 0$ limit[‡]. Thus our simple Galerkin approximation reproduces the exact expression for the contribution of the concentration fluctuations to the equilibrium dynamic structure factor.

Integration over ω of Eqs. (9.30) and (9.31) gives the Galerkin approximation to the static partial structure factors:

$$S_{cc}^{(\mathrm{E})}(q) = \frac{k_B \bar{T}_0}{\rho} \left(\frac{\partial c}{\partial \mu}\right)_T, \qquad S_{TT}^{(\mathrm{E})}(q) = \frac{k_B \bar{T}_0^2}{\rho c_P}\frac{8}{\pi^2}, \qquad (9.32)$$

to be compared with Eq. (5.34). Again we reproduce the expression for the intensity of the equilibrium concentration fluctuations exactly, while the expression for the intensity of the temperature fluctuations is recovered except for an overall factor $8/\pi^2$ which is close to unity. We conclude that the Galerkin approximation (9.9) gives a reasonable description of the structure factor of a fluid mixture.

9.5 Nonequilibrium fluctuations for positive separation ratio

Having checked the quality of the Galerkin approximation (9.9), we now proceed to evaluate and discuss the various nonequilibrium partial static

---

[‡]Provided that the concentration derivative of the chemical potential in Eq. (5.33a) is expressed in terms of ϵ_D

structure factors, obtained by following the general procedure described in Sect. 9.3.

In a one-component fluid a Rayleigh stationary bifurcation is the only instability mechanism. In the case of a binary mixture, the situation is complicated by the possible presence of an oscillatory Hopf instability for negative separation ratio ψ, as anticipated in the discussion of Sect. 5.1.1. However, when $\psi > 0$, the preferred instability mechanism continues to be stationary (Lhost et al., 1991; Schechter et al., 1974) and we believe that the conclusions of Chapter 10 regarding the relationship between experimental structure factor and equal-time autocorrelation functions continue to hold. For $\psi < 0$, this relationship may require further clarification that lies outside the scope of our present monograph.

As was already noticed in the "bulk" discussion below Eq. (5.32), complications arise because of the explicit appearance of Ra in the expressions of the decay rates when gravity is not neglected. As a consequence, the partial dynamic structure factors can no longer be split into a sum of an equilibrium plus a nonequilibrium part, as was the case in Eq. (5.32) where gravity was neglected. However, as earlier noticed in Chapter 5 for the "bulk" binary fluid, upon integration over the frequency ω, the *static* partial structure factors do indeed admit an additive decomposition into an equilibrium and a nonequilibrium contributions. Since we are here mainly interested in the structure factor as measured in shadowgraph experiments[‡], and shadowgraph experiments have so far primarily probed static fluctuations (see, however, the discussion in Sect. 10.1.3), we shall concentrate here only on the static structure factor, as was done for one-component fluid layers in Chapter 7.

9.5.1 Temperature fluctuations

Substituting the correlation matrix defined by Eq. (9.26) into Eq. (9.29), and integrating over the frequency ω, we find that in the small-scattering-angle approximation ($q_{\parallel} \simeq q$ and $q_{\perp} \simeq 0$) the partial static structure factor $S_{TT}(\tilde{q})$ can be expressed as:

$$S_{TT}(\tilde{q}) = S_{TT}^{(E)} + S_{TT}^{(NE)}(\tilde{q}, Ra), \qquad (9.33)$$

where $S_{TT}^{(E)}$ represents the short-ranged equilibrium contribution already reported in Eq. (9.32), while $S_{TT}^{(NE)}(\tilde{q}, Ra)$ is to be interpreted as the long-ranged nonequilibrium contribution. Such a nonequilibrium contribution to

---

[‡]Small-scattering-angle light scattering cannot probe wave numbers small enough for finite-size effects to be significant

the partial $S_{TT}(\tilde{q})$ structure factor can be expressed as:

$$S_{TT}^{(\mathrm{NE})}(\tilde{q}, Ra) = \frac{U_{TT}(\tilde{q}, Ra)}{\left[1 - \dfrac{Ra}{R_\mathrm{s}(\tilde{q})}\right]\left[1 - \dfrac{Ra}{R_\mathrm{o}(\tilde{q})}\right]} \frac{Ra}{B(\tilde{q})}, \tag{9.34}$$

which is more easily interpreted physically when it is decomposed in the following form (Ortiz de Zárate et al., 2004):

$$S_{TT}^{(\mathrm{NE})}(\tilde{q}, Ra) = \frac{A_{TT}^{(\mathrm{s})}(\tilde{q}, Ra)}{1 - \dfrac{Ra}{R_\mathrm{s}(\tilde{q})}} + \frac{A_{TT}^{(\mathrm{o})}(\tilde{q}, Ra)}{1 - \dfrac{Ra}{R_\mathrm{o}(\tilde{q})}}. \tag{9.35a}$$

with:

$$A_{TT}^{(\mathrm{s})}(\tilde{q}, Ra) = \frac{R_\mathrm{o}(\tilde{q})}{R_\mathrm{o}(\tilde{q}) - R_\mathrm{s}(\tilde{q})} \frac{U_{TT}(\tilde{q}, Ra)}{B(\tilde{q})} Ra \tag{9.35b}$$

and

$$A_{TT}^{(\mathrm{o})}(\tilde{q}, Ra) = \frac{R_\mathrm{s}(\tilde{q})}{R_\mathrm{o}(\tilde{q}) - R_\mathrm{s}(\tilde{q})} \frac{U_{TT}(\tilde{q}, Ra)}{B(\tilde{q})} Ra. \tag{9.35c}$$

In the denominators of the two terms in the LHS of Eq. (9.35a), we recognize the two (linear) instability conditions $Ra = R_\mathrm{s}(\tilde{q})$ and $Ra = R_\mathrm{o}(\tilde{q})$ with $R_\mathrm{s}(\tilde{q})$ and $R_\mathrm{o}(\tilde{q})$ earlier specified by Eqs. (9.22) and (9.23). Therefore, we physically interpret $A_{TT}^{(\mathrm{s})}(\tilde{q}, Ra)$ as representing the amplitude of nonequilibrium temperature fluctuations that are associated with the stationary (Rayleigh) instability, while $A_{TT}^{(\mathrm{o})}(\tilde{q}, Ra)$ represents the amplitude of nonequilibrium temperature fluctuations that are associated with the oscillatory (Hopf) instability. Of course, as is evident from Eqs. (9.35b) and (9.35c), these amplitudes vanish for $Ra = 0$ (equilibrium). We observe how the two instability mechanisms cooperate to cause an enhancement of the fluctuations when a binary mixture is driven outside equilibrium.

For the case of the $S_{\zeta\zeta}$ partial structure factor, a set of equations with the same structure as Eqs. (9.33) and (9.35) can be obtained, with the equilibrium contribution in (9.33) replaced by the corresponding $S_{\zeta\zeta}^{\mathrm{E}}$; and where we can identify two nonequilibrium amplitudes $A_{\zeta\zeta}^{(\mathrm{s})}(\tilde{q}, Ra)$ and $A_{\zeta\zeta}^{(\mathrm{o})}(\tilde{q}, Ra)$ associated with the stationary and with the oscillatory instability, respectively. These nonequilibrium amplitudes may, in turn, be expressed in terms of a single function $U_{\zeta\zeta}(\tilde{q}, Ra)$, similarly to Eqs. (9.35b) and (9.35c) for the corresponding temperature-fluctuations amplitudes. A similar decomposition is possible for the cross $S_{\zeta T}$ partial structure factor. Equations (9.33) and (9.35a) are a compact and convenient way of expressing the various nonequilibrium partial structure factors, allowing a clear and simple physical interpretation.

Notice that when Eq. (9.22) holds, a divergence in the first term in the RHS of Eq. (9.35a) appears. Hence, our expression for the nonequilibrium partial structure factor $S_{TT}^{\text{NE}}(\tilde{q})$ will be only valid for Ra values up to the same $Ra_{\text{c,s}}$ obtained from the linear instability condition. In addition, the second term of the RHS of Eq. (9.35a) will also be divergent when the Hopf instability condition (9.23) holds. As discussed earlier, for $\psi > 0$ the Hopf instability is secondary with respect to the (primary) stationary instability. Thus, the divergence associated with the Hopf mode occurs at Ra values larger than the $Ra_{\text{c,s}}$ obtained from (9.22). Thus, for $\psi > 0$ the Ra-interval of validity of Eq. (9.35a) is solely determined by the Rayleigh instability, in spite of the presence of a second instability mechanism. Again, as in Chapter 7 for a one-component fluid, we find that (linear) instabilities correspond to divergences in a (linear) theory of nonequilibrium fluctuations. Additionally, we remark that all partial structure factors (velocity, temperature, concentration and cross-terms) diverge simultaneously for the same value of Ra.

An explicit expression for the coefficient U_{TT} in Eq. (9.34), from which the nonequilibrium Rayleigh and Hopf amplitudes can be obtained from Eqs. (9.35b) and (9.35c), has been derived by Ortiz de Zárate et al. (2004) in the small Dufour-effect-ratio approximation. The equation for U_{TT} is rather long and complicated, and we refer the interested reader to the original publication for the explicit formulas. We find it more convenient here to present explicit expressions only for some limiting cases, while the complete result for the nonequilibrium temperature fluctuations will be discussed only numerically.

The first limiting case we shall discuss here is that of large wave numbers q, for which the intensity of the nonequilibrium temperature fluctuations becomes proportional to $\tilde{q}^{-4}$ (Ortiz de Zárate et al., 2004):

$$S_{TT}^{(\text{NE})}(\tilde{q}) \xrightarrow{q \to \infty} S_{TT}^{(\text{NE},\infty)} \frac{1}{\tilde{q}^4} + \mathcal{O}(\tilde{q}^{-6}), \qquad (9.36)$$

with proportionality constant:

$$\frac{S_{TT}^{(\text{NE},\infty)}}{S_{TT}^{(\text{E})}} = \frac{P_1^2 \, Pr}{Pr + 1} \left[\left(1 + \psi - \frac{2\sqrt{2}P_0}{\pi P_1} \frac{2 + Le}{1 + Le} \psi \right) Ra + \frac{Ra^2 \mathcal{R}_T}{Pr} \right], \qquad (9.37)$$

where $S_{TT}^{(\text{E})}$ is the equilibrium partial structure factor calculated within the zeroth-order Galerkin approximation, Eq. (9.32). The dimensionless parameter $\mathcal{R}_T$ was defined in Eq. (9.28). To obtain a more compact expression we have added in Eqs. (9.36) and (9.37) the contributions from the stationary and from the oscillatory fluctuations. Notice that, as was the case for $Ra = 0$ in Eq. (9.30), the $\mathcal{O}(\epsilon_{\text{D}}^{-1})$ term in the correlation matrix (9.27) does not contribute to the TT-partial structure factor. At large q, boundary effects are negligible, so that Eq. (9.36) is to be compared with the exact result obtained in Sect. 5.3 without using any Galerkin approximation. If we integrate Eq. (5.35b) for the "bulk" $S_{TT}^{(NE)}(\omega, q)$ over the frequency ω, we indeed obtain a contribution proportional to q^{-4} like (9.36). Explicitly, in the limit $\epsilon_D \to 0$ one finds (Law and Nieuwoudt, 1989):

$$\frac{S_{TT}^{(NE,\infty)}}{S_{TT}^{(E)}} = \frac{c_p (\nabla T_0)^2}{T a_T (a_T + \nu)} L^4, \tag{9.38}$$

where the factor L^4 arises from having expressed (9.36) in terms of dimensionless wave number $\tilde{q}$. We note that Eq. (9.38) equals the term proportional to Ra^2 in Eq. (9.37) except for the factor $P_1^2 \simeq 1$. The term proportional to Ra in Eq. (9.37) arises from gravity, which was neglected in Sect. 5.3. We recall that a similar behavior appeared in the case of a one-component fluid, where the term proportional to Ra in Eq. (4.30) arose from gravity (Law and Sengers, 1989). Furthermore, if we take $\psi = 0$ in Eq. (9.37) (corresponding to the one-component limit), we reproduce Eq. (4.30) except for the same factor P_1^2 close to unity. These observations support again the validity of our Galerkin approximation.

A second interesting case is the limit of small wave numbers q, where the nonequilibrium enhancement becomes proportionally to $\tilde{q}^2$ (Ortiz de Zárate et al., 2004). We recall from Chapter 5 that in this $q \to 0$ limit the "bulk" (*i.e.*, in the absence of boundary conditions) nonequilibrium partial structure factor $S_{TT}^{(NE)}$ reaches, as a consequence of the gravity (Segrè and Sengers, 1993), a finite nonzero value. Thus, the boundary conditions (vanishing temperature fluctuations at the walls) cause the enhancement of the nonequilibrium temperature fluctuations to vanish at small q. This feature is similar to the case of a one-component fluid discused in Eq. (7.23).

A third interesting limiting case is that of $\psi = 0$, in which case our current system reduces to the Rayleigh-Bénard problem for a one-component fluid, which was analyzed in Sect. 7.4 in terms of a different set of Galerkin test functions. The result of substituting $\psi = 0$ in Eq. (9.33) can be simplified to:

$$S_{TT}(\tilde{q}) = S_{TT}^{(E)} \left[1 + \frac{Ra}{Pr + \dfrac{C_1(\tilde{q})}{C_2(\tilde{q})}} \frac{Pr + \mathcal{R}_T Ra}{\dfrac{C_1(\tilde{q})B(\tilde{q})}{P_1^2} - Ra} \right], \tag{9.39}$$

to be compared with Eq. (7.40). We first find that the result for the equilibrium structure factor obtained by taking $Ra = 0$ in Eq. (9.39) is 19% lower than the exact "bulk" result, $S_E(\gamma - 1)/\gamma$; while with the Galerkin test functions used in Sect. 7.4 (Ortiz de Zárate and Sengers, 2002; Oh

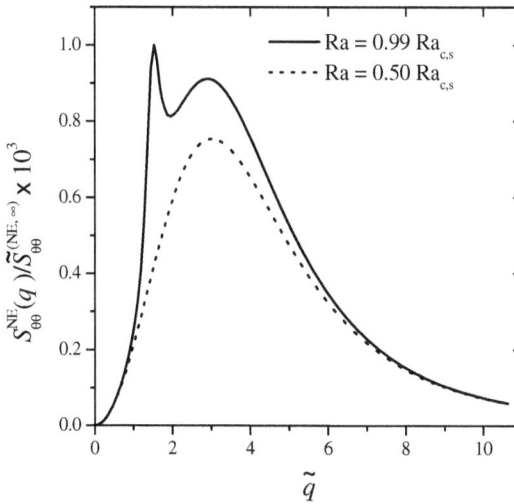

Figure 9.1: Appearance of a bimodal distribution for the amplitude of nonequilibrium temperature fluctuations as the stationary instability is approached. The broad peak corresponds to the Hopf fluctuations, while the sharper peak to the left for larger Ra corresponds to the Rayleigh fluctuations. Parameter values are quoted in the text.

et al., 2004), the approximation for the equilibrium structure factor was 17% lower. Secondly, the critical Rayleigh number obtained from Eq. (9.39) is $Ra_c = 1728$ (Niederländer et al., 1991), which agrees better with the correct value $Ra_c = 1708$ than the value $Ra_c = 1750$ obtained from the Galerkin test functions employed in Sect. 7.4. The limit for large q obtained from Eq. (9.39) is 21% lower than the true asymptotic limit, to be compared with a 20% difference obtained from the Galerkin test functions used in Sect. 7.4. It is difficult to decide which of the two sets of Galerkin test functions represent best the nonequilibrium structure factor of a one-component fluid layer subjected to an uniform temperature gradient. Since Eq. (9.39) gives a significantly better approximation for Ra_c it can be expected to yield a good approximation for the nonequilibrium structure factor at values of q close to q_c.

From the asymptotic behaviors discussed before, we conclude that the nonequilibrium partial structure factor $S_{TT}^{(NE)}$ for arbitrary ψ exhibits a crossover from a $\tilde{q}^{-4}$ behavior at large $\tilde{q}$ to a $\tilde{q}^2$ behavior at small $\tilde{q}$. This implies that there will be at least one local maximum for some nonzero wave number $\tilde{q}_m$, as was the case for a one-component fluid layer discussed in Chapter 7. Most interestingly, however, we find that, depending on ψ and Ra, the amplitude of the nonequilibrium temperature fluctuations may display two maxima, implying a bimodal distribution of the these fluctuations as a function of $\tilde{q}$. The two maxima correspond to the two competing instability mechanisms (Rayleigh and Hopf). This is clearly shown in Fig. 9.1, where for two values of Ra we have plotted as a function of $\tilde{q}$ the amplitude of the nonequilibrium temperature fluctuations, normalized by the coefficient multiplying $\tilde{q}^{-4}$ in the asymptotic expansion for large $\tilde{q}$. The data in Fig. 9.1 correspond to $Le = 32$, $Pr = 4.16$ and $\psi = 0.025$. In the expression for the dimensionless correlation matrix (9.27), we used parameter values $\epsilon_D = 10^{-4}$ and $\mathcal{R}_T = 5/\epsilon_D$; which are quite reasonable values for ordinary liquid mixtures. The solid curve in Fig. 9.1 corresponds to $Ra = 419$, which is $Ra = 419 \simeq 0.99 Ra_{c,s}$ for the quoted values of Pr, Le and ψ, while the dotted curve corresponds to $Ra = 210 \simeq 0.5 Ra_{c,s}$. In Fig. 9.1 the appearance of a bimodal structure factor as the instability is approached is evident. For the larger value of Ra, the right broader maximum (at higher values of $\tilde{q}$) arise from Hopf fluctuations, while the left sharper maximum arise from Rayleigh fluctuations. For the data in Fig. 9.1, as $Ra \rightarrow Ra_{c,s}$, a divergence will appear at the position $\tilde{q}_{c,s}$ of the left maximum. Notice that for the lower Ra value, the broader maximum corresponding to Hopf fluctuations masks the second maximum corresponding to Rayleigh fluctuations.

9.5.2   Concentration fluctuations

In the case of a fluid mixture, we need to consider not only nonequilibrium temperature fluctuations, but also nonequilibrium concentration fluctuations induced by the Soret effect. In practice, concentration fluctuations are usually easier to observe than temperature fluctuations (Li et al., 1998, 2000; Vailati and Giglio, 1996, 1997b). Hence, an evaluation of the cc partial static structure factor is even more interesting than that of the temperature partial structure factor discussed in the previous section. In this chapter we are using the variable ζ instead of concentration. Since $\delta c = \beta^{-1}[\zeta - \psi \alpha \delta T]$, to calculate the concentration partial structure factor we need to consider the autocorrelation of the temperature, the autocorrelation of the ζ variable, and the cross-correlation between the temperature and the ζ-fluctuations. This results in a quite tedious calculation. Some simplification arises because, as discussed in Sect. 9.4, to obtain a consistent approximation only the terms proportional to ϵ_D^{-1} and to $\mathcal{R}_T$ in the correlation matrix (9.27) are to be considered. Similarly to Eq. (9.33), the final result obtained for the S_{cc} partial structure factor admits an additive decomposition in an equilibrium plus a nonequilibrium part. Moreover, the nonequilibrim partial structure factor $S_{cc}^{(NE)}(\tilde{q}, Ra)$ can be expressed like in Eqs. (9.35), so as to identify the independent contributions from the Hopf and the Rayleigh instability mechanisms. As in Eqs. (9.35b)-(9.35c), the amplitudes of these Rayleigh and Hopf nonequilibrium fluctuations can be expressed in terms of a single function $U_{cc}(\tilde{q}, Ra)$. The explicit expression for this $U_{cc}(\tilde{q}, Ra)$

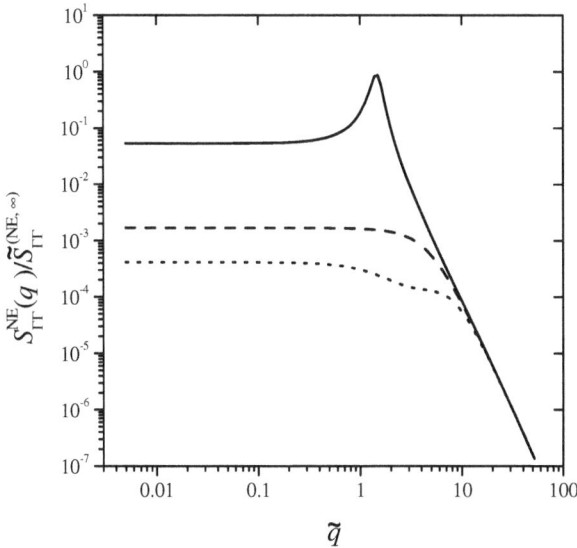

Figure 9.2: Normalized amplitude of nonequilibrium concentration fluctuations as a function of $\tilde{q}$, for three values of Ra. Solid curve is for $Ra = 419$, which is close to the $Ra_{c,s}$. Dashed curve is for $Ra \approx 0$, which corresponds to thermal nonequilibrium in microgravity. Dotted curve is for large and negative Rayleigh number $Ra = -1500$, as have been used in some experiments (Vailati and Giglio, 1996).

function becomes very long and complicated, so that we shall discuss here the results for the amplitude of nonequilibrium concentration fluctuations in graphical form, as we did for the nonequilibrium temperature fluctuations in Sect. 9.5.1.

In Fig. 9.2 we have plotted the nonequilibrium partial static structure factor $S_{cc}^{(NE)}(\tilde{q}, Ra)$ as a function of $\tilde{q}$ for three values of Ra, retaining for all other physical properties the same values adopted in Fig. 9.1 for the nonequilibrium temperature partial structure factor. As in Fig. 9.1, we have divided $S_{cc}^{(NE)}$ by the proportionality factor $S_{cc}^{(NE,\infty)}$ for large $\tilde{q}$, see Eq. (9.36). Then, the three sets of data displayed in Fig. 9.2 have a common large-q limit, simply given by $\tilde{q}^{-4}$. The solid curve corresponds to the same set of parameters for which the temperature fluctuations were displayed in Fig. 9.1 as a solid curve. The dotted curve also corresponds to the same set of parameters, except the Rayleigh number, for which a large and negative value was chosen, namely $Ra = -1500$, in accordance with the overstable situation investigated in some experiments (Vailati and Giglio, 1996). The dashed curve corresponds to a Ra number close to zero, showing nonequilibrium concentration fluctuations in the absence of gravity.

From an examination of the data displayed in Fig. 9.2, as well as from some additional numerical and analytical studies, we arrive at the following conclusions:

1. The presence of two maxima which was so evident in Fig. 9.1 for the nonequilibrium temperature fluctuations is masked in the concentration fluctuations. Notice that the solid curve in Fig. 9.2 is for the same Ra number as the solid curve in Fig. 9.1. We believe this feature is mainly a consequence of the fact that, for usual binary fluids, the Le number is large, so that the contribution of Hopf fluctuations to concentration fluctuations is negligible. This point will be discussed in more detail in Sect. 9.6.

2. Unlike the amplitude of the nonequilibrium temperature fluctuations, the amplitude of the nonequilibrium concentration fluctuations does not go to zero at $\tilde{q} \to 0$. Instead, it reaches a constant limit at $\tilde{q} \to 0$ for any value of the Ra number. This behavior is a direct consequence of the different nature of the boundary conditions: a zero derivative instead of a zero value in Eq. (9.1). When for mathematical simplicity the unrealistic case of two free and permeable walls (implying no concentration fluctuations at the boundaries) is considered, then the partial structure factor $S_{cc}^{\mathrm{NE}}(\tilde{q})$ does vanish at $\tilde{q} \to 0$ (Sengers and Ortiz de Zárate, 2001).

3. As a consequence of #2, below a certain value of Ra the maximum enhancement is reached at $\tilde{q} = 0$. This seems evident in the plots for negative or small Ra in Fig. 9.2. Note that this is never true for the temperature fluctuations.

The constant limit at $q \to 0$ of the nonequilibrium enhancement of concentration fluctuations can be, after some long algebra, evaluated within the Galerkin approximation (9.10) employed in this chapter. We obtain:

$$\frac{S_{cc}^{\mathrm{NE}}(\tilde{q}, Ra)}{S_{cc}^{(\mathrm{E})}} \xrightarrow{\tilde{q} \to 0} \frac{2\sqrt{2}P_1}{\pi} RaLe \frac{\psi P_0 + \dfrac{2\sqrt{2}P_1}{\pi} \dfrac{\mathcal{R}_T \epsilon_{\mathrm{D}} Ra}{Pr}}{\Lambda^4 - \dfrac{2\sqrt{2}P_1 P_0}{\pi} \psi RaLe}. \tag{9.40}$$

Notice that, as explained above, only the terms proportional to $\epsilon_{\mathrm{D}}^{-1}$ and to $\mathcal{R}_T$ in the correlation matrix (9.27) have been considered. It is also worth noticing the presence of a divergence in Eq. (9.40) for $\psi LeRa = 2\sqrt{2}P_1 P_0/\pi\Lambda^4 \simeq 678.5$, which cancels the denominator. However, it can be demonstrated that such a divergence always occurs for a Ra number larger than the critical $Ra_{\mathrm{c,s}}$, and thus has no relevance in practice. We shall further comment on this point later.

We conclude this section by noting in Fig. 9.2, for the case of large and negative Ra number, the presence of a small dip in the amplitude of the nonequilibrium fluctuations for $\tilde{q}$ values close to the ones where, for positive Ra, there appears the maximum associated with the stationary convective instability. It is difficult to assess whether such a small dip is real or is an artifact of the Galerkin approximation (Ortiz de Zárate et al., 2004).

9.6 Large-Lewis-number approximation for positive separation ratio

In the previous section, we have shown how one can derive solutions of the linearized random Boussinesq equations for the nonequilibrium structure factor of binary fluid mixtures with arbitrary values of Le, Pr and Ra. However, even with our simple Galerkin approach, such expressions are rather long and complicated; thus a practical simplification would be useful. To obtain such a simplification, we take advantage of the separation in order of magnitude of the various diffusivities, since for commonly used liquid mixtures $\nu \gg a \gg D$ (Segrè et al., 1993a). As a consequence, like in Sect. 5.2, we shall adopt a large-Le approximation (Velarde and Schechter, 1972), that has been shown to describe the most important concentration fluctuations in the bulk. We start by performing a large-Le expansion in the definitions (9.22) and (9.23) of $R_{\rm s}(\tilde{q})$ and $R_{\rm o}(\tilde{q})$. Then, it can be readily demonstrated that:

$$\frac{R_{\rm o}(q)}{R_{\rm o}(q) - R_{\rm s}(q)} = 1 + D_2(q)\frac{1}{Le} + \mathcal{O}\left(\frac{1}{Le^2}\right),$$

$$\frac{R_{\rm s}(q)}{R_{\rm o}(q) - R_{\rm s}(q)} = D_2(q)\frac{1}{Le} + \mathcal{O}\left(\frac{1}{Le^2}\right), \tag{9.41}$$

where

$$D_2(q) = \frac{\pi\sqrt{2}P_1}{4P_0\psi C_1(q)}\left[1 + \psi - \frac{2\sqrt{2}P_0\ \psi C_1(q)}{\pi P_1[PrC_2(q) + C_1(q)]}\right]. \tag{9.42}$$

We recognize in the LHS of Eqs. (9.41) the factors appearing in the general expressions (9.35b)-(9.35c) for the amplitude of the nonequilibrium Rayleigh and Hopf fluctuations, respectively. We therefore conclude from Eqs. (9.41) that, when Le is large, Hopf fluctuations can be neglected when compared to Rayleigh fluctuations. As a consequence, the large Lewis-number approximation neglects the oscillatory instability (Velarde and Schechter, 1972). Next, by performing the same kind of large-Le expansion for the functions $U_{TT}(\tilde{q}, Ra)$, etc. appearing in the general expression (9.34) of the nonequilibrium partial structure factors, it can be shown that:

$$U_{TT}(\tilde{q}, Ra) = \mathcal{O}(1), \quad U_{\zeta\zeta}(\tilde{q}, Ra) = \mathcal{O}(Le), \quad U_{T\zeta}(\tilde{q}, Ra) = \mathcal{O}(1). \tag{9.43}$$

In these expansions only the terms proportional to $\epsilon_{\rm D}^{-1}$ and to $\mathcal{R}_T$ in the correlation matrix (9.27) have been considered. From Eqs. (9.41) and (9.43), we conclude that the dominant contribution, in the large-Lewis-number limit, to the nonequilibrium concentration fluctuations is given by the stationary (Rayleigh) part of the $S_{\zeta\zeta}^{\rm (NE)}$ partial structure factor. Neglecting any other contribution, using Eq. (9.41) and the $\mathcal{O}(Le)$ expansion of $U_{\zeta\zeta}(\tilde{q}, Ra)$, we obtain a rather simple expression for the amplitude of the nonequilibrium concentration fluctuations, namely:

$$
\frac{S_{cc}^{NE}(\tilde{q}, Ra)}{S_{cc}^{(E)}} = \left\{ \left[\frac{\mathcal{R}_T \epsilon_D}{Pr} - P_0^2 \psi^2 \frac{D_2(\tilde{q})}{B(\tilde{q})} \right] \frac{2\sqrt{2} P_1}{\pi} Ra + P_0 \psi \right\}
$$
$$
\times \frac{2\sqrt{2} P_1 Le Ra}{\pi B(\tilde{q}) \left[1 - \dfrac{Ra}{R_s(\tilde{q})} \right]}, \quad (9.44)
$$

where $D_2(\tilde{q})$ was defined in Eq. (9.42). It should be noticed that, since in the large-Le approximation the contribution from temperature fluctuations can be neglected, the ratio given by Eq. (9.44) can be identified with $A_c(q)$ in Chapter 5.

The approximation (9.44) exhibits the correct asymptotic behaviors for both large and small $\tilde{q}$. Thus, it reaches a finite nonzero limit at $\tilde{q} \to 0$ and it is proportional to $\tilde{q}^{-4}$ for large $\tilde{q}$. The small $\tilde{q}$ limit of Eq. (9.44) continues to be given by Eq. (9.40), since terms of lower order in Le only contribute in $\tilde{q}^2$ order in that limit. The proportionality coefficient of the large $\tilde{q}$ limit, similar to the one defined in Eq. (9.36) for the temperature fluctuations, is easily evaluated as:

$$
\frac{S_{cc}^{(NE,\infty)}}{S_{cc}^{(E)}} = \left\{ \frac{\mathcal{R}_T \epsilon_D}{Pr} \frac{2\sqrt{2} P_1}{\pi} Ra + P_0 \psi \right\} \frac{2\sqrt{2} P_1}{\pi} Le Ra, \quad (9.45)
$$
$$
= \left\{ \frac{8 P_1^2}{\pi^2} \left(\frac{\partial \mu}{\partial c} \right) \frac{(\nabla c_0)^2}{\nu D} + \frac{2\sqrt{2} P_0 P_1}{\pi} \frac{\beta g \nabla c_0}{\nu D} \right\} L^4.
$$

Comparing Eq. (9.45) with the prefactor in Eq. (5.19) we observe that the only difference is the presence of two numbers, $8 P_1^2/\pi^2$ and $2\sqrt{2} P_0 P_1/\pi$, which are close to unity. This results further shows the appropriateness of the Galerkin approximation. This feature of Eq. (9.44) is especially important, because the "bulk" limit, Eq. (5.19), has been experimentally verified for negative Ra and relatively large q values, see Chapter 10.

With the help of Eq. (9.45), we observe that the small-q limit of $S_{cc}^{NE}(\tilde{q}, Ra)$, which in the large-$Le$ approximation continued to be given by Eq. (9.40), can be more conveniently expressed as:

$$
A_c(\tilde{q}) = \frac{S_{cc}^{NE}(\tilde{q}, Ra)}{S_{cc}^{(E)}} \xrightarrow{\tilde{q} \to 0} \frac{S_{cc}^{(NE,\infty)}}{S_{cc}^{(E)} \Lambda^4 \left(1 - \dfrac{\psi Ra Le}{678.5} \right)}. \quad (9.46)
$$

We stress that the large-Le expansion leading to Eq. (9.44) is completely systematic, except for the fact that we did not expand the Rayleigh instability condition in the denominator. We have proceeded in this way because, as will be discussed later, it greatly improves the approximation, in particular for Ra values close to the instability. A more systematic large-Le expansion is obtained by assuming

$$1 - \frac{Ra}{R_s(\tilde{q})} \simeq 1 - \frac{P_0^2}{B(\tilde{q})}\, \psi LeRa \qquad (9.47)$$

in the denominator of Eq. (9.44) (Ortiz de Zárate et al., 2004). Notice that, in spite of Le being large, the concentration Rayleigh number (5.16) can be of $\mathcal{O}(1)$ (Ortiz de Zárate et al., 2004). It is interesting to note that both the large $\tilde{q}$ limit, Eq. (9.45) and the small-$\tilde{q}$ limit, Eq. (9.46) are independent of whether the systematic approximation (9.47) is adopted in the denominator or not.

A consequence of having adopted Eq. (9.47) will be that the maximum (as a function of $\tilde{q}$) of the nonequilibrium structure factor will always be at $\tilde{q} = 0$. As discussed elsewhere (Schechter et al., 1974), this is actually the case when a large-Lewis-number approximation is applied to the starting Boussinesq equations. A second consequence will be that the critical Rayleigh number, *i.e.*, the Ra value at which nonequilibrium fluctuations diverge, will be specified by the condition $\psi LeRa = 678.5$ that cancels the denominator of Eq. (9.46). However, one should notice that the structure factor, as given by (9.44), diverges (for $\psi > 0$) for the same condition (9.22) obtained for the linear Rayleigh instability within our Galerkin approximation (Lhost et al., 1991), which usually occurs at a $q_{c,s} \neq 0^{\ddagger}$. This is the main reason why we prefer not to apply a large-Le expansion to the Rayleigh instability condition in the denominator of Eq. (9.44). This detail demonstrates the advantage of having obtained Eq. (9.44) from the full set of binary Boussinesq equations (9.24), instead of having adopted the $Le \to \infty$ approximation in the starting equations.

In this respect we should mention that the random Boussinesq equations at large Le, Eqs. (5.9), can be solved for realistic boundary conditions by expanding the solution in the set of corresponding hydrodynamic modes, as done in Chapter 6 for a one-component fluid. Then, by a method similar to that of Sect. 7.3, the exact small-q limit of the nonequilibrium enhancement of concentration fluctuations can be evaluated. The result is:

$$A_c(\tilde{q}) \xrightarrow{\tilde{q} \to 0} \left\{ \left(\frac{\partial \mu}{\partial c}\right) \frac{(\nabla c_0)^2}{\nu D} + \frac{\beta g \nabla c_0}{\nu D} \right\} \frac{L^4}{720 \left(1 - \dfrac{\psi RaLe}{720}\right)}, \qquad (9.48)$$

to be compared with the Galerkin approximation (9.46). Again we find a very good accordance between the Galerkin approximation and the known exact limiting behavior (9.48). For the prefactor we refer to the comments

---

‡The comments after Eq. (9.22) about the value of $\tilde{q}_{c,s}$ apply here also.

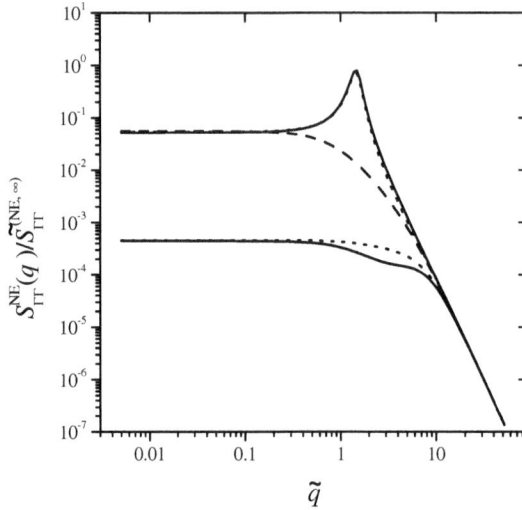

Figure 9.3: Large Lewis number approximation (9.44) to the nonequilibrium amplitude of concentration fluctuations. Solid curves represent full Galerkin results, while dashed curves represent such large Le number approximation (see text for further comments).

after Eq. (9.45). For the critical concentration Rayleigh number in the strict large-Lewis-number approximation, *i.e.*, with the substitution (9.47) in the denominator of Eq. (9.44); we obtained 678.5 with the Galerkin approximation, that compares well with the exact result 720.

To illustrate the appropriateness of the approximate expression (9.44), we show in Fig. 9.3 a plot of Eq. (9.44) for the same parameter values as for the plots in Fig. 9.1 and 9.2, that is, $Le = 32$, $Pr = 4.15$, and for two of the values of the Rayleigh number considered in Fig. 9.2: $Ra = 419$, which is close to the Rayleigh instability, and for $Ra = -1500$, which is a Rayleigh number of the order employed in some experiments (Li et al., 2000). As in previous plots, we have normalized the nonequilibrium structure factors by dividing with the corresponding proportionality constant of the large $\tilde{q}$ asymptotic expansion, $S_{cc}^{(\text{NE},\infty)}$. Consequently, all plots in the figure share the same asymptotic limit $(1/\tilde{q}^4)$ for large $\tilde{q}$. The solid curves in Fig. 9.3 correspond to the full Galerkin expression for the amplitude of the non-equilibrium concentration fluctuations; they are actually the same curves plotted in Fig. 9.2. The dotted curves correspond to the large-Le approximation: Eq. (9.44) divided by Eq. (9.45). The dashed curve corresponds to the the strict large-Lewis-number approximation, *i.e.*, with the substitution (9.47) in the denominator of Eq. (9.44). We only plotted the strict large-Lewis-number approximation for $Ra = 419$, because for the negative $Ra = -1500$ is completely undistinguishable from the complete Eq. (9.44). It is evident that the large-Lewis-number approximation (9.44) represents the full structure factor quite well; actually for Ra close to $Ra_{c,s}$ it is almost undistinguishable from the one obtained in the full Galerkin approximation. It is also obvious that the substitution of Eq. (9.47) in the denominator of Eq. (9.44) gives very bad results for Ra numbers close to the instability. However, for negative Ra such a substitution does not affect appreciably to the goodness of the large-Lewis-number approximation (9.44).

To finalize this section, we wish to recall that, in deducing Eq. (9.44), in accordance to Eq. (9.41), we have completely neglected the oscillatory instability. Consequently, Eq. (9.44) will only be valid for $\psi > 0$. In the case $\psi < 0$ it so happens that, although the factor determining the amplitude of the Hopf fluctuations, second line of (9.41), is indeed $\mathcal{O}(Le^{-1})$ in our large-Le expansion, the corresponding denominator is almost zero in the neighborhood of some $\tilde{q}_{c,s}$, so that the actual contribution of the Hopf fluctuations may be very large. The approximation (9.44) will not be good for $\psi < 0$ and for $\tilde{q}$ values close to the $\tilde{q}_{c,o}$ of the Hopf instability. We should also emphasize that neglecting Hopf fluctuations, as done in Eq. (9.44), is only justified for the nonequilibrium concentration fluctuations. For the temperature fluctuations it turns out that the amplitude of the Rayleigh and the Hopf contributions are of the same order in Le. Therefore, both have to be taken into account simultaneously, and this is the reason why the amplitude of the concentration fluctuations shown in Fig. 9.2 for $Ra = 419$ does not exhibit the bimodal distribution found in Fig. 9.1 for the temperature fluctuations at the same Ra.

9.7 Concluding remarks

In this chapter, we have employed a simple Galerkin approximation scheme to calculate the nonequilibrium contribution to the various thermodynamic fluctuations for the linearized binary Boussinesq equations using realistic boundary conditions. The same set of Galerkin test functions has been previously employed by Lhost et al. (1991) to successfully analyze the linear instability problem. Since the conclusions of our work may be hidden by the long and cumbersome algebra, we summarize here our main findings.

First of all, we found that the nonequilibrium fluctuations in any thermodynamic variable for the binary Boussinesq problem may be classified as arising from one of the two instability mechanisms present. So it is possible to distinguish between nonequilibrium fluctuations associated with the Rayleigh stationary instability and nonequilibrium fluctuations associated with the Hopf oscillatory instability, see Eq. (9.35a).

The amplitude of nonequilibrium temperature fluctuations, as a function of the wave number q of the fluctuations, decays as q^{-4} for large q (in accordance with the well-known behavior of the "bulk" structure factor studied in Chapter 5) and goes to zero as q^2 for extremely small q, as previously found for the case of a pure fluid in Chapter 7. Between these two limiting behaviors (and close to the stationary instability for $\psi > 0$) the nonequilibrium enhancement of temperature fluctuations may present two maxima, reflecting the presence of two competing instability mechanisms.

Due to the different kinds of boundary condition (null derivative *versus* null function), the nonequilibrium enhancement of concentration fluctuations does not go to zero at $q \to 0$, but it reaches a constant nonzero limit, given by Eq. (9.40) in our Galerkin approximation. We conclude that the physical nature of boundary conditions determines the behavior at $q \to 0$ of the corresponding nonequilibrium fluctuations and, in principle, may be experimentally investigated. A consequence of the different $q \to 0$ behavior of the concentration fluctuations is that, for $\psi > 0$, the maximum enhancement of these fluctuations is usually at $q = 0$, except for Ra numbers very close to the stationary instability.

Finally, we have obtained a simple approximation for the amplitude of the nonequilibrium concentration fluctuations taking advantage of the fact that the Le number is large for common binary mixtures or solutions. We have shown that, for $\psi > 0$, such an approximation is equivalent to neglecting both temperature fluctuations and the fluctuations associated with the Hopf instability. As further discussed in Chapter 10, our simple and systematic $Le \to \infty$ approximation compares well with the full nonequilibrium enhancement of the concentration fluctuations for reasonable values of the various thermophysical properties involved.

Chapter 10

Experiments on nonequilibrium fluctuations in the Rayleigh-Bénard problem

As already mentioned in Chapter 3, thermal fluctuations in fluids in thermodynamic equilibrium have been traditionally investigated by light scattering. The same light-scattering techniques can also be used to investigate fluctuations in fluids in nonequilibrium states. Furthermore, since the intensity of fluctuations in nonequilibrium fluids is significantly enhanced as compared to equilibrium fluids, it has also become possible to probe nonequilibrium fluctuations with shadowgraphy, an experimental technique previously used to visualize convection patterns in nonequilibrium fluids (de Bruyn et al., 1996). The main aim of this chapter is to review these experimental techniques and the results obtained for nonequilibrium fluctuations in the Rayleigh-Bénard problem.

We have divided the material to be presented in this chapter in two main parts. In the first part, Sect. 10.1, we discuss the experimental techniques that have thus far been employed to measure nonequilibrium fluctuations. The discussion of light scattering in Sect. 10.1.1 is brief, since excellent monographes exist on the topic (Berne and Pecora, 1976; Boon and Yip, 1980; Chu, 1991). Although these books deal with light scattering in equilibrium fluids, many of the issues apply equally well to light scattering in nonequilibrium fluids. In Sect. 10.1.2 we give a more detailed discussion of quantitative shadowgraphy, since this method is a newer experimental technique and the reader may be less acquainted with its fundamentals.

The remainder of the chapter reviews the actual experimental results that have been obtained for thermal fluctuations in fluids and fluid mixtures in nonequilibrium steady states. Section 10.2 deals with nonequilibrium light-scattering experiments in one-component liquids, including experiments for both the Rayleigh and the Brillouin components of the fluctuation spectrum. In Sect. 10.3 we discuss shadowgraph experiments in one-component fluids. Since most shadowgraph experiments yield the intensity of the fluctuations, they pertain to the Rayleigh component of the spectrum (recall from Eq. (4.45) that the total intensity of the Brillouin components of the spectrum is unaffected by the presence of a temperature or concentration gradient). However, the method can also be used for studying the time dependence of the fluctuations by measuring the shadowgraph image as a functions of the exposure time; this topic is addressed in Subsects. 10.1.3 (theory) and 10.3.2 (experiments). In Sect. 10.4 we review nonequilibrium light-scattering experiments in binary mixtures that have been used to probe nonequilibrium fluctuations induced by the Soret effect. Finally, in Sect. 10.5 we review shadowgraph experiments in binary mixtures that have been mainly used to measure nonequilibrium fluctuations in free-diffusion processes.

10.1   Experimental techniques

10.1.1   Small-angle light scattering

Light scattering results from the interaction between light (electromagnetic radiation) and matter. In a classical picture, when a electromagnetic wave propagates in a dielectric and transparent medium, the electric field accelerates the electrons causing the atoms to radiate. The light scattered by the medium will be the sum of all such electromagnetic radiation. If the illuminated (scattering) volume is divided into many volume elements of equal size[‡], the scattered field is the superposition of the fields radiated from each one of these volume elements. If all the "points" (volume elements) inside the scattering volume are optically identical (*i.e*, they have the same refractive index) there will be no scattered light in any other direction than the forward direction. This is so because the scattered fields from different "points" will differ only in phase. If the scattering volume contains many scatterers, the scattered field of any given scatterer can always be statistically paired with the scattered field of another scatterer which will be identical in amplitude but opposite in phase, so that the fields cancel, leaving light propagating only in the forward direction. As a consequence, scattered light arises from differences in the refractive index, *i.e* from *fluctuations* of the refractive index inside the scattering volume.

[‡]Large compared to the cube of the incident wavelength

To obtain an expression for the intensity of scattered light, we consider a plane wave incident in an nonmagnetic nonconducting nonabsorbing medium. Let the incident electric field be:

$$\mathbf{E}(\mathbf{r}, t) = \hat{\mathbf{p}}_0 E_0 \, \exp \mathrm{i}(\mathbf{q}_\mathrm{i} \cdot \mathbf{r} - \omega_\mathrm{i} t), \tag{10.1}$$

where $\hat{\mathbf{p}}_0$ is the (unitary) polarization vector, E_0 the field amplitude (so that the intensity of the incident light is $I_0 = |E_0|^2$), $\mathbf{q}_\mathrm{i}$ the wave vector of the incident beam and ω_i its frequency. The magnitude of $\mathbf{q}_\mathrm{i}$ is $q_\mathrm{i} = 2\pi n_0/\lambda_0 = q_0 n_0$, where n_0 the (average) refractive index of the medium and where λ_0 and q_0 are the wavelength and the wave number of the incident light in vacuo, respectively. By using standard electrodynamics (Berne and Pecora, 1976) it can be demonstrated that the spectrum of light $I_\mathrm{s}(\omega)$ scattered by a volume ΔV at a distance R depends on the scattering angle as:

$$I_\mathrm{s}(\omega) = \frac{I_0 n_0 q_\mathrm{i}^4}{4\pi^2 R^2} \frac{1}{\Delta\tau} \int_{-\Delta\tau/2}^{\Delta\tau/2} dt \int_{-\Delta\tau/2}^{\Delta\tau/2} dt' \int_{\Delta V} d^3\mathbf{r} \int_{\Delta V} d^3\mathbf{r}'$$
$$\times \langle \delta n(\mathbf{r}, t) \cdot \delta n(\mathbf{r}', t') \rangle \, e^{\, \mathrm{i}\,[\mathbf{q} \cdot (\mathbf{r} - \mathbf{r}') + \omega(t - t')]}, \tag{10.2}$$

where δn are the refractive-index fluctuations, so that n will be represented as $n(\mathbf{r}, t) = n_0 + \delta n(\mathbf{r}, t)$. In Eq. (10.2) $\Delta\tau$ represents the collecting time during which photons are counted for obtaining a spectrum; in practice it is large so that one can take the limit $\Delta\tau \to \infty$ as will be discussed below. The spectrum $I_\mathrm{s}(\omega)$ as given by Eq. (10.2) depends on the scattering angle through the scattering vector $\mathbf{q}$. The scattering vector is defined as the difference between the wave vector $\mathbf{q}_\mathrm{i}$ of the incident light and the wave vector of the scattered light $\mathbf{q}_\mathrm{s}$ in the medium (Berne and Pecora, 1976):

$$\mathbf{q} = \mathbf{q}_\mathrm{i} - \mathbf{q}_\mathrm{s}. \tag{10.3}$$

The wave vector $\mathbf{q}_\mathrm{s}$ of the scattered light is defined by the scattering experiment geometry. As an illustration we give in Fig. 10.1 a schematic representation of a typical light-scattering experiment in a fluid layer. As further discussed below, because of the long-ranged nature of the nonequilibrium fluctuations, one needs to perform the light-scattering experiments at small scattering angles. As shown in this figure, the scattering vector $\mathbf{q}_\mathrm{s}$ is pointed in the direction from the center of the scattering volume to the position of the light detector, and its magnitude q_s is the wave number of the scattered light. The change of wavelength in the scattering process is very small, so that $q_\mathrm{i} \simeq q_\mathrm{s}$. Then the magnitude q of the scattering vector is related to the incident wave number q_i and the scattering angle θ through the Bragg-Williams condition (Berne and Pecora, 1976):

$$q = 2q_\mathrm{i} \, \sin(\theta/2). \tag{10.4}$$

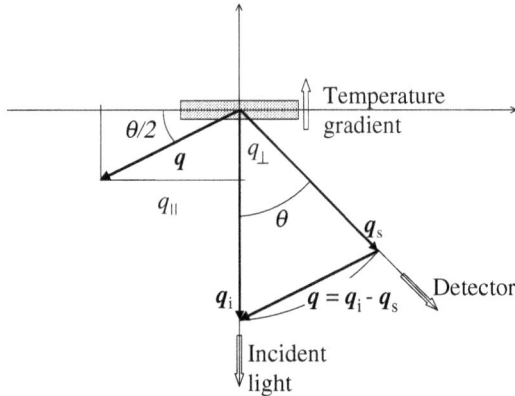

Figure 10.1: Schematic representation of a small-angle nonequilibrium light-scattering experiment. $\mathbf{q}_i$ is the wave vector of the incident light, $\mathbf{q}_s$ is the wave vector of the scattered light. The magnitude $q = |\mathbf{q}_i - \mathbf{q}_s|$ of the scattering wave vector is related to the scattering angle θ by Eq. (10.4). For clarity, the size of the scattering angle θ has been exaggerated.

The scattering angle θ is the angle between the wave vectors $\mathbf{q}_i$ and $\mathbf{q}_s$. In Eq. (10.2) ω represents the difference between the frequency of the scattered light ω_s and that of the incident light ω_i.

We do not give here a derivation of Eq. (10.2) based on Maxwell's equations; instead we refer the reader to excellent books where detailed accounts can be found (Berne and Pecora, 1976). We do mention that Eq. (10.2) is based on the assumption that only single-scattering events contribute to the total scattered light intensity, *i.e.*, multiple-scattering events have been neglected. The presence of q_i^4 as a prefactor in Eq. (10.2) is noteworthy: light with a shorter wavelength is scattered more strongly than light with a larger wavelength. Thus, Ar (green) lasers have an advantage over He-Ne (red) lasers as a light source, especially for one-component fluids where the intrinsic intensity of hydrodynamic fluctuations is small. It also explains why the sky is blue: when we look at the sky in a direction other than the sun, we are receiving light scattered by the atmosphere, which is enriched at shorter wavelengths in accordance with the q_i^4 prefactor in Eq. (10.2).

Equation (10.2) relates the spectrum of light scattered by a fluid to the autocorrelation function $\langle \delta n(\mathbf{r}, t) \cdot \delta n(\mathbf{r}', t') \rangle$ of the refractive-index fluctuations. This autocorrelation function can, in turn, be related to the hydrodynamic fluctuations through the use of an "equation of state" $n(\rho, T, c)$ see, *e.g.*, Eq. (10.13) below. In the remainder of this section we shall shall discuss the relationship between the scattering spectrum (10.2) and the various

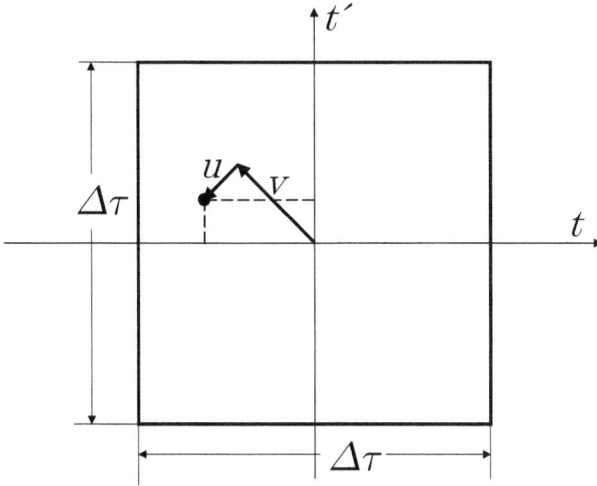

Figure 10.2: Derivation of Eqs. (10.7) and (10.20), see text.

structure factors introduced in previous chapters.

First of all, we note that one usually calculates from fluctuating hydrodynamics the double Fourier transform $\langle \delta n^*(\mathbf{q}, \omega) \cdot \delta n(\mathbf{q}', \omega') \rangle$ of the autocorrelation function for the index of refraction. In all cases studied in the previous chapters the two-frequencies autocorrelation function is proportional to a delta function due to time-translation symmetry; see Eqs. (3.23), (4.19), (4.38), (5.13) or (6.25). In general we may write:

$$\langle \delta n^*(\mathbf{r}, \omega) \cdot \delta n(\mathbf{r}', \omega') \rangle = F(\omega, \mathbf{r}, \mathbf{r}') \, 2\pi \, \delta(\omega - \omega'), \tag{10.5}$$

or

$$\langle \delta n(\mathbf{r}, t) \cdot \delta n(\mathbf{r}', t') \rangle = \frac{1}{2\pi} \int_{-\infty}^{\infty} F(\xi, \mathbf{r}, \mathbf{r}') \, e^{-i\xi(t - t')} \, d\xi, \tag{10.6}$$

where we prefer to designate the integration variable by ξ to avoid confusion in the subsequent development. Next, substituting Eq. (10.6) into Eq. (10.2), and performing the integrations over t and t', we obtain:

$$I_{\mathrm{s}}(\omega) = \frac{I_0 n_0 q_{\mathrm{i}}^4}{4\pi^2 R^2} \int_{\Delta V} d\mathbf{r} \int_{\Delta V} d\mathbf{r}' \, F(\omega, \mathbf{r}, \mathbf{r}') \, e^{\, i\mathbf{q} \cdot (\mathbf{r} - \mathbf{r}')}, \tag{10.7}$$

where we have assumed that the collecting time $\Delta \tau$ is large enough, so that:

$$\frac{1}{\Delta\tau} \iint_{-\Delta\tau/2}^{\Delta\tau/2} dt\, dt'\, e^{-\mathrm{i}t(\xi-\omega)}\, e^{\,\mathrm{i}t'(\xi-\omega)} \simeq 2\pi\,\delta(\xi-\omega). \tag{10.8}$$

This formula can be derived by performing a change of variables in the double integral over t and t' in Eq. (10.2) (see Fig. 10.2):

$$\begin{aligned} u &= \frac{1}{\sqrt{2}}(t'+t), \\ v &= \frac{1}{\sqrt{2}}(t'-t), \end{aligned} \tag{10.9}$$

which amounts to a $\pi/4$ rotation in the $\{t, t'\}$ plane. The original interval of integration then transforms into $v \in [-\Delta\tau/\sqrt{2}, \Delta\tau/\sqrt{2}]$ and, for a given value of v, $u \in [-(\Delta\tau/\sqrt{2}) + |v|, (\Delta\tau/\sqrt{2}) - |v|]$. The jacobian of the transformation (10.9) is unity. With this change of variables, integration over variable u in Eq. (10.8) is readily performed. If we denote by X the integral in (10.8), we obtain:

$$X(x, \Delta\tau) = \sqrt{2} \int_{-\frac{\Delta\tau}{\sqrt{2}}}^{\frac{\Delta\tau}{\sqrt{2}}} dv\, e^{\,\mathrm{i}\sqrt{2}\,vx} - \frac{2}{\Delta\tau} \int_{-\frac{\Delta\tau}{\sqrt{2}}}^{\frac{\Delta\tau}{\sqrt{2}}} dv\, |v|\, e^{\,\mathrm{i}\sqrt{2}\,vx}, \tag{10.10}$$

with $x = (\xi - \omega)$. The first of the integrals in (10.10) converges in the $\Delta\tau \to \infty$ limit to a delta function $2\pi\,\delta(x)$, while the second integral goes to zero in the same limit. The limit $\Delta\tau \to \infty$ is to be understood in the sense that:

$$\operatorname*{Lim}_{\Delta\tau\to\infty} \int_{-\infty}^{\infty} f(x)\, X(x, \Delta\tau)\, dx = 2\pi\, f(0), \tag{10.11}$$

for any function $f(x)$ for which the integral (10.11) converges. For instance, it is easy to verify that (10.11) is true for a test function $f(x) = 1/(x^2 + 1)$. That (10.11) holds for arbitrary $f(x)$ can be verified by representing $f(x)$ in terms of its Fourier transform, but we are not giving here further details.

Equation (10.7) shows that the spectrum of scattered light is proportional to the spectrum of the fluctuations. In particular, the total intensity of light scattered by the fluid will be obtained by integration over the frequency ω of the spectrum, as was done in deriving Eqs. (3.25) and (4.28).

An important modern technique is measuring light scattering with photon-correlation spectroscopy. A digital correlator counts repeatedly the number of photons registered by the detector over a sample time and computes the average over all sample times yielding a correlation function of the refractive-index fluctuations directly at times t and t'. The time-dependent correlation function only depends on the time difference $t - t'$, as shown by Eq. (10.6). Details on the photon-counting technique may be found in the books of Berne and Pecora (1976) and of Chu (1991), together with other topics like homodyne/heterodyne light scattering that we do not discuss here.

Note that Eq. (10.8) and, hence, Eq. (10.7), hold for frequencies $\omega \gg \Delta\tau^{-1}$, a condition always satisfied in practice, since the frequencies involved are of the order of GHz (see Fig. 3.1). In the case of photon-counting experiments, the time, over which measurements are accumulated for obtaining the time correlation function, needs to be much larger than the decay times of the exponentials appearing in the autocorrelation of the refractive-index fluctuations. This requires good laser stability.

It is evident from Eq. (10.7) that the scattering spectrum, or the time correlation function in photon-counting experiments, depends on the scattering volume ΔV and on the geometry of the experiment through the scattering vector $\mathbf{q}$. In the case of one-component fluids, the refractive-index fluctuations are related to the fluctuations of the density and the fluctuations of the temperature through the (local) "equation of state" $n = n(\rho, T)$. For usual fluids it turns out that:

$$\left(\frac{\partial n}{\partial \rho}\right)_T \gg \left(\frac{\partial n}{\partial T}\right)_\rho, \tag{10.12}$$

so that the autocorrelation function of the refractive-index fluctuations will be simply related to the autocorrelation function of the density fluctuations by:

$$\langle \delta n^*(\omega, \mathbf{q}) \cdot \delta n(\omega', \mathbf{q}') \rangle = \left(\frac{\partial n}{\partial \rho}\right)_T^2 \langle \delta\rho^*(\omega, \mathbf{q}) \cdot \delta\rho(\omega', \mathbf{q}') \rangle. \tag{10.13}$$

For a one-component fluid in equilibrium the double Fourier transform of the density autocorrelation function is proportional to a delta function $\delta(\mathbf{q} - \mathbf{q}')$: see Eq. (3.23). If one substitutes Eq. (3.23) into Eq. (10.2), performs the integrations over t and t', as in Eq. (10.7), and assumes that the scattering volume is large enough so that the 3-dimensional version of Eq. (10.8) applies to the spatial integrations, one finds:

$$\frac{I_s(\omega)}{\Delta V} = \frac{I_0 n_0 q_i^4}{4\pi^2 R^2} \left(\frac{\partial n}{\partial \rho}\right)_T^2 \rho m_0 \, S(\omega, \mathbf{q}). \tag{10.14}$$

Therefore, as anticipated in Chapter 3, the spectrum of light scattered by the fluid is indeed proportional to the dynamic structure factor $S(\omega, \mathbf{q})$ introduced in Eq. (3.23). The total intensity of light scattered by the fluid will be given by the integral over the frequency of the spectrum. Using the definition (3.24) of the static structure factor $S(\mathbf{q})$, we find:

$$\frac{I_s}{\Delta V} = \frac{I_0 n_0 q_i^4}{2\pi R^2} \left(\frac{\partial n}{\partial \rho}\right)_T^2 \rho m_0 \, S(\mathbf{q}). \tag{10.15}$$

In the Boussinesq approximation the density depends only on temperature in accordance with Eq. (4.3). In that case the theory yields the autocorrelation function of the temperature fluctuations, which is related to the structure factor by Eq. (4.19).

In discussing the nonequilibrium effects on the Brillouin doublet we found in Sect. 4.3 that the Fourier-transformed autocorrelation function of the density fluctuations is not simply proportional to a delta function $\delta(\mathbf{q} - \mathbf{q}')$, but that it has a more complicated structure, as given by Eq. (4.38). To demonstrate expression (4.41) for the dynamic structure factor let us consider first the term in (4.38) that is proportional to $\delta(\boldsymbol{\xi} - \boldsymbol{\xi}' - \mathbf{k})$, where we renamed the wave vector $\mathbf{q}$ of the fluctuations as $\boldsymbol{\xi}$ to avoid confusion in the following development. Applying a double inverse Fourier transform in $\boldsymbol{\xi}$ and $\boldsymbol{\xi}'$ to the second term in the RHS of Eq. (4.38), and integrating the resulting expression over $\boldsymbol{\xi}'$, we observe that the expression of the (real-space) density autocorrelation function will contain a nonequilibrium term given by:

$$\langle \delta\rho^*(\omega, \mathbf{r})\delta\rho(\omega', \mathbf{r}')\rangle_{\text{NE},1} = i\, \frac{\Delta T}{(2\pi)^3}\, 2\pi\, \delta(\omega - \omega')$$

$$\times \int_{\mathbb{R}^3} d\boldsymbol{\xi}\, N(\omega, \boldsymbol{\xi}, \boldsymbol{\xi} - \mathbf{k})\, e^{-i\boldsymbol{\xi}\cdot\mathbf{r}}\, e^{i(\boldsymbol{\xi} - \mathbf{k})\cdot\mathbf{r}'}, \quad (10.16)$$

where the function $N(\omega, \mathbf{q}, \mathbf{q}')$ was defined by Eq. (4.40). Next, substituting Eq. (10.16) into Eq. (10.7) and handling the time integrals as explained in the derivation of Eq. (10.8), we see that the light-scattering spectrum $I_s(\omega)/\Delta V$ will contain a nonequilibrium contribution proportional to the integral:

$$\left(\frac{I_s(\omega)}{\Delta V}\right)_{\text{NE},1} \propto i\, \frac{\Delta T}{(2\pi)^3} \int_{\mathbb{R}^3} d\boldsymbol{\xi}\, N(\omega, \boldsymbol{\xi}, \boldsymbol{\xi} - \mathbf{k})\, X(\boldsymbol{\xi}, \mathbf{k}, \mathbf{q}) \qquad (10.17)$$

with

$$X(\boldsymbol{\xi}, \mathbf{k}, \mathbf{q}) = \frac{1}{\Delta V} \iint_{\Delta V \times \Delta V} d\mathbf{r}\, d\mathbf{r}'\, e^{-i\boldsymbol{\xi}\cdot\mathbf{r}}\, e^{i(\boldsymbol{\xi} - \mathbf{k})\cdot\mathbf{r}'}\, e^{i\mathbf{q}\cdot(\mathbf{r} - \mathbf{r}')}. \quad (10.18)$$

Performing in the integral (10.18) a 3-dimensional version of the change of variables (10.9) (illustrated in Fig. 10.2), and integrating over a 3-dimensional variable $\mathbf{u}$ by assuming ΔV to be a cube, we can express (10.18) as:

$$X(\boldsymbol{\xi}, \mathbf{k}, \mathbf{q}) = \qquad\qquad\qquad\qquad\qquad\qquad\qquad\qquad (10.19)$$

$$\frac{1}{\Delta V} \int_{\sqrt{2}\Delta V} d\mathbf{v}\, e^{i\sqrt{2}\mathbf{v}\cdot(\boldsymbol{\xi} - \mathbf{q} - \frac{1}{2}\mathbf{k})} \prod_i \frac{2\sqrt{2}}{k_i}\, \sin\left[\frac{k_i}{2}(\Delta L - \sqrt{2}|v_i|)\right],$$

where the product is over the three spatial dimensions. The factor $\sqrt{2}\Delta V$ under the integral means that the three dimensions of the cube $\Delta V = \Delta L^3$

have to be scaled by a factor $\sqrt{2}$. Next, we recall from Sect. 4.3 that we are really interested only in the linear **k** terms. Thus, we expand the products of sines under the integral in powers of **k**, and then we reproduce in the RHS of Eq. (10.19) the 3-dimensional version of Eq. (10.10). Using exactly the same arguments as before, we find that for small **k**:

$$X(\boldsymbol{\xi}, \mathbf{k}, \mathbf{q}) \simeq (2\pi)^3 \, \delta(\boldsymbol{\xi} - \mathbf{q} - \tfrac{1}{2}\mathbf{k}). \tag{10.20}$$

Finally, substitution of Eq. (10.20) into Eq. (10.17) reproduces Eq. (4.41).

We conclude this section discussing in some detail the experimental arrangement for measuring the nonequilibrium dynamic structure Eq. (6.28) of fluid layers. (Law et al., 1990; Segrè et al., 1992; Li et al., 1994a, 1998; Vailati and Giglio, 1996, 1997b). The scattering medium in such experiments is a thin horizontal fluid layer bounded by two parallel plates whose temperatures can be controlled independently so as to establish a temperature gradient across the fluid layer. The temperature gradient can be parallel or antiparallel to the direction of gravity. The horizontal plates are furnished with windows allowing a laser beam to propagate through the fluid in the direction parallel to the gravity and to the temperature gradient. A schematic representation of such an experiment was shown in Fig. 10.1.

Combining Eq. (10.13) with the equation of state (4.3) in the Boussinesq approximation, we see that the autocorrelation of the refractive-index fluctuations will be proportional to the autocorrelation function of the temperature fluctuations. In the actual experiments the scattering angle is very small and one may assume that the entrance pinhole of the detector is indeed receiving light from all points in the beam trajectory across the medium. Thus the scattering volume will be a cylinder of length L with cross-section ΔS, the thickness of the laser beam[‡]. Substituting Eq. (6.25) for the autocorrelation function of the temperature fluctuations into Eq. (10.7), performing the spatial integral over the cross-section ΔS of the scattering volume by using the 2-dimensional version of Eq. (10.8), one may verify that the spectrum of light, $I_s(\omega)/\delta V$, scattered with scattering vector $\mathbf{q} = \{\mathbf{q}_\parallel, q_\perp\}$ is indeed proportional to $S(\omega, \mathbf{q})$, as defined by Eq. (6.28).

In addition, we note that in an actual light-scattering experiment the magnitude of the components $q_\parallel$ and $q_\perp$ of the scattering vector are not independent variables, since they are related to the scattering angle, θ, by:

$$\begin{aligned} q_\parallel &= q\cos(\theta/2) = 2q_i \sin(\theta/2)\cos(\theta/2), \\ q_\perp &= q\sin(\theta/2) = 2q_i \sin^2(\theta/2), \end{aligned} \tag{10.21}$$

see Fig. 10.1. Hence, for the interpretation of small-angle light-scattering experiments one may use in practice the approximation $q_\parallel \simeq q$, $q_\perp \simeq 0$, so that the nonequilibrium structure factor $S(q_\parallel, q_\perp)$, defined by Eq. (6.28),

---

[‡]Use of a gaussian profile for the laser beam will make no difference

depends only on the magnitude q of the scattering wave vector $\mathbf{q}$, which is the experimentally relevant quantity. This approximation has been amply used in this book.

From Fig. 7.2 we see that to measure the maximum enhancement of nonequilibrium fluctuations one would need scattering vectors whose magnitude is of the order $\tilde{q}_c$, which is of the order of the inverse thickness of the layer; this implies values of the scattering angle θ such that $Lq_0\theta \simeq \tilde{q}_c$. However, the vacuo wave number of incident light is typically $k_0 \simeq 10^5$ cm^{-1}, so that for reasonable layer thicknesses it is actually impossible to probe fluctuations with wave numbers $\tilde{q}$ that are of the order of $\tilde{q}_c$ by light scattering. The reason is that the presence of stray light from optical surfaces determines a minimum angle below which it is impossible to perform measurements because of very poor signal to noise ratio. However, it is possible to the nonequilibrium fluctuations accurately at larger wave numbers and to verify the q^{-4} dependence of the dynamic structure factor in the limit $\tilde{q} \to \infty$ predicted by Eqs. (7.12) and (7.27). This $\tilde{q} \to \infty$ limit refers to large q values but always within the hydrodynamic range, *i.e.*, corresponding to wavelengths much larger that the molecular size. In this book we always deal with structure factors in the hydrodynamic range, which for molecular fluids correspond to the range observed by light-scattering or shadowgraph techniques. For researchers working in neutron or X-ray scattering our $q \to \infty$ limit corresponds to their $q \to 0$ limit.

Before closing this section, we remark that the spectrum of scattered light is proportional to $S(\omega, \mathbf{q})$, as given by Eq. (6.28), only for $q \gg L_s^{-1}$, where L_s is the thickness of the scattering volume, or, in the geometry we are considering, the thickness of the laser beam, see Eq. (10.10). Otherwise, the approximations leading to (10.8) are not completely justified. As a consequence, for extremely small q, effects related to the small thickness of the beam, not discussed here, may appear and the application of light-scattering becomes difficult at such extremely small scattering angles. Fortunately, there is another experimental technique, namely, quantitative shadowgraphy to be discussed below, that is more suitable for probing fluctuations at extremely small wave numbers q.

10.1.2   Quantitative shadowgraph analysis

Quantitative shadowgraph analysis is a promising alternative experimental technique for measuring the intensity of nonequilibrium fluctuations (Wu et al., 1995; de Bruyn et al., 1996; Bodenschatz et al., 2000; Vailati and Giglio, 1997a; Brogioli et al., 2000b). Instead of a laser beam, an extended uniform monochromatic light source is employed to illuminate the fluid layer. Then many shadowgraph images of a plane perpendicular to the temperature gradient are obtained with a CCD detector, which registers the spatial distribution of the intensity $I(\mathbf{x}_\parallel)$, where $\mathbf{x}_\parallel$ is a two-dimensional

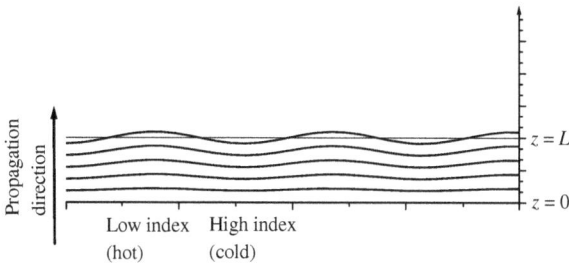

Figure 10.3: Surfaces of constant phase of a plane light wave propagating through the shadowgraph medium, adapted from Trainoff (1999).

position vector in the imaging plane.

Recently, Trainoff and Cannell (2002) have presented a detailed theoretical analysis of the quantitative shadowgraph method based on physical optics. They studied not only shadowgraph images produced by deterministic patterns above the convective threshold, but also shadowgraph images produced by fluctuations below the onset of convection. For fluctuations below the threshold, Trainoff and Cannell (2002) assumed a sinusoidal dependence of the index-of-refraction fluctuations on the vertical z-coordinate. To generalize their result and to make contact with the static structure factor observed by light scattering as given by Eq. (6.31), we present here a derivation of the shadowgraph intensity $I(\mathbf{x}_\parallel)$ produced by arbitrary nonequilibrium fluctuations in fluids in quiescent states below the convective threshold.

For this purpose, we consider a plane wave that propagates perpendicularly to the horizontal xy-plane. Inside the fluid medium, the imposed temperature gradient ∇T_0 induces an index-of-refraction gradient ∇n_0 in the z-direction. In addition, there are small fluctuations in the index of refraction associated with the temperature fluctuations. Hence, the local index of refraction of the medium can be represented by:

$$n(\mathbf{r}, t) = n_0 + \nabla n_0 z + \left(\frac{\partial n}{\partial T} \right)_P \delta T(\mathbf{x}_\parallel, z, t), \qquad (10.22)$$

where we use the subscript $\parallel$ to emphasize that the corresponding vectors are two dimensional in the horizontal xy-plane. Note that we are assuming a completely arbitrary dependence of the fluctuations on the spatial coordinates x and y, as well as z, only limited by the BC of vanishing temperature fluctuations at $z = 0$ and $z = L$. For the propagation of light inside the fluid we use the paraxial approximation; that is, we neglect beam bending and assume that light rays inside the fluid are parallel to the vertical z-direction (Trainoff and Cannell, 2002; Trainoff, 1999). We also neglect any possible dispersion effects of the medium on the propagating light. If we arbitrarily take the phase of the electromagnetic wave at the entrance plane as zero, the phase accumulated at the exit plane $\Phi(\mathbf{x}_\parallel, t)$ will be determined by (see Fig. 10.3):

$$
\begin{aligned}
\Phi(\mathbf{x}_\parallel, t) &= q_0 \int_0^L n(\mathbf{x}_\parallel, z, t)\, dz \\
&= \left(q_0 n_0 L + \tfrac{1}{2} q_0 \nabla n_0 L^2\right) + k_0 \left(\frac{\partial n}{\partial T}\right)_P \int_0^L \delta T(\mathbf{x}_\parallel, z, t)\, dz \\
&= \Phi_0 + \delta\Phi(\mathbf{x}_\parallel, t).
\end{aligned}
\tag{10.23}
$$

where, since the fluctuations are small, $\delta\Phi \ll \Phi_0$. We note that Trainoff and Cannell (2002) adopted for the propagation of light inside the shadowgraph medium an eikonal approximation, which is more complete, since it includes beam bending and incorporates not only a phase modulation but also an amplitude modulation for the exit field. However, for small fluctuations below threshold, as also mentioned by Trainoff and Cannell (2002), one may to first order consider the phase change only and ignore amplitude effects. In this limit, the eikonal approximation is equivalent to the paraxial approximation and Eq. (10.23) is just the generalization for arbitrary z-dependence of Eq. (52) in the paper by Trainoff and Cannell (2002). From Eq. (10.23), it follows that the complex amplitude $E(\mathbf{r}, t)$ of the electromagnetic field at the exit plane, $z = L$, is given by:

$$
E(\mathbf{x}_\parallel, L, t) = E_0 \exp\left\{ \mathrm{i}[\Phi_0 + \delta\Phi(\mathbf{x}_\parallel, t)] \right\},
\tag{10.24}
$$

where E_0 is the amplitude at the entrance of the cell. To propagate the exit field from the $z = L$ plane up to an imaging plane located at an arbitrary value of $z > L$, we use the Fresnel approximation (Born and Wolf, 1989), so that:

$$
\begin{aligned}
&E(\mathbf{x}_\parallel, z, t) \\
&= \mathrm{i}\frac{e^{-\mathrm{i}q_0 z}}{\lambda_\mathrm{a} z} \int_A d^2 \mathbf{x}_{\parallel 1} E_0\, e^{\mathrm{i}[\Phi_0 + \delta\Phi(\mathbf{x}_{\parallel 1}, t)]}\, e^{\frac{-\mathrm{i}q_0}{2z}(\mathbf{x}_\parallel - \mathbf{x}_{\parallel 1})^2},
\end{aligned}
\tag{10.25}
$$

where λ_a is the wavelength of the light in the air behind the cell. We shall assume that its corresponding wave number $q_0 = 2\pi/\lambda_\mathrm{a}$ can be identified with the vacuum wave number appearing in Eq. (10.23). In Eq. (10.25), the symbol A under the integral sign indicates that the two-dimensional integral extends over the finite optical aperture of the system, which has a surface area A. We note that, for the particular case where $\delta\Phi(\mathbf{x}_{\parallel 1}, t)$ is a periodic function of $\mathbf{x}_{\parallel 1}$, this problem has been studied recently by Berry and Bodenschatz (1999), who have demonstrated that an initial periodic phase grating is multiply reconstructed by Talbot interference. Next, we use the fact that the fluctuations are small, so that we may expand Eq. (10.25) up to linear order in the fluctuations:

$$E(\mathbf{x}_\parallel, z, t) \simeq iE_0 \frac{e^{-i(q_0 z - \Phi_0)}}{\lambda_a z} \int_A d^2 \mathbf{x}_{\parallel 1} \left[1 + i\, \delta\Phi(\mathbf{x}_{\parallel 1}, t) \right] e^{\frac{-iq_0}{2z}(\mathbf{x}_\parallel - \mathbf{x}_{\parallel 1})^2}$$

$$= E_0 e^{-i(q_0 z - \Phi_0)} \left\{ 1 - \frac{q_0}{2\pi z} \int_A d^2 \mathbf{x}_{\parallel 1}\, \delta\Phi(\mathbf{x}_{\parallel 1}, t) e^{\frac{-iq_0}{2z}(\mathbf{x}_\parallel - \mathbf{x}_{\parallel 1})^2} \right\},$$

$$(10.26)$$

where we have assumed that the optical aperture of the system is large enough so that the integral of the nonfluctuating part can be extended over the entire horizontal xy-plane ($\mathbb{R}^2$); this is equivalent to neglecting diffraction effects at the edge of the optical aperture. To obtain the intensity in the imaging plane, we evaluate the modulus squared of the complex amplitude, so that:

$$I(\mathbf{x}_\parallel) = \qquad\qquad\qquad\qquad\qquad\qquad\qquad\qquad\qquad (10.27)$$

$$I_0 \left\{ 1 - \frac{q_0}{2\pi z} \frac{1}{\Delta t} \int_0^{\Delta t} dt \int_A d^2 \mathbf{x}_{\parallel 1}\, \delta\Phi(\mathbf{x}_{\parallel 1}, t)\, e^{\frac{-iq_0}{2z}(\mathbf{x}_\parallel - \mathbf{x}_{\parallel 1})^2} - \text{c.c.} \right\},$$

where $I_0 = |E_0|^2$ is the intensity of the incident light and Δt is the response time of the detector, which is the time that the device is counting photons in a single measurement. The term c.c. in Eq. (10.27) represents the complex conjugate of the integral. Terms quadratic in the fluctuations have been neglected. Equation (10.27) represents the intensity actually measured by the CCD detector in a single shadowgraph picture. Since $\langle \delta\Phi \rangle = 0$, the coefficient I_0 can be obtained by averaging the observed $\mathcal{I}(\mathbf{x}_\parallel)$ over many shadowgraph pictures. From Eq. (10.27), it follows that the shadowgraph signal $\mathcal{I}(\mathbf{x}_\parallel)$ is given by:

$$\mathcal{I}(\mathbf{x}_\parallel) = \frac{I(\mathbf{x}_\parallel) - I_0}{I_0} \qquad\qquad\qquad\qquad\qquad\qquad (10.28)$$

$$= -\frac{q_0}{2\pi z} \frac{2}{\Delta t} \int_0^{\Delta t} dt \int_A d^2 \mathbf{x}_{\parallel 1} \delta\Phi(\mathbf{x}_{\parallel 1}, t)\, \cos\left[\frac{q_0}{2z}(\mathbf{x}_\parallel - \mathbf{x}_{\parallel 1})^2 \right],$$

where we have used that $\delta\Phi(\mathbf{x}_{\parallel 1}, t)$ is a real quantity. In quantitative shadowgraph analysis of fluctuating patterns one determines the spatial power spectrum of $\mathcal{I}(\mathbf{x}_\parallel)$ averaged over fluctuations (Wu et al., 1995; Brogioli et al., 2000b), which can be expressed as:

$$\langle |\mathcal{I}(\mathbf{q}_\parallel)|^2 \rangle = \int d^2 \mathbf{x}_\parallel \int d^2 \mathbf{x}_\parallel'\, \langle \mathcal{I}^*(\mathbf{x}_\parallel) \cdot \mathcal{I}(\mathbf{x}_\parallel') \rangle\, \exp\left[i \mathbf{q}_\parallel (\mathbf{x}_\parallel - \mathbf{x}_\parallel') \right], \quad (10.29)$$

where the integrals are extended over $\mathbb{R}^2$. We note that, with the definition (10.29), the spatial power spectrum $\langle |\mathcal{I}(\mathbf{q}_\parallel)|^2 \rangle$ has dimensions of length to the fourth power. With the aid of Eq. (10.28), $\langle|\mathcal{I}(\mathbf{q}_\|)|^2\rangle$ can be related to the autocorrelation function $\langle\delta\Phi^*(\mathbf{x}_{\|1},t)\cdot\delta\Phi(\mathbf{x}'_{\|1},t')\rangle$ of the phase accumulutated by the light as it crosses the medium, which in turn can be related to the structure factor of the fluid, $S(\omega,q_\|,z,z')$, as defined by Eq. (6.25). Fourier transforming Eq. (10.23) for $\delta\Phi(\mathbf{x}_{\|1},t)$ in the horizontal xy-plane and in the time t and using Eq. (6.25), we obtain:

$$\langle\delta\Phi^*(\mathbf{x}_{\|1},t)\cdot\delta\Phi(\mathbf{x}'_{\|1},t')\rangle = \frac{1}{(2\pi)^3}\left(\frac{\partial n}{\partial T}\right)_P^2 \frac{m_0 q_0^2}{\rho\alpha_p^2}\int_0^L dz \int_0^L dz' S(\omega,q_\|,z,z')$$

$$\times \int d^2\mathbf{q}_\| \, e^{-i\mathbf{q}_\|(\mathbf{x}_{\|1}-\mathbf{x}'_{\|1})}\int_{-\infty}^{\infty} d\omega \, e^{-i\omega(t-t')}, \quad (10.30)$$

where the integral in the two-dimensional Fourier variable $\mathbf{q}_\|$ is again over the entire $\mathbb{R}^2$ plane. Integration over t and t', as required for the calculation of the spatial autocorrelation of the shadowgraph signal, yields:

$$\frac{1}{\Delta t^2}\int_0^{\Delta t} dt \int_0^{\Delta t} dt' \langle\delta\Phi^*(\mathbf{x}_{\|1},t)\cdot\delta\Phi(\mathbf{x}'_{\|1},t')\rangle$$

$$\simeq \frac{1}{(2\pi)^2}\left(\frac{\partial n}{\partial T}\right)_P^2 \frac{L m_0 q_0^2}{\rho\,\alpha_p^2}\int d^2\mathbf{q}_\| \, S(q_\|,0)\, e^{-i\mathbf{q}_\|(\mathbf{x}_{\|1}-\mathbf{x}'_{\|1})}, \quad (10.31)$$

where we have assumed that the response time Δt of the detector is very short compared to the characteristic evolution time of the hydrodynamic fluctuations, so that, after integrating over t and t', we can take $\Delta t \simeq 0$. In deriving Eq. (10.31), we have also used the definition of the static structure factor, Eq. (6.29), and the definition, Eq. (6.31), of the quantity $S(q_\|,q_\perp)$ probed in light-scattering experiments. We note that the validity of Eq. (10.31) is not affected by the geometry of any scattering volume, as is the case in small-angle light-scattering experiments. Combining Eqs. (10.28) and (10.31), we obtain:

$$\langle\mathcal{I}^*(\mathbf{x}_\|)\cdot\mathcal{I}(\mathbf{x}'_\|)\rangle = \left(\frac{q_0^2}{\alpha_p\pi z}\right)^2\frac{m_0 L}{\rho(2\pi)^2}\left(\frac{\partial n}{\partial T}\right)_P^2\int d^2\mathbf{q}'_\| \, S(q'_\|,0) \quad (10.32)$$

$$\times\int_A d^2\mathbf{x}_{\|1}\int_A d^2\mathbf{x}'_{\|1} e^{-i\mathbf{q}'_\|(\mathbf{x}_{\|1}-\mathbf{x}'_{\|1})}\cos\left[\frac{q_0}{2z}(\mathbf{x}_\|-\mathbf{x}_{\|1})^2\right]\cos\left[\frac{q_0}{2z}(\mathbf{x}'_\|-\mathbf{x}'_{\|1})^2\right],$$

where we have designated the Fourier variable as $\mathbf{q}'_\|$, to avoid confusion in the subsequent development. Upon substituting Eq. (10.32) into Eq. (10.29), we obtain a complicated expression for $\langle|\mathcal{I}(\mathbf{q}_\|)|^2\rangle$, with several integrations to be performed. We first integrate over the variables $\mathbf{x}_\|$ and $\mathbf{x}'_\|$, whose domain is $\mathbb{R}^2$. These two integrals can be performed exactly yielding:

$$\langle |\mathcal{I}(\mathbf{q}_\parallel)|^2 \rangle = \frac{L m_0 q_0^2}{\rho \alpha_p^2 \pi^2} \left(\frac{\partial n}{\partial T} \right)_P^2 \sin^2 \left(\frac{q_\parallel^2 z}{2 q_0} \right) \int d^2 \mathbf{q}_\parallel' \; S(q_\parallel', 0)$$

$$\times \int_A d^2 \mathbf{x}_{\parallel 1} \; e^{i \mathbf{x}_{\parallel 1} \cdot (\mathbf{q}_\parallel - \mathbf{q}_\parallel')} \int_A d^2 \mathbf{x}_{\parallel 1}' \; e^{-i \mathbf{x}_{\parallel 1}' \cdot (\mathbf{q}_\parallel - \mathbf{q}_\parallel')}. \quad (10.33)$$

Next, we perform the integral over $\mathbf{x}_{\parallel 1}'$, assuming that the optical aperture of the system is large enough so that the integral can be extended from A to the entire $\mathbb{R}^2$ plane, so as to obtain $(2\pi)^2 \; \delta(\mathbf{q}_\parallel - \mathbf{q}_\parallel')$. Then, we perform the integral over $\mathbf{q}_\parallel'$ and subsequently the integral over $\mathbf{x}_{\parallel 1}$, to arrive at the final result (Brogioli et al., 2000b):

$$\langle |\mathcal{I}(\mathbf{q}_\parallel)|^2 \rangle = \frac{4 m_0 V q_0^2}{\rho \; \alpha_p^2} \left(\frac{\partial n}{\partial T} \right)_P^2 \; \sin^2 \left(\frac{q_\parallel^2 z}{2 q_0} \right) \; S(q_\parallel, 0), \quad (10.34)$$

where $V = LA$ is the sample volume illuminated by the light. In Eq. (10.34), the sine term plays the role of optical transfer function (Brogioli et al., 2000b); its periodicity shows the phenomenon of Talbot interference. Equation (10.34) is the generalization of Eq. (56) in the work of Trainoff and Cannell (2002) for an arbitrary z-dependence of the index-of-refraction fluctuations. We again use the notation $S(q_\parallel, q_\perp)$ for the structure factor in accordance with Eq. (6.28) as previously adopted for the analysis of light-scattering experiments. Trainoff and Cannell (2002) have also evaluated small modifications to Eq. (10.34) due to experimental effects such as inhomogeneities in the illumination, angular spread in the incident beam or finite spectral bandwidth of the light source.

Equation (10.34) has been derived under the assumption of homogeneous illumination with a plane wave front completely parallel to the (transparent) fluid walls. In practice there are likely to be some small optical misalignments so that it is more convenient to determine the shadowgraph transfer function experimentally by replacing the nonequilibrium fluid by a uniformly scattering media or by other tricks. We note that Eq. (10.34) is rotationally invariant, since it depends only on the modulus of the vector $\mathbf{q}_\parallel$. Hence, from a practical point of view it is advantageous to perform azimuthal averages of the measurements, so as to cancel errors resulting from inhomogeneities. For these reasons Eq. (10.34) is usually represented by (Brogioli et al., 2000b; Wu et al., 1995; de Bruyn et al., 1996):

$$\overline{\langle |\mathcal{I}(\mathbf{q}_\parallel)|^2 \rangle} = \check{T}(q_\parallel) \; S(q_\parallel, 0). \quad (10.35)$$

The azimuthal average of $\langle |\mathcal{I}(\mathbf{q})|^2 \rangle$ is indicated by the overline. The symbol $\check{T}(q_\parallel)$ represents an optical transfer function, which now is a quantity to be determined experimentally (Brogioli et al., 2000b; de Bruyn et al.,

1996). Anyway, by applying a two-dimensional Fourier transformation to the shadowgraph signal (de Bruyn et al., 1996; Bodenschatz et al., 2000; Brogioli et al., 2000b), one can determine the structure factor of the fluid as a function of the wave number q at $q_\perp = 0$. Hence, both small-angle light scattering and shadowgraphy yield the structure factor $S(q_\parallel = q, q_\perp = 0)$. For light scattering, q is the scattering wave number as given by Eq. (10.4), whereas for shadowgraphy q is the modulus of the two-dimensional Fourier vector in the imaging plane.

10.1.3   Dependence of shadowgraph images on collecting times

An important experimental aspect of the shadowgraph technique, which has not been analyzed by Trainoff and Cannell (2002), is the effect of the camera-exposure time Δt, to be referred to as the collecting time. In deriving Eq. (10.31) we assumed that the collecting time Δt was very short compared to the evolution time of the hydrodynamic fluctuations. However, as discussed in Sect. 8.1, close to a nonequilibrium instability, a slowing down in the evolution of thermal fluctuations becomes important, so that the assumption $\Delta t \simeq 0$ may no longer be justified. It is easy to incorporate the effects of a nonzero Δt to the shadowgraph theory developed in the previous section. Instead of assuming $\Delta t \simeq 0$ in Eq. (10.31), we retain the dependence of the integral on Δt. Then, our final result for the shadowgraph signal, Eq. (10.35), instead of being proportional to $S(q_\parallel, 0)$, will be proportional to the integral:

$$\overline{\langle |\mathcal{I}(\mathbf{q})|^2 \rangle} = \check{T}(q)\, \frac{1}{\Delta t^2} \int_0^{\Delta t} dt \int_0^{\Delta t} dt'\, S(q, |t - t'|), \tag{10.36}$$

where $S(q, \tau)$ is the same Fourier transform of the dynamic structure factor discussed in Eq. (6.28), but in the small-angle approximation ($q_\parallel = q$, $q_\perp = 0$). Indeed, applying double Fourier transformations in frequency to Eq. (6.26) and using Eqs. (10.22) and (10.23), it may be shown that $S(q, q_\perp \simeq 0, |t - t'|)$ will be proportional to the autocorrelation function $\langle \delta\Phi(t) \cdot \delta\Phi(t') \rangle$ of the phase accumulated by the plane wave as it crosses the shadowgraph medium.

As was discussed in Sect. 8.1, for positive Ra but below Ra_c the time correlation function $S(q, \tau)$ in the small-angle approximation can be expressed as a series of exponentials. Then the influence of collecting times on the shadowgraph signal $\overline{\langle |\mathcal{I}(\mathbf{q})|^2 \rangle}$ can be in principle evaluated by substituting Eq. (8.2) into Eq. (10.36). Shadowgraph experiments require significant thermal-noise levels, so that in practice they are performed close to the convective instability. In the vicinity of an instability we can adopt a most-unstable-mode approximation for the structure factor, as discussed in

Sect. 8.2. We then obtain:

$$
\overline{\langle |\mathcal{I}|^2\rangle} = \breve{T}\,\frac{1}{\Delta t^2}\int_0^{\Delta t} dt \int_0^{\Delta t} dt' \left\{ A_1^{(-)}\,e^{-\Gamma_1^{(-)}\,|t-t'|} + A_1^{(+)}\,e^{-\Gamma_1^{(+)}\,|t-t'|}\right\}
$$

$$
= \breve{T}\left\{ \frac{2A_1^{(-)}}{\Delta t^2}\,\frac{\Delta t\,\Gamma_1^{(-)} - 1 + e^{-\Gamma_1^{(-)}\Delta t}}{[\Gamma_1^{(-)}]^2} \right.
$$

$$
\left. + \frac{2A_1^{(+)}}{\Delta t^2}\,\frac{\Delta t\,\Gamma_1^{(+)} - 1 + e^{-\Gamma_1^{(+)}\Delta t}}{[\Gamma_1^{(+)}]^2} \right\}. \tag{10.37}
$$

In the limit $\Delta t \to 0$, Eq. (10.37) reduces to Eq. (10.35), as it should.

An analysis of Eq. (10.37) reveals that the effect of the collecting time on the shadowgraph signal does become important. Defining $S(q, \Delta t)$ as the quantity between brackets in Eq. (10.37), and comparing with the result (10.35) for zero collecting time, one can show (Oh et al., 2004) that when $\Delta t \neq 0$ the height of the maximum is smaller, while the position of the maximum is displaced towards slightly larger values of $\tilde{q}_{\mathrm{m}}$. Most notably, taking into account that in the large q-limit $\Gamma_N^{(\pm)}(q) \propto q^2$, while $A_N^{(\pm)}(q) \propto q^{-4}$ (see Sect. 6.3.2), one finds that $S(q, \Delta t)$ for a nonzero collecting time Δt becomes proportional to q^{-6} in the large-q limit, in contrast to the q^{-4} proportionality for zero collecting time (Oh et al., 2004).

10.2 Nonequilibrium light-scattering experiments in one-component fluids

The effect of the presence of a temperature gradient is expected to cause a wave-number-dependent enhancement of the Rayleigh component in the light-scattering spectrum of a fluid, described in Sect. 4.2, and the appearance of an asymmetry in the intensity of the two Brillouin components, described in Sect. 4.3. In this chapter we review the experimental attempts that have been made to investigate thermal nonequilibrium fluctuations in one-component fluids.

10.2.1 Propagating modes

As mentioned in Chapter 6, when a one-component fluid is driven out of thermal equilibrium, propagating thermal modes are expected to appear at large negative values of the Rayleigh number, *c.f.*, condition (6.33) for a fluid between two free boundaries. Such propagating modes have been observed by Boon et al. (1979) and by Lallemand and Allain (1980) from forced-Rayleigh-scattering experiments in acetone at $Ra \simeq -69\,000$. Although the existence of propagating modes was verified experimentally, quantitative agreement with the theory was not achieved, in part because of problems with beam bending inside the measurement cell caused by the refractive-index gradient and in part because of diffraction effects on the detection beam. An additional difficulty is that the experiments must be done at small values of the scattering angle q because, as shown in Fig. 6.1 propagating modes (*i.e.*, nonzero imaginary part of the decay rates) only appear at moderately small values of the wave number, since for large q the decay rates always reach the (real) asymptotic limits $a_T q^2$ or νq^2.

10.2.2 Brillouin doublet

That the imposition of a temperature gradient should cause an asymmetry in the intensity of the Brillouin components has been expected for a long time and detection of such an asymmetry was the goal of early attempts to observe the scattering of light in a nonuniformly heated body. For instance, Landsberg and Shubin (1939) investigated the scattering in a nonuniformly heated quartz single crystal illuminated with white light, and Chistyi (1977) used a yttrium aluminium garnet in the presence of a steady temperature gradient. None of these earlier attempts found any deviation of the intensity distribution from that for a uniformly heated body. This is probably due to the fact that, before the invention of the laser, experiments lacked the required accuracy to resolve this phenomenon.

For nonequilibrium light-scattering experiments to be successful, it is desirable to use a fluid layer with a small height L, so at to avoid any convective disturbances due to residual temperature inhomogeneities in the horizontal direction However, the size of the mean free path of phonons is significant and for small thickness layers, the Brillouin scattering may be affected by phonons bouncing from the hot wall versus those bouncing from the cold wall. On the other hand, since the asymmetry, Eq. (4.45), is inversely proportional to q^2, small scattering angles are again required in the experiments. These two facts further complicate the experimental verification of the asymmetry of the Brillouin spectrum (4.45).

An asymmetry in the Brillouin peaks has been observed by Beysens et al. (1980) who used an Ar laser as light source and a Fabry-Pérot spectrometer to analyze light scattered by water at 313 K. The experiments were performed with a $\mathbf{q} \parallel \nabla T_0$ geometry, for which the laser beam propagates perpendicularly to the temperature gradient to measure the scattering at small scattering angles. A problem with such an arrangement is beam bending causing uncertainties in the value of the scattering wave number q. The experimental results of Beysens et al. (1980) indicated that the asymmetry $\epsilon_B(\mathbf{q})$ exists, and it is indeed proportional to $\nabla T_0/q^2$ as predicted by Eq. (4.45). However, the results were quantitatively off by a factor of 3, which the authors attributed to the presence of beam bending and to interaction of the phonons with the container walls (Beysens, 1983).

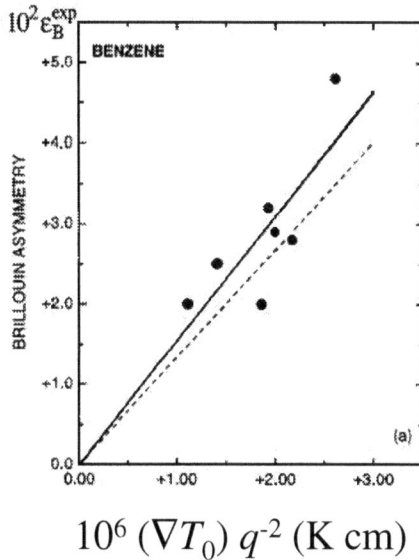

$$10^6 \, (\nabla T_0) \, q^{-2} \, (\text{K cm})$$

Figure 10.4: Intensity asymmetry ϵ_B of Brillouin scattering for benzene at 320 K as a function of $\nabla T_0/q^2$. Symbols represent experimental data. The solid line is a linear fit to the experimental data, while the dotted line is predicted theoretically from Eq. (4.45). Adapted from Suave and de Castro (1996).

Subsequent experimental evidence for an asymmetry in Brillouin scattering induced by a stationary temperature gradient has been reported by Kiefte et al. (1984) from light-scattering experiments in liquid oxygen at low temperatures, water at 300 K and fused quartz at 315 K. Again a quantitative verification of Eq. (4.45) was hampered by finite-size effects related to sound reflections at the walls of the container. The most accurate nonequilibrium Brillouin-scattering measurements have so far been obtained by Suave and de Castro (1996) with benzene and isopropyl alcohol. The scattering cell of Suave and de Castro (1996) was larger so that the effects of sound reflection at the walls are smaller than in the previous studies (Beysens et al., 1980; Kiefte et al., 1984). As an example of their experimental results, we show in Fig. 10.4 the asymmetry ϵ_B in the intensity of the Brillouin components as a function of $\nabla T_0/q^2$ for benzene at an average temperature of 320 K. The experimental data were obtained at q values ranging from 4000-6300 cm^{-1} at various stationary temperature gradients. The solid line represents a linear fit to the experimental data, while the dotted line represents the prediction from Eq. (4.45) and literature values for the sound velocity c and the sound-attenuation coefficient D_V of benzene. Although the experimental slope is 15% larger than the predicted theoretically, the difference can be easily explained by uncertainties in the experimental q related to beam bending.

After a first unsuccessful attempt (Hattori et al., 1996), Hattori et al. (1998) have also obtained quantitative experimental evidence for the asymmetry in the Brillouin doublet using a different technique, namely, optical beating spectroscopy. The liquids employed were ethanol and water, again at room temperature. The authors found that, to explain their experimental results, non-uniformity in the temperature gradient had to be taken into account in the theory.

10.2.3   Rayleigh scattering

Since the nonequilibrium enhancement in the intensity of the Rayleigh line is predicted to be proportional to q^{-4} in accordance with Eq. 4.27, one needs not only small scattering angles, but the angles should be measured with a high accuracy. Hence, possible beam-bending effects on the measurement of the scattering angle become even more serious for Rayleigh scattering than for Brillouin scattering. Fortunately, since the nonequilbrium enhancement has a maximum when $\nabla T_0 \perp \mathbf{q}$, it is possible to avoid any beam bending by letting the laser beam propagate in the direction parallel or antiparallel to the temperature gradient. This is the optimal arrangement schematically shown in Fig. 10.1.

The first experimental attempt to determine the effect of a temperature gradient on Rayleigh scattering in a nonequilibrium fluid was reported by Wegdam et al. (1985). In their experiments, an inhomogeneous temperature distribution was created in carbon tetrachloride by generating a interference pattern inside the measurement cell with a powerful (infrared) CO_2 laser. An additional Ar laser was used for the light-scattering measurements. In these experiments the temperature gradient is not uniform, so that it is dubious whether the results of Sect. 4.2 apply quantitatively. The experimental setup deviates from the optimal configuration of Fig. 10.1. Wegdam et al. (1985) did observe an enhancement of the Rayleigh line proportional to q^{-4}, but they found a mysterious fast component in qualitative disagreement with the theoretical prediction. They attributed this fast component erroneously to viscous fluctuations, but it was probably caused by the presence of residual dust particles in the liquid

A more detailed nonequilibrium Rayleigh-scattering experiment was performed by Law et al. (1988) in pure toluene at an average temperature of 40°C. The experimental cell corresponded to a typical Rayleigh-Bénard design, where a temperature gradient was imposed by maintaining the two horizontal walls at different temperatures. An Ar laser was employed as the light source, and photon-counting spectroscopy was used to analyze the

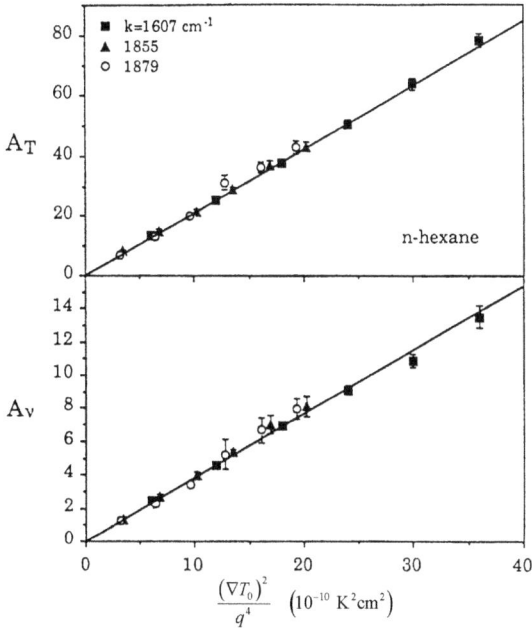

Figure 10.5: Enhancement of nonequilibrium thermal A_T and viscous A_ν fluctuations in liquid n-hexane at $25°C$ as a function of $(\nabla T_0)^2/q^4$. Symbols indicate experimental data and the solid lines the values predicted theoretically from Eq. (4.27). Adapted from Li et al. (1994b).

scattered light. Furthermore, cross correlation was employed, in order to obtain a noiseless time correlation function up to times of the order of the decay time of viscous fluctuations: $1/\nu q^2$. Beam bending was avoided in the experimental design of Law et al. (1988) with the incident laser beam propagating vertically, parallel to the temperature gradient, through windows in the top and bottom plates, corresponding to the optimal arrangement shown in 10.1. As mentioned earlier, measurement of the scattering intensity at small angles then corresponds to the $\mathbf{q} \perp \nabla T_0$ configuration, for which the nonequilibrium effects in the Rayleigh line are largest.

With this experimental setup Law et al. (1988) were able to verify that the nonequilibrium time-dependent correlation function consists of two exponentials with decay times $a_T q^2$ and νq^2, in accordance with Eq. (4.26). The amplitude of the component with decay rate $a_T q^2$ exhibits an important nonequilibrium enhancement, proportional to $(\nabla T_0)^2/q^4$ in accordance to Eq. (4.27), while the amplitude of the mode with decay rate νq^2 is negative, causing a reduction of the time correlation function for very small times (of the order of μs). However, limited accuracy in the knowledge of the temperatures at the boundaries of the cell wall and, hence, of the effective temperature gradient in the liquid inside the optical cell, precluded a complete quantitative agreement between the experimental enhancement and Eq. (4.27).

With an improved optical-cell arrangement Law et al. (1990) repeated the experiment with toluene at an average temperature of 25°C and then obtained complete quantitative agreement with the theoretical predictions. Later, Segrè et al. (1992) and Li et al. (1994a) obtained more experimental data for toluene as well as for n-hexane at 25°C, using the same experimental setup previously employed by Law et al. (1990) except that a new multiple-tau correlator provided better resolution in the measurement of experimental correlation functions. The better resolution also enabled them to replace the cumbersome Ar laser by a convenient (and stable) He-Ne laser. As an example of the experimental results, we present in Fig. 10.5 the (positive) enhancement of the nonequilibrium thermal fluctuations and the (negative) enhancement of the viscous fluctuations as a function of $(\nabla T_0)^2/q^4$. The symbols represent experimental data, and the solid lines the values predicted theoretically from Eq. (4.27) and the known thermophysical properties of n-hexane at 25°C. It is important to note that the agreement between experiment and theory shown in Fig. 10.5 is obtained without using any adjustable parameters in the analysis of the experimental data.

The experiments of Segrè et al. (1992) and Li et al. (1994a) correspond to Ra numbers from $-25,000$ to $-300,000$, and dimensionless wave numbers ranging from $\tilde{q} = 640$ down to $\tilde{q} = 345$. It should be noticed that, in spite of the large and negative Ra values, in these experiments propagating modes are not expected. For instance, when $Ra = -300,000$ we evaluated that, for free boundaries, the "window" of wave numbers for which the decay rates are propagating ends at $\tilde{q} \simeq 20.68$, well below the experimental $\tilde{q}$. For rigid boundaries the "window" of propagating modes can only be evaluated numerically. However, the maximum $\tilde{q}$ for which decay rates have nonzero imaginary part is expected to move to lower $\tilde{q}$ values. Indeed, propagating modes were not observed in the actual experiments. We conclude that the experiments of Segrè et al. (1992) and Li et al. (1994a) indeed probed what we have referred to in Chapter 4 as "bulk" nonequilibrium fluctuations far from the Rayleigh-Bénard convective instability and should be interpreted by Eqs. (4.26)-(4.27), as done in Fig. 10.5. With light scattering it becomes difficult to cover scattering wave numbers low enough to study the influence of gravity or finite-size effects on the nonequilibrium structure factor. At extremely small q, nonequilibrium fluctuations can be investigated by shadowgraph techniques, as discussed in the next section. However, shadowgraph has the disadvantage over light scattering of being less accurate, and the level of agreement between experiment and theory shown in Fig. 10.5 has not yet been surpassed.

10.3 Shadowgraph experiments in one-component fluids

In the past decade, shadowgraphy has emerged as an alternative powerful experimental technique to study nonequilibrium fluctuations. It was originally developed for measuring convection patterns in dissipative nonequilibrium systems (Ahlers, 1991; de Bruyn et al., 1996; Bodenschatz et al., 2000). More recently, the technique has also been used to study fluctuations close to but below the onset of convection The shadowgraphy technique is particularly suited to study fluctuations at much smaller wave numbers q than those accessible with light scattering. The intrinsic accuracy of quantitative shadowgraphy for measuring nonequilibrium fluctuations in quiescent fluids, however, is lower than that of light scattering and requires large thermal noise. For this reason shadowgraph measurements in one-component fluids have been performed in highly compressible fluids. A problem with highly compressible fluids is that the Boussinesq approximation may no longer be adequate.

Experiments thus far reported in the literature have probed the divergence and the critical slowing down of the thermal fluctuations near the onset of Rayleigh-Bénard convection.

10.3.1 Divergence of critical fluctuations near the onset of convection

Wu et al. (1995) have used the shadowgraph technique to measure the nonequilibrium structure factor in a layer of compressed supercritical carbon dioxide in which a temperature gradient was established so that Ra was close to (but below) the critical Rayleigh number. The combination of a highly compressible fluid and closeness to the convective instability made the nonequilibrium fluctuations to be detectable in shadowgraph pictures. The measurement cell had a transparent top plate and a mirror bottom plate, and it was illuminated from above by a collimated monochromatic diode, producing a homogeneous illumination (Wu et al., 1995).

As an illustration, we show in Fig. 10.6 two shadowgraph pictures (left column) together with the corresponding shadowgraph signal (right column), *i.e.*, the modulus squared of the Fourier transform of the pictures in the left column. Due to the finite size of the pictures, the raw shadowgraph signal contains a central peak which has been deleted in Fig. 10.6. The pictures in the right column of Fig. 10.6 show a circular structure in accordance with the fact that the nonequilibrium structure factor depends only on the magnitude q of the vector $\mathbf{q}$, as a consequence of the translational invariance below the instability[‡]. Azimuthal averaging will produce a function with a

---

[‡]Above the instability, translational invariance is broken and the ring in Fig. 10.6

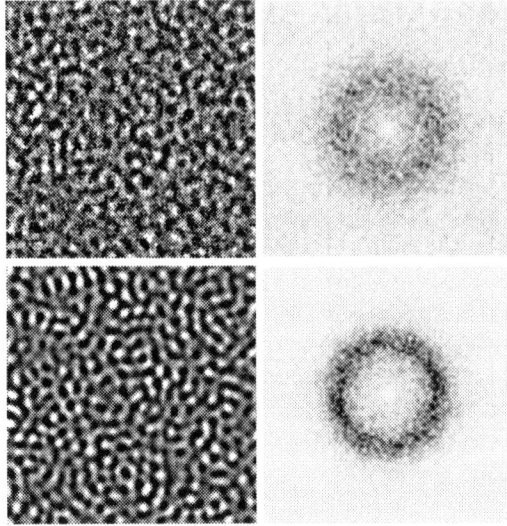

Figure 10.6: Shadowgraph images of a Rayleigh-Bénard cell filled with supercritical CO_2. The left column are actual pictures, while the right column are the modulus squared of the two-dimensional Fourier transform of the pictures in the right column (background subtracted). Top pictures are for $\epsilon = -0.6$ and bottom pictures for $\epsilon = -0.14$. Courtesy of Guenter Ahlers.

single peak at q values around the radius of the ring, similar to the curves presented in Fig. 7.3. As was discussed in Chapter 7, the appearance of a peak in the nonequilibrium structure factor is a consequence of the competition between the enhancement produced by the imposed temperature gradient and a suppression due to the finite size of the fluid layer. The peak, for $Ra \lesssim Ra_c$ appears at $\tilde{q} \simeq \tilde{q}_c$. As mentioned above, with shadowgraphy one can probe fluctuations at much smaller wave numbers than with light scattering, so shadowgraph techniques can be used to examine effects induced by the presence of boundaries in the system as the appearance of a maximum (ring) in the right column of Fig. 10.6 shows.

By analyzing a large series of shadowgraph pictures like the ones presented in Fig. 10.6, Wu et al. (1995) were able to study the divergence of the power of thermal fluctuations as the convective instability is approached, a topic discussed theoretically in Sect. 8.4. They confirmed that the mean square fluctuation amplitude indeed increases inversely proportional to $\sqrt{-\epsilon}$, in accordance with the predicted asymptotic behavior in Eq. (8.19). The proportionality factor was of the same order of magnitude as predicted by the theory, but the accuracy was limited, due to both non-Boussinesq effects and experimental uncertainties in the optical transfer

---

collapses in two peaks (Oh and Ahlers, 2003)

function (Wu et al., 1995; Oh et al., 2004).

Similar results, *i.e.*, a $1/\sqrt{-\epsilon}$ proportionality in the intensity of the fluctuations below the convection threshold has been observed by Quentin and Rehberg (1995) for a liquid mixture.

We mention that, in spite of inconclusive earlier experimental measurements (Ahlers et al., 1981), the results reviewed in this and the previous section show that, at least for the RB problem, simple "thermal" noise (as discussed in this book) is enough to describe fluctuations below the convective instability. Hence, it does not appear necessary lo look for "fancy" sources of noise, such as multiplicative, correlated or colored noise (García Ojalvo et al., 1993; García Ojalvo and Sancho, 1999), to explain these large fluctuations.

10.3.2 Critical slowing down of fluctuations near the onset of convection

As shown in Sect. 8.1, the theory predicts that the decay rate of the fluctuations will decrease as the Rayleigh-Bénard instability is approached, a phenomenon known as critical slowing down near the onset of convection. This phenomenon has been observed experimentally by several investigators and different techniques.

Using forced Rayleigh scattering Allain et al. (1978) measured the decay of imposed horizontal spatially periodic temperature profiles with various wave numbers q. The decay time of these imposed deviations from the steady state increased as the temperature gradient approached the value associated with the critical Rayleigh number Ra_c. Furthermore, the decay times became larger for periodic temperature profiles with q closer to q_c. A critical slowing down of nonequilibrium fluctuations has also been observed by Sawada (1978) with an acoustic method. In this experiment, first a Rayleigh-Bénard convection pattern was established and then the cell was turned over so as to change the sign of the Rayleigh number Ra. The experimental observations were interpreted with an amplitude equation that did not include any wave-number dependence. The single decay time extracted from the data did indeed increase as the onset of convection was approached, but numerical agreement with the amplitude equation was poor. Using neutron scattering, Pedersen and Riste (1980) and Otnes and Riste (1980) observed the critical slowing down of nonequilibrium fluctuations close to the onset of convection in a liquid crystal. The results of these pioneering papers on fluctuations below the instability are rather qualitative. Behringer and Ahlers (1977) and Westried et al. (1978) have presented more quantitative experimental studies showing the slowing down of the dynamics of convection patterns as the Rayleigh-Bénard instability is approached from *above*.

A quantitative study of the critical slowing down of fluctuations below

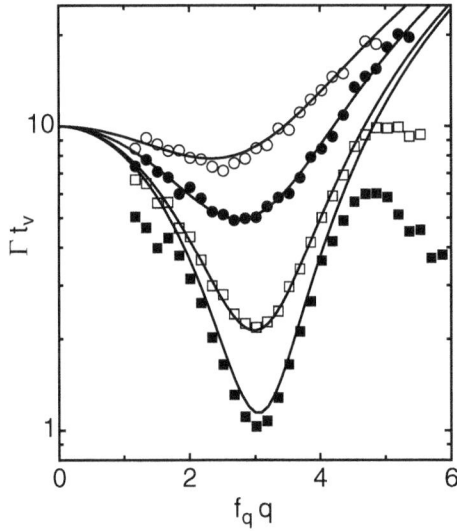

Figure 10.7: Dimensionless decay rate Γt_v, where Γ is the experimental decay rate and t_v the vertical relaxation time, as a function of $\tilde{q}$. The symbols represent experimental data. The solid curves are obtained from the Galerkin approximation (7.33). From Oh et al. (2004).

the convection threshold has been performed recently by Oh et al. (2004) with an experimental setup similar to that used by Wu et al. (1995), described in the previous section. As was done by Wu et al. (1995), Oh et al. (2004) performed the experiments in a supercritical fluid not far from the critical point, this time sulfur hexafluoride, to enhance the thermal noise. The decay rate of the thermal fluctuations was evaluated by taking advantage of the dependence of the shadowgraph signals on the collecting time Δt, as explained in Sect. 10.1.3. By taking the ratio of shadowgraph signals obtained with different Δt, one can eliminate any experimental uncertainties in the optical transfer function $\check{T}(q)$, see Eq. (10.37). As an example, Fig. 10.7 shows the experimental decay rate Γ, made dimensionless by multiplying with a vertical thermal relaxation rate $t_\nu \simeq L^2/a_T$, as a function of the experimental dimensionless wave number $\tilde{q}$. Since the measurements were obtained close to the convective instability, the product Γt_v can be identified with the product of the dimensionless slowest decay rate $\tilde{\Gamma}_1^{(-)}(q)$ and the Prandtl number: $Pr\,\tilde{\Gamma}_1^{(-)}(q)$. A problem with experiments in a fluid near the critical point is that the thermophysical properties and, hence, the Rayleigh number Ra, varies strongly with the vertical position in the fluid layer, so that non-Boussinesq effects will be large. To account for non-Boussinesq effects Oh et al. (2004) used an effective Rayleigh number obtained by multiplying the Boussinesq estimate for Ra by an adjustable scale factor. The solid curves in Fig. 10.7 represent Eq. (7.33) in terms of an effective Rayleigh number (see Oh et al. (2004) for details).

The theoretical prediction that the intensity of the thermal fluctuations will diverge and that the decay rate of the fluctuations will vanish upon approaching the threshold for convection and further elucidated in Chapter 8, is reminiscent of what happens to thermal fluctuations near second-order phase transitions in equilibrium thermodynamics. The theory of thermal fuctuations in this book is based on a linear coupling of the fluctuations and, hence, is a mean-field theory in the language of Swift and Hohenberg (1977). Nonlinear coupling between the fluctuations will change the transition between the hydrodynamically quiescent state and the convective state ultimately from a second-order to a first-order transition (Hohenberg and Swift, 1992). Experimental evidence for this phenomenon has recently been reported by Oh and Ahlers (2003), again by using shadowgraph techniques. A theoretical treatment of the effects of nonlinear couplings between the fluctuations is outside the scope of the present volume.

10.4 Light scattering in binary systems

As reviewed in Chapter 5, the Rayleigh component of the dynamic structure factor of a binary mixture contains two modes. The decay rates $\Gamma_\pm(q)$ of these modes are given by the second line of Eq. (5.26). When the Dufour effect ratio ϵ_D defined by Eq. (2.76) is small, as is the case for most common liquid mixtures, the decay rates are $\Gamma_+ \simeq a_T q^2$ for the temperature fluctuations, and $\Gamma_- \simeq D q^2$ for the concentration fluctuations. For most liquid mixtures the decay rate $D q^2$ of the concentration fluctuations is much larger than the decay rate $a_T q^2$ of the temperature fluctuations. Under the same $\epsilon_D \simeq 0$ condition, the relative intensity of the temperature and the concentration fluctuations is given by the quantity $\mathcal{R}$ defined by Eq. (5.38). Usually $\mathcal{R}$ is much larger than unity, in particular for mixtures whose components have significantly different refractive indices, so that the intensity of the concentration fluctuations is much larger than that of the temperature fluctuations. Hence, in many light-scattering experiments in liquid mixtures only concentration fluctuations have been observed in practice.

In Sect. 10.4.1 we review the experimental attempts to measure nonequilibrium fluctuations in binary liquid mixtures and in Sect. 10.4.2 we review a experimental results for nonequilibrium concentration fluctuations in polymer solutions. All these experiments deal with nonequilibrium fluctuations induced by the imposition of a temperature gradient. In Sect. 10.5, we consider nonequilibrium concentration fluctuations induced by concentration gradients resulting from free-diffusion processes.

10.4.1 Nonequilibrium fluctuations in liquid mixtures

Experimental measurements of the nonequilibrium enhancement of the concentration fluctuations induced by the Soret effect in binary liquid mixtures have been obtained at the University of Maryland (Segrè et al., 1993a; Li et al., 1994a,b, 1995) and at the University of Milan (Vailati and Giglio, 1996, 1997b).

At the University of Maryland, Sengers and coworkers have performed nonequilibrium Rayleigh-scattering experiments in mixtures of liquid toluene and liquid n-hexane subjected to a temperature gradient far from the critical point of demixing and for large negative values of Ra. The separation ratio of this mixture is positive, so that the experiments were performed in the region of overstability, corresponding to the fourth quadrant in Fig. 5.1. They were able to determine not only the enhancement of the scattering due to the nonequilibrium concentration fluctuations, but also the contributions from the nonequilibrium viscous and temperature fluctuations. For this mixture, the Dufour effect ratio ϵ_D is sufficiently small so that, provided buoyancy is neglected, the nonequilibrium Rayleigh component of the structure factor should be given by the sum of three lorentzians in accordance with Eq. (5.41). The time-dependent correlation function was determined by photon-correlation spectroscopy. The experimental correlation functions could be readily analyzed in terms of three exponentials, since in this liquid mixture the three decay rates in the small-ϵ_D approximation (*i.e.*, $\Gamma_+ = a_T q^2$, $\Gamma_- = D q^2$ and $\Gamma_\nu = \nu q^2$) are well separated in time scale[‡]. Segrè et al. (1993a) determined the amplitudes A_c, A_T and A_ν of the three components of the nonequilibrium fluctuations of toluene+n-hexane mixtures with three different concentrations at various values of the wave number q and of the temperature gradient ∇T_0. In the experiments the wave number q was large enough so that gravity effects could be neglected. Just as in the nonequilibriumn Rayleigh-scattering experiments, the experimental arrangement corresponds to $\mathbf{q} \perp \nabla T_0$, so that at the small scattering angles q can again be identified with $q_\parallel$, the configuration for which which Eq. (5.41) was derived. It thus follows from Eqs. (5.42) that the three independent enhancements should be proportional to $(\nabla T_0)^2/q^4$:

$$A_\nu, A_T, A_c \propto \frac{\nabla T_0^2}{q^4}. \tag{10.38}$$

As an example, we show in Fig. 10.8 the experimental amplitudes, obtained for an equimolar mixture of toluene and n-hexane, as a function of $(\nabla T_0)^2/q^4$ (Li et al., 1994b). The experiments confirm that the nonequilibrium enhancements of the fluctuations are indeed proportional to $(\nabla T_0)^2$ and inversely proportional to q^4 (Segrè et al., 1993a; Li et al., 1994a,b, 1995).

[‡]The Lewis number of the mixture is $Le \simeq 35$

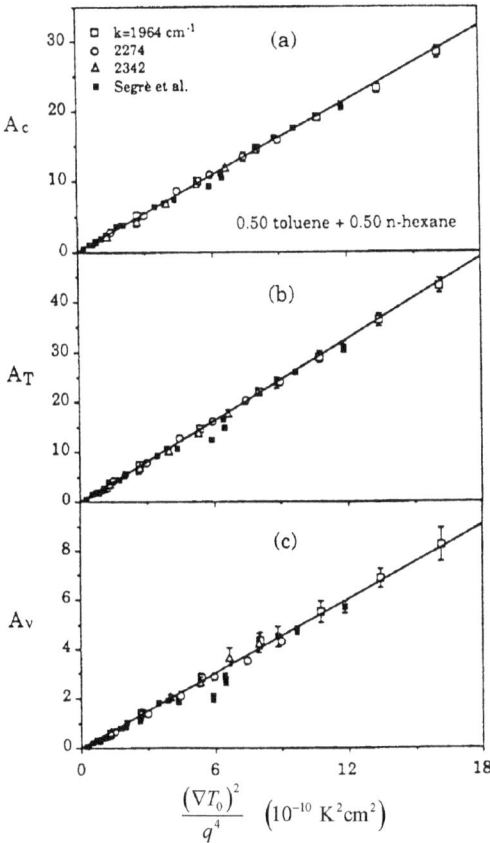

Figure 10.8: Nonequilibrium enhancement A_c of the concentration, A_T of the temperature and A_ν of the viscous fluctuations, as a function of $\nabla T_0^2 q^{-4}$ for an equimolar mixture of toluene in n-hexane. The solid line in (a) represent a linear fit to the data; while in (b) and (c) solid lines are obtained theoretically from Eqs. (5.42) and an effective S_T value deduced from (a). Adapted from Li et al. (1994b).

For a quantitative interpretation of the observed nonequilibrium enhancements one needs reliable values for the derivative $(\partial\mu/\partial c)_T$ (usually expresed in terms of the osmotic compressibility), the kinematic viscosity ν, the mass diffusion coefficient D and the Soret coefficient S_T. At the time of the experiments, no accurate values were available for the Soret coefficient of mixtures of toluene and n-hexane. The solid lines in Fig. 10.8 represent a fit of Eqs. (5.42) to the experimental amplitudes with values for the various thermophysical properties deduced from the literature except for the Soret coefficient S_T which was treated as an adjustable constant. Quantitative agreement of the theoretical prediction with all three experimental amplitudes, A_c, A_T and A_ν, was obtained in terms of the same effective value for S_T. However, when the Soret coefficient S_T was subsequently measured by Zhang et al. (1996) and by Köhler and Müller (1995), the experimental results obtained for S_T turned out to differ by about 25% from the effective values of S_T deduced from the nonequilibrium concentration fluctuations (Segrè et al., 1993a; Li et al., 1994a,b, 1995). The reason for this difference

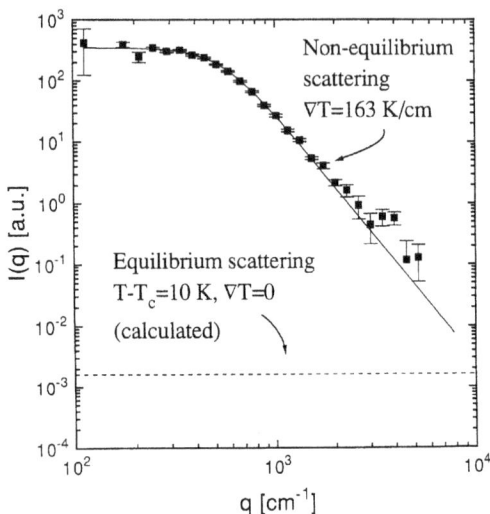

Figure 10.9: Intensity of light scattered by a mixture of aniline and cyclohexane as a function of the scattering wave number. The symbols represent experimental data obtained with $\nabla T_0 = 163$ K cm^{-1}. The solid curve indicates a fit to (10.40). Reproduced with permission from Vailati and Giglio (1996).

has not yet been explained (Segrè et al., 1993a; Li et al., 1994a,b, 1995). As we shall see in the next subsection, quantitative agreement has been obtained for nonequilibrium concentration fluctuations in a polymer solution for which the Soret coefficient is two orders of magnitude larger than for a liquid mixture like toluene+n-hexane.

Vailati and Giglio (1996, 1997b) have measured nonequilibrium concentration fluctuations induced by the Soret effect in a mixture of liquid aniline and cyclohexane. To enhance the intensity of the concentration fluctuations, the light-scattering experiments were performed in the homogeneous phase, but rather close to the critical consolute point. This arrangement assures the absence of significant contributions from temperature or viscous fluctuations to the total intensity of the scattered light. Moreover, since at a consolute point the diffusion coefficient vanishes, a liquid mixture near a consolute point has a very large Lewis number, so that the light-scattering intensity should be given by Eq. (5.19). Vailati and Giglio (1996, 1997b) only measured the intensity of the concentration fluctuations and not its decay rate. However, they were able to cover scattering angles smaller than those accessible in the earlier experiments of Li et al. (1994b,a) and, hence, to probe wave numbers sufficiently small for buoyancy effects due to gravity to become apparent. To simplify the analysis of the experimental data,

(Vailati and Giglio, 1996) introduced a "roll-off" scattering wave number defined as:

$$q_{\mathrm{RO}}^4 = -\frac{\beta\ g\ \nabla c_0}{\nu\ D} = -\psi\frac{Ra}{Le}\frac{1}{L^4}. \tag{10.39}$$

The experiments, as those of Sengers and coworkers, were performed for $Ra < 0$ and $\psi > 0$, so that the right-hand side of (10.39) is positive and q_{RO} is a real quantity. In terms of q_{RO} we obtain from (5.19):

$$A_c \propto \frac{\nabla T_0^2}{q^4 + q_{\mathrm{RO}}^4}. \tag{10.40}$$

For large q, Eq. (10.40) reduces to (10.38). However, in the limit $q \to 0$, the q^{-4} divergence in Eq. (10.38) is quenched and Eq. (10.40) predicts that a constant limiting value for $\tilde{S}_{\mathrm{NE}}$ will be reached (Segrè and Sengers, 1993).

The intensity of the scattered light, measured by Vailati and Giglio (1996) at a temperature gradient ∇T_0=167 K/cm is shown in Fig. 10.9 as a function of the wave number q of the fluctuations. The intensity of the fluctuations in thermal equilibrium was too weak to be observed and had to be estimated. The solid curve represents a fit to Eq. (10.40) and the crossover from a q^{-4} divergence to the gravitationally induced saturation at very small q is clearly observed. From the fit of Eq. (10.40) to the experimental data, Vailati and Giglio (1996) deduced $q_{\mathrm{RO}} = 537$ cm^{-1}, which is comparable to the value $q_{\mathrm{RO}} = 410$ cm^{-1} estimated from (10.39) using literature values for the thermophysical properties of this mixture. The intensity of the nonequilibrium fluctuations is plotted in Fig. 10.9 in arbitrary units. The authors did not have the resolution to determine the absolute magnitude of the nonequilibrium enhancement of the concentration fluctuations with sufficient accuracy to make a reliable check of the validity of the prefactor in Eqs. (5.20) or (5.21).

10.4.2 Concentration fluctuations in dilute polymer solutions

For dilute polymer solutions in a good solvent, the diffusion coefficient D is much smaller than for binary mixtures while the Soret coefficient S_T is much larger. The combination of these two facts results in the presence of larger concentration fluctuations that evolve very slowly and, consequently, can be easily detected. In addition, since the temperature and and viscous fluctuations are weak and have decay times much shorter than that of the concentration fluctuations, the experimental correlation functions obtained can be analyzed in terms of a single exponential, given by Eq. (5.17) in the limit of large Lewis number Le.

Experimental values for the nonequilibrium enhancement A_c and for the decay rate Γ_D of the concentration fluctuations have been obtained for dilute

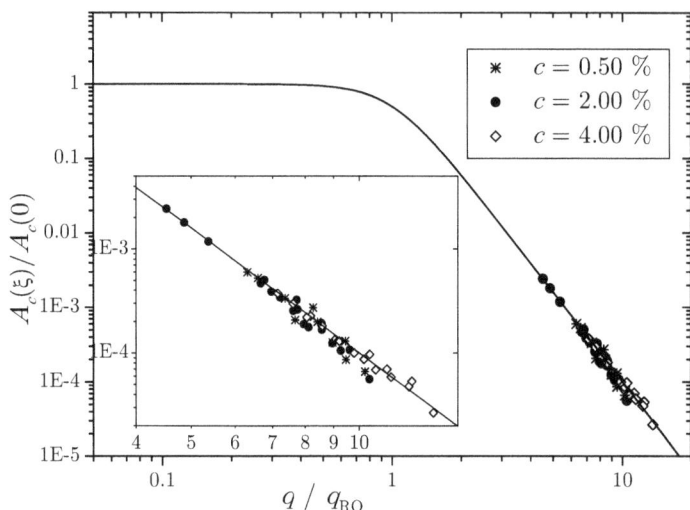

Figure 10.10: Scaled nonequilibrium enhancement of the concentration fluctuations as a function of the scaled scattering wave number q/q_{RO} for solutions of polystyrene in toluene at several concentrations. The solid curve represents the universal function (10.41). The insert shows the range were the actual experimental data have been obtained (Sengers and Ortiz de Zárate, 2002).

solutions of polystyrene ($M_W = 96\,400$) in toluene from low-angle dynamic light-scattering measurements at the University of Maryland (Li et al., 1998, 2000; Sengers et al., 2000). For this mixture, in the concentration range of the samples investigated, the separation ratio ψ is positive, varying from $\psi = 0.121$ for $c = 0.25\%$ to $\psi = 1.101$ for $c = 4.00\%$ (Sengers et al., 2000). Hence, the polymer solution is indeed in a stable thermally conducting state when heated from above (*i.e.*, for $Ra < 0$). The experiments were performed at negative Rayleigh numbers down to $Ra = -25000$, well inside the stability zone. The Lewis number for this polymer solution is $Le \simeq 1500$, so that the assumption of large Lewis number is justified indeed. The light-scattering experiments were performed at scattering angles for which gravity effects can be neglected,so that the noncritical enhancement $A_c(q)$ and the decay rate Γ_D of the concentration fluctuations will be given by Eq. (5.20).

The measured decay rates of the nonequilibrium concentration fluctuations were the same, within experimental accuracy, as those of the equilibrium concentration fluctuations, *i.e.*, $\Gamma_D = Dq^2$. With the wave number q deduced from a geometric measurement of the scattering angle, the values obtained for the diffusion coefficient D from the experimental decay rates agreed with the values of D measured by an independent beam-bending technique (Li et al., 2000).

By comparing the amplitudes of the observed time-dependent correlation functions in thermal equilibrium and out of thermal equilibrium, experimental values were obtained for the enhancement $A_c(q)$ of concentration fluctuations at 25°C (Li et al., 2000). An elegant comparison between theory and experiment can be formulated in terms of a scaled universal representation of $A_c(q)$ adopted by Brogioli et al. (2000b) for studying concentration fluctuations in free-diffusion processes. For this purpose we note that (5.19), with the definition (10.39) for q_{RO} and the small-angle approximation $q_\parallel \simeq q$, can be rewritten as:

$$\frac{A_c(\xi)}{A_c(0)} = \frac{1}{1 + \xi^4},$$

(10.41)

where $\xi = q/q_{\mathrm{RO}}$ is a scaled wave number and where

$$A_c(0) = 1 + \left(\frac{\partial \mu}{\partial c}\right)_T \frac{(\nabla c_0)^2}{\nu D\, q_{RO}^4},$$

(10.42)

is the enhancement in the $q \to 0$ limit.

The scale factors $A_c(0)$ and q_{RO} in (10.41) depend on the thermophysical properties of the solution and on the value of the stationary concentration gradient ∇c_0, which in turn depends on the applied stationary temperature gradient ∇T_0 through the Soret coefficient S_T in accordance with Eq. (5.3). For a given ∇T_0, q_{RO} and $A_c(0)$ can be calculated from the literature values for the various thermophysical properties collected by Li et al. (2000). If we then scale the experimental data for $A_c(q)$ by $A_c(0)$ and the wave number q by q_{RO}, a universal curve should be obtained according to Eq. (10.41). The results are shown in Fig. 10.10. The inset in Fig. 10.10 shows in more detail the data in the actual range of experimental wave numbers. To make Fig. 10.10 more legible, only experimental data corresponding to three polymer concentrations have been plotted, but there is no difference when the data obtained for all different polymer solutions are included. It should be emphasized that the curve in Fig. 10.10, unlike the ones in Fig. 10.8 or in Fig. 10.9, does not represent a fit to the experimental data, but represents the universal function defined in Eq. (10.41) with the experimental data scaled by the known thermophysical properties for the polymer solutions. Although the experiment does not cover small enough wave numbers for the saturation effect due to gravity to be observed, Fig. 10.10 shows excellent agreement between theory and experiment. In fact, the light-scattering data obtained for polystyrene solutions in toluene represent the first accurate confirmation of the theory for nonequilibrium concentration fluctuations, including the prefactor in (5.20).

10.5   Shadowgraphy in binary mixtures

Giglio and coworkers at the University of Milan have used the shadowgraph technique to measure the nonequilibrium concentration fluctuations induced in liquid mixtures by concentration gradients associated with free diffusion (Brogioli et al., 2000b,a). Specifically, Brogioli et al. (2000b) studied free-diffusion in concentrated aqueous solutions of several solutes, including glycerol and the polymer poly-ethylene-glycol. The initial state was prepared by filling the cell with the lighter liquid and carefully injecting the denser liquid from below, so that a sharp meniscus was formed. From this initial state free diffusion proceeds until an uniform concentration is reached inside the cell.

As was mentioned in Sect. 5.4, nonequilibrium concentration fluctuations in free diffusion exhibit some of the same characteristic features as nonequilibrium concentration fluctuations induced when the concentration gradient results from an imposed temperature gradient. Specifically the contribution from the nonequilibrium concentration fluctuations to the scattered-light intensity are described by Eq. (5.19), if one replaces the stationary concentration gradient ∇c_0 by $\nabla c(z,t)$. Thus the actual concentration gradient will depend on the time t and on the vertical coordinate z. This concentration gradient $\nabla c(z,t)$ can, in turn, be calculated by solving the free-diffusion equation with a step function as initial condition. Averaging over the vertical variable z one obtains a time-dependent vertically averaged structure factor, which is the quantity measured in the shadowgraph experiments. The averaged structure factor has a expression similar to Eq. (10.41), with a time-independent amplitude, but with a time-dependent roll-off wave vector $q_{RO}(t)$ (Vailati and Giglio (1998)).

Brogioli et al. (2000b) confirmed the presence of a dependence of the nonequilibrium structure factor on q^{-4} and a rollover due to gravitational effects, in excellent quantitative agreement with theory, both for polymer solutions and for liquid mixtures. We may thus conclude that the theoretical prediction (5.19) for the nonequilibrium concentration fluctuations has been confirmed experimentally. This quantitative confirmation of the accuracy of fluctuating hydrodynamics to describe nonequilibrium concentration fluctuations for free-diffusion processes makes additional quantitative measurements for liquid mixtures when the concentration gradient is induced by thermal diffusion desirable, see Sect. 10.4.1.

More recently, Brogioli et al. (2003) have modified the shadowgraph technique by setting up a schlieren blade. With this modified setup they measured nonequilibrium fluctuations in free diffusion of urea in water. These new results obtained with the modified shadowgraph technique confirm again the main features of the theory for nonequilibrium fluctuations in the absence of finite-size effects, namely a q^{-4} dependence of the intensity of fluctuations for large q that crosses over to a constant limit at $q \to 0$.

Chapter 11

Other nonequilibrium fluctuations

In chapters 4-9 we have focussed our attention on nonequilibrium fluctuations in fluids and fluid mixtures subjected to a stationary temperature gradient, to which we have referred as the Rayleigh-Bénard (RB) problem. Nonequilibrium fluctuations in the RB problem have received considerable attention in the literature so that much information is available for this case. Hence, the RB problem may be considered as representative for studying fluctuations in nonequilibrium systems. As we have seen in Chapter 10, many theoretical predictions concerning the long-ranged nature of nonequilibrium fluctuations in the RB problem have been verified experimentally.

In the present chapter we discuss some other cases for which nonequilibrium fluctuations have been studied in the literature at different levels of detail. Our main goal in this chapter is to show how the physically most relevant features encountered in the study of nonequilibrium fluctuations in the RB problem also appear in other nonequilibrium situations. Specifically, we shall see that a generic scale invariance leading to an algebraic spatial decay of the hydrodynamic equal-time correlation functions is a universal feature of nonequilibrium fluctuations.

Our account of nonequilibrium fluctuations in other systems will be much more brief than our treatment of nonequilibrium fluctuations in the RB problem in the previous chapters. Section 11.1 deals with fluctuations in one-component fluids under a uniform shear flow, sometimes referred to as plane Couette flow. In Sect. 11.2 we consider the problem of interfacial nonequilibrium fluctuations, and in Sect. 11.3 we review nonequilibrium fluctuations in nematic liquid crystals. We conclude the chapter by discussing nonequilibrium fluctuations in reaction-diffusion problems in Sect. 11.4. Fluctuations in reaction-diffusion processes have received considerable attention in the literature, but their study is usually performed in the framework of a so-called chemical master equation. This approach is slightly different from what we have been considering so far in this book, since it is not a thermodynamic method (temperature plays no role). Nevertheless we include this case in some detail, because it illustrates the generic scale invariance leading to an algebraic spatial dependence of the nonequilibrium hydrodynamic correlation functions. Moreover, the results of the last section can be interpreted thermodynamically as isothermal fluctuations in chemical reactions between extremely dilute species.

11.1 Nonequilibrium fluctuations in isothermal one-component fluids under shear

Many authors have investigated nonequilibrium fluctuations in a uniformly sheared one-component fluid, sometimes referred to as plane Couette flow (Machta et al., 1980; Tremblay et al., 1981; Lutsko and Dufty, 1985, 2002; Wada and Sasa, 2003). An excellent review of the earlier work has been presented by Tremblay (1984). In most cases this system has been investigated in the isothermal approximation, which means that temperature fluctuations are neglected. In the absence of temperature fluctuations in the system, they do not contribute to density fluctuations, so that this approach is equivalent to taking $\alpha_p = 0$, an assumption earlier adopted in Sect. 4.3 to evaluate modifications to the Brillouin lines of a fluid subjected to a uniform temperature gradient. The assumption of a perfectly uniform temperature (no temperature fluctuations), in conjunction with the approximation that $\alpha_p \simeq 0$, causes the heat equation to be irrelevant for this problem (Tremblay et al., 1981). Note that with this assumption viscous heating is neglected. Hence, the set of relevant hydrodynamics equations reduces to the continuity Eq. (2.6) and the Navier-Stokes Eq. (2.51). The equation of state will be simply: $d\rho = \varkappa_T \rho \, dp$.

The difference with the development in Sect. 4.3 is that now the system is brought out of equilibrium by a uniform shear. In particular, we consider a fluid layer confined between two horizontal plates located at $z = -\frac{1}{2}L$ and $z = +\frac{1}{2}L$. The upper plate moves in the direction of the x axis with a velocity $\mathbf{v}(z = L/2) = v_0 \, \hat{\mathbf{x}}$, where $\hat{\mathbf{x}}$ is the unit vector in the direction of the x-axis. The lower plate also moves in the x direction with a velocity opposite to that of the upper plate, thus $\mathbf{v}(z = -L/2) = -v_0 \, \hat{\mathbf{x}}$. The stationary solution of the hydrodynamic equations (2.6) and (2.51) with the boundary conditions (BC) mentioned above is $\rho(\mathbf{r}, t) = \rho_0$ (uniform) and:

$$\mathbf{v}(\mathbf{r}, t) = \mathbf{v}_0 = \frac{2v_0}{L} z \, \hat{\mathbf{x}}. \tag{11.1}$$

The quantity $\dot{\gamma}_0 = 2v_0/L$ is referred to as the (uniform) shear rate (units: s^{-1}). Because of the symmetry of the BC, the fluid velocity at the midplane $z = 0$ is zero. Tremblay et al. (1981) have considered the slightly more general case of a nonzero velocity at the midplane, but this only adds more algebra to our problem, while the physical consequences remain the same. Thus, we focus here on the more transparent case with symmetric BC. Indeed, by substituting Eq. (11.1) into the hydrodynamic equations (2.6) and (2.51), one readily verifies that (11.1) with uniform density (and pressure) is a solution of these hydrodynamic equations while also satisfying the appropriate BC. The solution mathematically represented by Eq. (11.1) thus corresponds to an isothermal fluid layer subjected to a uniform shear rate. The x-axis is the direction of the velocity, the y-axis is the direction of the vorticity and the z-axis is the direction of the gradient (Batchelor, 1967; Tritton, 1988). The solution (11.1) to the hydrodynamic equations is usually referred to as *laminar flow*.

We want to study fluctuations around the stationary nonequilibrium solution described by Eq. (11.1). To obtain the fluctuating-hydrodynamics equations for this problem, we substitute into the hydrodynamic equations the stationary solution for the fields plus some spatiotemporal fluctuation, namely: $\rho(\mathbf{r}, t) = \rho_0 + \delta\rho(\mathbf{r}, t)$ and $\mathbf{v}(\mathbf{r}, t) = \mathbf{v}_0 + \delta\mathbf{v}(\mathbf{r}, t)$. In addition, we add a stochastic part to the dissipative fluxes. Since we are assuming here iso-entropic evolution (no temperature fluctuations), the only random dissipative flux to be considered is the random stress tensor $\delta\Pi(\mathbf{r}, t)$. The correlation functions among the various components of $\delta\Pi$ is given by the FDT for a one-component fluid, Eq. (3.6c). If one further assumes that fluctuations are small, one may linearize in the fluctuating fields, so as to obtain (Tremblay et al., 1981):

$$\partial_t(\delta\rho) + \dot{\gamma}_0 z \, \partial_x(\delta\rho) + \rho_0 \, \boldsymbol{\nabla} \cdot \delta\mathbf{v} = 0,$$

$$\partial_t(\delta\mathbf{v}) + \dot{\gamma}_0 z \, \partial_x(\delta\mathbf{v}) + \dot{\gamma}_0 \, \delta v_z \, \hat{\mathbf{x}} = \frac{-\boldsymbol{\nabla}\delta\rho}{\rho_0^2 \varkappa_T} + \nu\nabla^2(\delta\mathbf{v})$$

$$+ \frac{1}{\rho_0}\left(\eta_\mathrm{v} + \frac{\eta}{3}\right)\boldsymbol{\nabla}(\boldsymbol{\nabla} \cdot \delta\mathbf{v}) + \frac{1}{\rho_0}\boldsymbol{\nabla}\,(\delta\Pi)\,.$$

$$(11.2)$$

Tremblay et al. (1981) solved the system of Eqs. (11.2) for the fluctuating fields by the usual "bulk" procedure of applying a full spatiotemporal Fourier transform, *i.e.*, without considering BC for the fluctuating fields. In addition, since in this case the system is isothermal, we can neglect any inhomogeneity in the strength of the noise, so that the first source of nonequilibrium effects in hydrodynamic fluctuations discussed in Sect. 3.5 is absent here. All nonequilibrium effects arise from mode coupling caused by the advection as evident from the terms proportional to $\dot{\gamma}_0$ in the LHS of Eq. (11.2). The calculation of the "bulk" structure factor using full spatial

Fourier transformations is complicated here by the explicit appearance of the vertical variable z in the LHS of Eqs. (11.2). This difficulty can be dealt with by using the same trick as in Sect. 4.3, namely, by replacing the z variable with $\sin kz$ and applying a k-expansion at the end of the calculation.

We shall not further discuss here the calculation of Tremblay et al. (1981), but refer the reader to the original paper for the details. We just mention that, to obtain a compact expression, at the end of the procedure a small $\dot{\gamma}_0$ limit is taken[‡]. The final result for the dynamic structure factor is given by Eq. (4.20) in Tremblay et al. (1981), which reads:

$$S(\omega, \mathbf{q}) = S_{\mathrm{E}}(\omega, q) \left[1 - \frac{2 D_V \omega^2 \, \dot{\gamma}_0 q_x q_z}{(\omega^2 - c^2 q^2)^2 + (\omega D_V q^2)^2} \right], \qquad (11.3)$$

where $S_{\mathrm{E}}(\omega, q)$ is the same equilibrium structure factor as in (4.39), which in this case contains only the Brillouin lines due to the iso-entropic $\alpha_p \simeq 0$ assumption. In comparing (11.3) with the result of Tremblay et al. (1981), one should note that we have adopted here symmetric BC, so that the stationary velocity at the midplane is zero; hence, we do not find a Doppler shift in the frequency. It is interesting to compare (11.3) with (4.42) obtained in Sect. 4.3 for the Brillouin lines in the presence of a stationary temperature gradient. Both formulas have a similar mathematical structure; however, in the numerator of the nonequilibrium enhancement we find a ω^2 term in Eq. (11.3), while in Eq. (4.42) we had a ω^3 term. This means that the enhancement of the Brillouin lines in the presence of a uniform shear is symmetric, in contrast to the case of a uniform temperature gradient, where we found an asymmetric enhancement at linear order in the gradient. As a consequence, the dimensionless static structure factor is modified in linear order in the shear rate. Actually, integrating (11.3) over the frequency, in accordance with (3.24), we obtain the dimensionless nonequilibrium static structure factor as:

$$S(\mathbf{q}) = S_{\mathrm{E}} \frac{1}{\gamma} \left[1 - \frac{\dot{\gamma}_0 \, q_x q_z}{D_V q^4} \frac{4 + \epsilon_1^2}{4 - \epsilon_1^2} \right], \qquad (11.4)$$

where $\epsilon_1 = D_V q^2 / cq$ is a small dimensionless parameter that can be neglected for most liquids: $\epsilon_1 \simeq 0$, see Eq. (3.20). In Eq. (11.4), the prefactor S_{E} represents the dimensionless amplitude of the density fluctuations for a fluid in equilibrium, as given by Eq. (3.28) and S_{E}/γ is the amplitude of the equilibrium Brillouin lines. Notice from Eq. (11.4) that the intensity of the Brillouin lines is affected in linear order in the nonequilibrium parameter $\dot{\gamma}_0$. Recall that for the case of a fluid layer subjected to a uniform temperature gradient, as elucidated in Sect. 4.3, the total integrated intensity of the Brillouin lines was unaffected at linear order in the gradient.

[‡]This is the only limit for which the stationary solution (11.1) is stable

Equation (11.4) shows that the static structure factor of a one-component isothermal fluid subjected to uniform shear exhibits an anisotropic enhancement. The equal-time density autocorrelation function continues to be related to $S(\mathbf{q})$ by Eq. (3.25), so that the equal-time density autocorrelation function will have a nonequilibrium long-ranged component. The long-ranged component is directly proportional to the nonequilibrium control parameter $\dot{\gamma}_0$, and for fluctuations with large wave number q, it becomes proportional to q^{-2}. As discussed in previous chapters, when a fluid layer is subjected to an uniform temperature gradient, the most important effect is a nonequilibrium enhancement of the Rayleigh line, which becomes proportional to q^{-4} in the large-q limit. As elucidated in Sect. 7.5 (see in particular Fig. 7.4), the q^{-4} dependence in Fourier space corresponds to a dependence of the equal-time density autocorrelation on r^1 in real space, showing the spatially long-ranged nature of the nonequilibrium enhancement. As discussed by Lutsko and Dufty (1985, 2002) for a fluid subjected to uniform shear, the q^{-2} dependence in Fourier space translates to a r^{-1} dependence in real space. However, while the algebraic r^{-1} dependence from linearized fluctuating hydrodynamics indicated long-ranged spatial correlations, it should be noted that Eq. (11.4) is only valid for $\dot{\gamma}_0/D_v q^2 \ll 1$. Lutsko and Dufty (2002) showed that a nonperturbative solution indicates that the spatial decay as r^{-1} crosses over to a stronger algebraic decay as $r^{-11/3}$. While the equal-time fluctuations in the presence of shear are long-ranged compared to molecular scales, they do not become macroscopic and finite-size effects do not play a significant role unlike the case of fluctuations in the presence of a temperature or concentration gradient.

It is obvious from the previous discussion that the pressure fluctuations will also have a nonequilibrium long-ranged component when a one-component fluid is under shear. For the case of an divergence-free (incompressible) fluid, this issue has been addressed by Wada and Sasa (2003) based on the results of Lutsko and Dufty (1985, 2002) for the nonequilibrium component of the velocity fluctuations. For an incompressible fluid there are no density fluctuations, but pressure fluctuations are related to velocity fluctuations through the Navier-Stokes Eq. (2.51). Furthermore, Wada (2004) studied nonequilibrium concentration fluctuations in incompressible fluid mixtures subjected simultaneously to shear and to a stationary concentration gradient. He found a crossover from the typical q^{-4} enhancement due to the presence of a stationary concentration gradient (see Chapter 5) to a weaker divergence for smaller q as a consequence of the presence of shear. Thus, shear induces a quench of the nonequilibrium fluctuations caused by a stationary concentration gradient (Wada, 2004).

Equation (11.4) was obtained under the assumption $\alpha_p \simeq 0$, which excludes the presence of a Rayleigh line in the spectrum of the fluctuations of the fluid. To investigate the effect of shear on the Rayleigh line one needs to relax the $\alpha_p \simeq 0$ approximation and must incorporate the heat equation

in the theory. Such an investigation has been carried out by García-Colín
and Velasco (1982) and Sahoo and Sood (1983a,b, 1984); it seems that
the Rayleigh line is unaffected by the presence of shear in the fluid layer
Sahoo and Sood (1984). Nonequilibrium fluctuations in the related pattern-
forming Taylor-Couette problem have been studied by Treiber (1996), who
used an amplitude equation formalism only applicable to nonequilibrium
fluctuations close to, but below, the Taylor instability. The case of ideal
gases under flow has been recently studied by Jou et al. (2005), borrowing
the concept of effective temperature introduced by Cugliandolo et al. (1997)
to describe systems with slow relaxation processes.

In spite of these theoretical studies, to our knowledge experimental data
for shear-induced long-ranged fluctuations in molecular fluids are not available. A notable exception is the effect of shear on critical fluctuations that
have been investigated in considerable detail both theoretically and experimentally as more recently reviewed by Onuki (2002). There also exists
an extensive literature on the effects of shear in complex fluids like poly-
mers, polymer solutions, micellar and amphiphilic systems, liquid crystals,
suspensions, etc. (Onuki, 1997), which we do not discuss in this book.

11.2   Nonequilibrium interface fluctuations

In a two-phase system, like a liquid in equilibrium with its vapor, there are
not only thermal fluctuations in the bulk phases, but also fluctuations of the
location of the interface, especially capillary waves that broaden the width
of the interface (Weeks, 1977; Rowlinson and Widom, 1982). Associated
with these capillary waves are propagating modes commonly referred to as
"ripples" or "ripplons". Just as propagating phonons lead to the presence of
two Brillouin components in the dynamic structure factor of the bulk fluid
in accordance with Eq. (3.27), so do the propagating ripplons lead to the
presence of two Brillouin components in the dynamic structure factor of the
interface. To a good approximation, the frequency shift of the Stokes and
anti-Stokes components from the propagating ripplons is given by (Landau
and Lifshitz, 1959):

$$\omega_{\rm c} = \pm \left(\sigma q^3/\rho\right)^{1/2}, \tag{11.5}$$

where σ is the surface tension; Eq. (11.5) is to be compared with a frequency
shift of $\pm cq^2$ in the bulk structure factor (3.27) of a fluid. The width of
the Brillouin components is determined by the damping rate of the ripplons
which is proportional to $(\eta/\rho)q^2 = \nu q^2$ (Levich, 1962). For a dense liquid
in equilibrium with a dilute vapor, ρ and η in the expression for the dynamic structure factor of the interface may be identified with the density
and the shear viscosity of the liquid. When the density ρ of the liquid is
known, dynamic surface light scattering from capillary waves can be used

to determine the surface tension and the viscosity of liquids (Fröba et al., 2003).

In addition to propagating ripplons, there are also nonpropagating diffusive interface fluctuations that cause the presence of a central or Rayleigh component in the spectrum of the interfacial dynamic structure factor (Bedeaux and Oppenheim, 1978). The Rayleigh-Brillouin triplet in the dynamic structure factor of a liquid-vapor interface has been observed experimentally by Mazur and Chung (1987).

Grant and Desai (1983) have extended the application of linear fluctuating hydrodynamics to a liquid-vapor interface in the presence of a temperature gradient along the interface. In the bulk fluid a temperature gradient causes an asymmetry in the intensity of the Brillouin components in the scattering spectrum from the phonons as was elucidated in Sect. 4.3. Similarly, a temperature gradient ∇T_0 causes an asymmetry in the intensity of the Brillouin components in the scattering spectrum from the ripplons:

$$\epsilon_B(\omega, \mathbf{q}) \simeq \overline{\epsilon_B}(\mathbf{q}) = \frac{3}{4} \frac{\sqrt{\sigma \rho}}{\eta} q^{-3/2} \, \hat{\mathbf{q}} \cdot \frac{\nabla T_0}{\bar{T}_0} \tag{11.6}$$

to be compared with the asymmetry of the Brillouin components in the structure factor of the bulk fluid, given by Eq. (4.45). An asymmetry of the Brillouin components from the ripplons on a liquid-vapor interface in the presence of a temperature gradient has been observed by Chung et al. (1988, 1990) in dynamic surface light-scattering experiments. The magnitude of the effect was found to be smaller than the value predicted by Eq. (11.6).

Nonequilibrium fluctuations at the interface between two dense fluid phases during diffusive mixing have been investigated experimentally by Cicuta et al. (2000). The complexity of these experiments, where light scattering by the interface is mixed with light scattered by the bulk fluid phases, makes the interpretation of the results difficult.

11.3 Nonequilibrium fluctuations in nematic liquid crystals

A nematic liquid crystal is a fluid phase in which non-spherical molecules tend to be locally aligned, but randomly distributed in position. The local alignment of the molecules defines a director field $\hat{\mathbf{n}}(\mathbf{r}, t)$, which varies smoothly over hydrodynamic space and time scales. The director field adds more hydrodynamic degrees of freedom to the system, so that new balance laws (in particular, conservation of angular momentum) appear in the thermo-hydrodynamical description of the system. To satisfy balance of torques acting on the director field, the stress tensor is required to have a nonzero antisymmetric part, $\Pi^{(a)} \neq 0$, in contrast with the case of an isotropic fluid examined in Chapter 3. The nonzero antisymmetric part of the stress tensor appears in the dissipation function Ψ of the system, where it plays the role of a new dissipative flux. As a consequence, to close the set of hydrodynamic equations as in Chapter 3, one needs to establish a new phenomenological relationship for $\Pi^{(a)}$, which leads to the introduction of additional viscosity coefficients that do not exist in the case of isotropic fluids. We see that the process of setting up fluctuating hydrodynamic equations is longer and more involved than for simple fluids.

In this brief section we shall not discuss the physics of nematic liquid crystals, or the details of the application of fluctuating hydrodynamics to liquid crystals. Instead, we shall provide the reader with some relevant references and mention the most significant physical conclusions reached in the literature. We focus on results dealing with fluctuations around nonequilibrium steady states in nematic crystals.

A systematic coverage of the physics of liquid crystals can be found in the book by de Gennes and Prost (1993), including fluctuations in equilibrium states. Other references on the topic are review articles by Stephen and Straley (1974) and by Val'kov et al. (1994); the latter more specifically devoted to a study of (equilibrium) fluctuations.

A first salient feature of fluctuations in nematic liquid crystals is that fluctuations in the director $\hat{\mathbf{n}}(\mathbf{r}, t)$ are very strong and completely dominate the scattering spectrum. Indeed, in the absence of any BC for the fluctuating director, the intensity of its *equilibrium* fluctuations turns out to diverge as q^{-2} for small wave numbers, $q \rightarrow 0$. These director fluctuations cause a strong turbidity in nematic liquid crystals, which have an opalescent appearance (Stephen and Straley, 1974). Thus the intensity of light scattered by a nematic medium is very large and can be readily observed with light scattering (Alms et al., 1977) or shadowgraphy (Rehberg et al., 1991b). A second salient feature of fluctuations in nematics is related to the intrinsic anisotropic character of the liquid crystals, which causes the structure of the light-scattering spectrum to depend on the polarization of incident and/or scattered light (Val'kov et al., 1994).

Regarding the physics of liquid crystals, an important characteristic is the strong coupling of the director with applied electric fields, since individual molecules tend to be aligned with such fields. This feature leads to a complete new phenomenology and introduces new complications in the theory such as a charge-balance equation. In particular, among the new nonequilibrium phenomena, convection (movement of the fluid liquid crystal) will appear when the intensity of an applied electric field is larger than a certain critical field[‡]. This new phenomenon is referred to as electrohydrodynamic convection (EHC) (Bodenschatz et al., 1988; Kramer and Pesh, 1995). Similarly to RB convection, above the threshold for EHC, stationary

---

[‡] Depending on whether the externally applied voltage is direct or alternating

patterns like rolls or hexagons may appear (Kramer and Pesh, 1995), as well as non-stationary patterns (Dennin et al., 1996). Of course, traditional RB convection, driven by temperature differences, also exist in liquid crystals. A review of the various types of convective instabilities in nematics, with an exhaustive comparison between RB convection and EHC has been presented by Kramer and Pesh (1995).

The theory of nonequilibrium fluctuations in liquid crystals has received more attention lately, in particular the case of nematics subjected to a stationary temperature gradient below the RB convection threshold (Pleiner and Brand, 1983a; Rodríguez and Camacho, 2002b), nematics subjected to a stationary pressure gradient (Rodríguez and Camacho, 2002a; Camacho et al., 2005), nematics in the presence of stationary shear flow (Pleiner and Brand, 1983b), and nematics below the threshold for EHC (Treiber and Kramer, 1994). The most interesting case is that of fluctuations below the threshold for EHC, because experimental information is available (see below). For this problem Treiber and Kramer (1994) developed an stochastic amplitude or envelope-equation formalism, which is equivalent to the linear Swift and Hohenberg (1977) approximation presented in Sect. 8.3 for RB convection in isotropic one-component fluids. This formalism is expected to describe appropriately nonequilibrium fluctuations close to the instability, but not close enough so that nonlinear effects become important. The most striking difference with RB convection in simple fluids is the breaking of rotational symmetry in the horizontal plane that happens in nematic EHC. As a consequence, the orientation of the convection rolls appearing above the EHC threshold has a preferred direction, determined by the BC for the director field (Treiber and Kramer, 1994).

Several experiments measuring nonequilibrium fluctuations below the onset of EHC have been performed in the past decade (Rehberg et al., 1991a,b; Winkler et al., 1992; Bisang and Ahlers, 1998; Scherer et al., 2000; Qiu and Ahlers, 2005). These experiments have confirmed most of the salient features of the Treiber and Kramer (1994) theory. For instance Rehberg et al. (1991b) experimentally verified that the intensity of the fluctuations increases as the onset of the EHC is approached following a power law $\propto (-\epsilon)^{-1/2}$. While in the case of RB convection the distance ϵ to the instability is measured in terms of a temperature, in the case of electroconvection it is measured in terms of a voltage. However, we note that the exponent of the power law is the same $-1/2$ as was found for the intensity of the fluctuations near the onset of RB convection in Sect. 8.4. Furthermore, the anisotropic enhancement of the fluctuations in the horizontal plane has also been observed (Bisang and Ahlers, 1998) by a shadowgraph technique similar to that described in Sect. 10.3 for RB convection. In addition, due to larger intrinsic noise, the nonlinear quench of the fluctuations when extremely close to the instability has been observed in EHC by Scherer et al. (2000). It is very difficult to observe such nonlinear effects near the onset of RB convection because thermal noise is in that case much weaker. The papers mentioned above deal with equal-time fluctuations, *i.e.*, only the total intensity of scattered light was measured. More recently Qiu and Ahlers (2005) have presented a shadowgraph experimental study of the *dynamics* of the fluctuations in a nematic liquid crystal near the onset of EHC.

While the attention has focused on nonequilibrium fluctuations in nematic liquid crystals, it would seem interesting to investigate nonequilibrium fluctuations in smectic and cholesteric phases of liquid crystals, because of a breakdown of conventional hydrodynamics in these systems (Mazenko et al., 1983).

11.4 Concentration fluctuations in reaction-diffusion problems

Concentration fluctuations in chemical reactions have been the subject of numerous studies over the years (van Kampen, 1982; Gardiner, 1985; Baras et al., 1996; Gillespie, 2000, 2001). Most of these studies are concerned with fluctuations in the homogeneous phase, *i.e.*, when the system is well stirred without any diffusion phenomena. There exists agreement in the literature that a proper theoretical framework for this problem can be based on the so-called chemical master equation (Gardiner, 1985). Then, the stationary solutions for the probability of the concentration fluctuations are Poisson distributions. However, starting from the chemical master equation, through a Kramers-Moyal approximation combined with the system-size expansion proposed by van Kampen (1976, 1982), it is possible to obtain a Fokker-Planck equation which is approximately equivalent to the original chemical master equation (Gardiner, 1985; Baras et al., 1996). The stationary solutions of the Fokker-Planck equation, or the equivalent Langevin equation, are gaussian. This fact has led to some debate about the general validity of the chemical Langevin equation (Gillespie, 2000, 2001; Zwanzig, 2001). For instance, it is known that the chemical Langevin equation fails when there is a bistability in the system of chemical reactions (Baras et al., 1996). We conclude that the Langevin equation is an approximation which is only valid when there is only one stable solution to the deterministic kinetic equations and when there are many particles per unit volume in the system. In this section we proceed by assuming that the approximations leading to the chemical Langevin equation are sufficiently well justified. We note that these approximations (many particles in small volume elements) are similar to the approximations that justify the hydrodynamic approach on which the previous chapters are based. Hence, we adopt here the Langevin equation to study the spatiotemporal evolution of the concentration fluctuations when both diffusion and chemical reactions are present.

Instead of discussing reaction-diffusion problems in a general context, we shall elucidate the features of fluctuations in these systems by treating two particular cases. Specifically, in Sect. 11.4.1, we consider the simple case of a chemical reaction between two active species in equilibrium (Berne and Pecora, 1976). For this case we shall calculate the static structure factor which will turn out to be short ranged in space. Then, in Sect. 11.4.2 we shall drive this chemical reaction to a nonequilibrium steady state by assuming that the inverse reaction proceeds by a different mechanism than the direct reaction. We shall see that the structure factor in such a nonequilibrium state is spatially long ranged, in contrast to when the chemical reactions are in equilibrium. This result is consistent with the general theme of this book. In the calculations of this section, we shall consider the system always in a quiescent $\mathbf{v} = 0$ state and we shall neglect any velocity or temperature fluctuations.

11.4.1 Equilibrium concentration fluctuations in reaction-diffusion

We first evaluate the contribution of the concentration fluctuations to the structure factor in a reaction-diffusion problem for which the chemical reactions are in equilibrium (Berne and Frisch, 1967; Blum and Salsbury, 1969; Berne and Pecora, 1976; Gardiner, 1985). We consider what is probably the simplest case, namely, a solution of an inert solvent with two active chemical species, A and B as solutes. These two species are in chemical equilibrium through the chemical reaction:

$$A \underset{k_2}{\overset{k_1}{\rightleftharpoons}} B, \tag{11.7}$$

where k_1 and k_2 are the kinetic chemical-reaction rates of the direct and of the inverse reaction, respectively, with units of inverse time. A detailed study of fluctuations for this chemical-kinetics problem can be found in the book of Berne and Pecora (1976), who have also discussed its equivalence to a dissociation problem in equilibrium. As explained above, a study of stochastic chemical reactions is usually based on a master-equation formalism (Gardiner, 1985). In this formulation, the relevant quantity is the probability $P(N_A, N_B; \mathbf{r}, t)$ of having, at a given time t, N_A molecules of species A and N_B molecules of species B inside an hydrodynamic volume element ΔV centered at a point $\mathbf{r}$. This probability is usually calculated subject to some initial values for the number of particles of the reactive species. When the system is in a quiescent state, if diffusion is neglected, the chemical master equation corresponding to (11.7) will be given by (Gardiner, 1985):

$$\frac{\partial P(N_A, N_B)}{\partial t} = k_1(N_A + 1) \ P(N_A + 1, N_B - 1) - k_1 N_A \ P(N_A, N_B)$$
$$+ k_2(N_B + 1) \ P(N_A - 1, N_B + 1) - k_2 N_B \ P(N_A, N_B), \tag{11.8}$$

where, to simplify the notation, we have dropped the spatiotemporal dependence of the probability $P(N_A, N_B)$. Following the usual procedures (Gardiner, 1985; Gillespie, 2000, 2001), one can obtain from Eq. (11.8) evolution equations for the mean values of the concentrations $\Delta V \, a(\mathbf{r}, t) = \langle N_A/N \rangle$ and $\Delta V \, b(\mathbf{r}, t) = \langle N_B/N \rangle$, where $N = N_A + N_B$ is the total number of active molecules in the volume element ΔV. The symbol $\langle \cdot \rangle$ means averaging over fluctuations, *i.e.*, averaging over the probability distribution $P(N_A, N_B; \mathbf{r}, t)$. When diffusion, not considered in Eq. (11.8), is included, one obtains the deterministic reaction-diffusion equation for a system of two active chemical species connected by the chemical reaction (11.7), namely:

$$
\begin{aligned}
\frac{\partial a}{\partial t} &= D_A \nabla^2 a - k_1 a + k_2 b, \\
\frac{\partial b}{\partial t} &= D_B \nabla^2 b + k_1 a - k_2 b.
\end{aligned}
\tag{11.9}
$$

The coefficients D_A and D_B are the diffusion coefficients of species A and of species B, respectively. We assume that both active chemical species diffuse independently so that $D_A \neq D_B$. Hence, in this section the concentrations of A and of B will be independent variables. It is important to note that the concentrations a and b in (11.9) are expressed as molecules (or moles) per unit volume, as is usual in chemical kinetics, while in the preceding chapters the concentrations were expressed in terms of mass fractions: c_A and c_B. The relationship between the two concentrations is:

$$
a = \frac{\rho}{m_A} \, c_A,
\tag{11.10}
$$

where ρ is the mass density of the (ternary) mixture and m_A the mass of the molecules A, with a similar relation for molecules B.

The homogeneous equilibrium solution of Eqs. (11.9) can be readily expressed as:

$$
\begin{aligned}
a_{\mathrm{eq}} &= \frac{k_2 n_p}{k_1 + k_2}, \\
b_{\mathrm{eq}} &= \frac{k_1 n_p}{k_1 + k_2},
\end{aligned}
\tag{11.11}
$$

where $n_p = a_{\mathrm{eq}} + b_{\mathrm{eq}}$ will be exclusively determined by the initial conditions, *i.e.*, by the total number of active molecules in the whole volume of the system. Thermodynamics tell us that an isolated system of A and B molecules with the chemical reaction (11.7), will reach a final equilibrium state where the concentrations of each kind of molecules will be given by Eqs. (11.11). The corresponding equilibrium constant K of the reaction is related to k_1 and k_2 by $K = k_2/k_1$, so that the law of mass action holds: $a_{\mathrm{eq}}/b_{\mathrm{eq}} = K$. The equilibrium solution is stable under fluctuations.

Next we consider concentration fluctuations around the equilibrium solution (11.11). As discussed by Gardiner (1985), the simplest procedure to study the spatiotemporal evolution of these fluctuations is the hydrodynamic approximation given by the chemical Langevin equation. As mentioned earlier, we assume that the system can be divided into cells or volume elements, sufficiently small for a local description, but that nevertheless contain a sufficiently large number of molecules so as to justify taking the hydrodynamic limit ($\Delta V \rightarrow \infty$, $N_A \rightarrow \infty$ and $a = N_A/\Delta V = $ constant). One can then perform a local expansion in the size of the cells, as suggested by van Kampen (1982). Such an expansion justifies the use of a chemical Langevin equation for the fluctuations (Gardiner, 1985). Thus, we assume that the evolution of the fluctuations around the equilibrium state can be described by the following set of linear stochastic differential equations (Gardiner, 1985):

$$\frac{\partial(\delta a)}{\partial t} = D_A \nabla^2(\delta a) - k_1(\delta a) + k_2(\delta b) - \frac{1}{m_A} \, \boldsymbol{\nabla} \cdot \delta \mathbf{J}^{(A)} + \delta\xi,$$
$$\frac{\partial(\delta b)}{\partial t} = D_B \nabla^2(\delta b) + k_1(\delta a) - k_2(\delta b) - \frac{1}{m_B} \, \boldsymbol{\nabla} \cdot \delta \mathbf{J}^{(B)} - \delta\xi,$$
(11.12)

where $\delta a(\mathbf{r}, t) = a(\mathbf{r}, t) - a_{\text{eq}}$ and $\delta b(\mathbf{r}, t) = b(\mathbf{r}, t) - b_{\text{eq}}$. In Eqs. (11.12), $\delta \mathbf{J}^{(A)}(\mathbf{r}, t)$ and $\delta \mathbf{J}^{(B)}(\mathbf{r}, t)$ represent the contribution to the fluctuations resulting from two independent random diffusive fluxes, as in Sect. 3.2.2 for binary mixtures. The prefactors m_A^{-1} and m_B^{-1} arise here, because we are now using concentrations in terms of molecules per unit volume, while in defining diffusion in previous chapters we expressed the concentrations in terms of mass fractions. An additional random force in Eq. (11.12) is $\delta\xi(\mathbf{r}, t)$, representing the contribution to the fluctuations arising from the chemical reactions. The term $\delta\xi(\mathbf{r}, t)$ can be interpreted as an stochastic degree of advancement of the chemical reaction (11.7), which is a new dissipative flux appearing in the nonequilibrium thermodynamics of chemical reacting systems (de Groot and Mazur, 1962). As a consequence of the stoichiometry of the particular chemical reaction (11.7) the random rate $\delta\xi(\mathbf{r}, t)$ appears with opposite signs in the evolution equations (11.12) for the concentration fluctuations δa and δb.

For the correlation functions between the random diffusive fluxes, we adopt the FDT as formulated by the authors who have studied diffusion in the context of a master equation (Gardiner, 1985), namely:

$$\langle \delta J_i^{(A)}(\mathbf{r}, t) \cdot \delta J_j^{(A)}(\mathbf{r}', t') \rangle = 2m_A^2 D_A a_{\text{eq}} \, \delta_{ij} \, \delta(\mathbf{r} - \mathbf{r}') \, \delta(t - t'),$$
$$\langle \delta J_i^{(B)}(\mathbf{r}, t) \cdot \delta J_j^{(B)}(\mathbf{r}', t') \rangle = 2m_B^2 D_B b_{\text{eq}} \, \delta_{ij} \, \delta(\mathbf{r} - \mathbf{r}') \, \delta(t - t'),$$
(11.13)

while the cross-correlation between the two random diffusive fluxes is zero. Of course, Eqs. (11.13) are compatible with the thermodynamic FDT presented in Sect. 3.2.2. Indeed, Eqs. (11.13) can be obtained from thermodynamics by first noticing that it implies the assumption that the particles perform independent random walks. This means that interactions between the molecules are neglected so that (11.13) actually only applies in the limit of infinite dilution. In this limit the mixture behaves as an ideal mixture for which the chemical potential of A species can be represented by:

$$\mu_A = \mu_A^\circ(p, T) + \frac{k_B T}{m_A} \ln x_A, \tag{11.14}$$

where x_A is the molecular fraction of A and where a factor m_A^{-1} appears because in this book chemical potentials have been defined per unit mass. If one assumes that the mole fraction of the solvent can be approximated by unity (infinite dilution), substitutes (11.14) or the equivalent equation for μ_B into Eqs. (3.9)-(3.10), one can verify with the help of (11.10) that the thermodynamic FDT, as presented in Sect. 3.2.2, indeed reduces to (11.13). In the limit of infinite dilution there is no interaction among the particles and no cross-diffusion effect is expected, neither in the FDT (11.13) nor in the mass-balance law (11.9).

The correlation function of the random rate of the chemical reaction is assumed to be local (short ranged) and can be obtained from the master equation (11.8) by the standard procedures described in detail in the literature (Gardiner and Chaturvedi, 1977; Gardiner, 1985; Baras et al., 1996). For the particular chemical reaction (11.7), one obtains:

$$\begin{aligned}
\langle \delta\xi(\mathbf{r}, t) \cdot \delta\xi(\mathbf{r}', t') \rangle &= 2k_1 a_{\text{eq}} \, \delta(\mathbf{r} - \mathbf{r}') \, \delta(t - t'), \\
&= [k_1 a_{\text{eq}} + k_2 b_{\text{eq}}] \, \delta(\mathbf{r} - \mathbf{r}') \, \delta(t - t').
\end{aligned} \tag{11.15}$$

Equation (11.15) can be derived not only by starting from the chemical master equation, but also from the nonequilibrium thermodynamics of a system with chemical reactions; see §10.2 in de Groot and Mazur (1962). The point is that the chemical kinetic relation in the RHS of (11.9), namely,

$$\xi = -k_1 a + k_2 b \tag{11.16}$$

between the rate ξ of the chemical reaction and the concentrations a and b can, in the limit of a dilute ideal solution and when concentrations are close to their equilibrium values, be interpreted as a linear phenomenological relation between a dissipative flux (the reaction rate) and the corresponding thermodynamic force (the affinity of the reaction). Then, as shown by Eq. (X.49) in the book of de Groot and Mazur (1962), the prefactor $2k_1 a_{\text{eq}}$ in the RHS of (11.15) is precisely $2k_B T$ times the corresponding phenomenological coefficient, as required by the generic thermodynamic FDT (3.3). It is important to note that, to interpret Eq. (11.16) as a linear phenomenological relation, the chemical reactions have to be very close to equilibrium, so that the affinity is much lower than $k_B T$ (de Groot and Mazur, 1962).

We now have all the information needed to solve the evolution equations (11.12) for the fluctuations around the equilibrium state. Applying a space and time Fourier transformation to Eqs. (11.12), the resulting set of algebraic equations may be written in the canonical form:

$$\mathsf{G}^{-1}(\omega, q) \begin{pmatrix} \delta a(\omega, \mathbf{q}) \\ \delta b(\omega, \mathbf{q}) \end{pmatrix} = \mathbf{F}(\omega, \mathbf{q}), \tag{11.17}$$

where ω is the frequency and $\mathbf{q}$ the wave number of the fluctuations. The inverse of the response function in this case is given by:

$$\mathsf{G}^{-1}(\omega, q) = \begin{pmatrix} \mathrm{i}\,\omega + D_A q^2 + k_1 & -k_2 \\ -k_1 & \mathrm{i}\,\omega + D_B q^2 + k_2 \end{pmatrix}, \tag{11.18}$$

while the noise $\mathbf{F}(\omega, \mathbf{q})$ is related to the Fourier transformed random dissipative fluxes by:

$$\mathbf{F}(\omega, \mathbf{q}) = \begin{pmatrix} \dfrac{-\mathrm{i}}{m_A}\, q_i J_i^{(A)}(\omega, \mathbf{q}) + \delta\xi(\omega, \mathbf{q}) \\ \dfrac{-\mathrm{i}}{m_B}\, q_i J_i^{(B)}(\omega, \mathbf{q}) - \delta\xi(\omega, \mathbf{q}) \end{pmatrix}. \tag{11.19}$$

Equation (11.18) can be readily solved for the Fourier transforms of the fluctuations $\delta a(\omega, \mathbf{q})$ and $\delta b(\omega, \mathbf{q})$. We are here interested in the autocorrelation function of the concentration fluctuations which can, in principle, be measured with light-scattering or shadowgraph experiments. For this purpose we need the correlation functions between the Fourier transforms of the random noise terms in the RHS of Eq. (11.18). They can be obtained by applying a double Fourier transformation to the FDT's (11.13) and (11.15). As usual, they are conveniently expressed in terms of a correlation matrix:

$$\langle F_\alpha^*(\omega, \mathbf{q}) \cdot F_\beta(\omega', \mathbf{q}') \rangle = C_{\alpha\beta}(q)\,(2\pi)^4\,\delta(\mathbf{q} - \mathbf{q}')\,\delta(\omega - \omega'), \tag{11.20}$$

where the correlation matrix for our current problem results in:

$$\mathsf{C}(q) = \begin{bmatrix} 2D_A q^2 a_{\mathrm{eq}} + k_1 a_{\mathrm{eq}} + k_2 b_{\mathrm{eq}} & -(k_1 a_{\mathrm{eq}} + k_2 b_{\mathrm{eq}}) \\ -(k_1 a_{\mathrm{eq}} + k_2 b_{\mathrm{eq}}) & 2D_B q^2 b_{\mathrm{eq}} + k_1 a_{\mathrm{eq}} + k_2 b_{\mathrm{eq}} \end{bmatrix}. \tag{11.21}$$

It is interesting to note that the temperature T or Boltzmann's constant k_B do not appear in Eq. (11.21). This justifies the assumption that, in the infinitely dilute limit and not far from equilibrium, reaction-diffusion problems are equivalent to random walks where the walkers have some probability of changing from A-type to B-type. These random walks problems are athermal, in the sense that temperature plays no role. Many studies on reaction-diffusion have been performed in this random-walk context (Gardiner, 1985).

In actual optical experiments one measures fluctuations in the refractive index of the mixture. Since we have here two independent concentrations, the structure factor will contain contributions from several partial structure factors, similar to Eq. (5.30) for the contributions of the independent temperature and concentration fluctuations to the structure factor of a binary mixture. However, for our purpose here and without loss of generality we may assume that fluctuations of the refractive index are caused only by the concentration fluctuations of species A. Then, a single structure factor $S(\omega, q)$ can be defined in terms of $\langle \delta a^*(\omega, \mathbf{q}) \, \delta a(\omega', \mathbf{q}') \rangle$ as:

$$\langle \delta a^*(\omega, \mathbf{q}) \, \delta a(\omega', \mathbf{q}') \rangle = \left(\frac{\partial a}{\partial n} \right)_{p,T}^2 S(\omega, q) \, (2\pi)^4 \, \delta(\omega - \omega') \, \delta(\mathbf{q} - \mathbf{q}'), \quad (11.22)$$

and we do not have to discuss other partial structure factors. In this equation $n(a, p, T)$ denotes the index of refraction of the solution as a function of the concentration of species A at a given pressure p and a given temperature T.

To calculate this structure factor we invert Eq. (11.18) to obtain the linear response function. Then, by using Eqs. (11.11), (11.21) and (11.22), and assuming $D_B \geq D_A$, we are able to represent the structure factor as:

$$S(\omega, q) = a_{\text{eq}} \left\{ \frac{A_+(q) \, \Gamma_+(q)}{\omega^2 + [\Gamma_+(q)]^2} + \frac{A_-(q) \, \Gamma_-(q)}{\omega^2 + [\Gamma_-(q)]^2} \right\}, \quad (11.23)$$

where the decay rates $\Gamma_\pm(q)$ of the fluctuations are

$$\Gamma_\pm(q) = \frac{1}{2} \left\{ (D_A + D_B) \, q^2 + k_1 + k_2 \right.$$
$$\left. \pm \sqrt{[(D_B - D_A)q^2 - k_1 + k_2]^2 + 4k_1 k_2} \right\} \quad (11.24)$$

and the amplitudes $A_\pm(q)$:

$$A_\pm(q) = 1 \mp \sqrt{1 + \frac{4k_1 k_2}{[\Gamma_+(q) - \Gamma_-(q)]^2}}. \quad (11.25)$$

Notice that the decay rates (11.24) are always real numbers, independent of the wave number q. Therefore, there are never propagating modes in reaction-diffusion problems.

Equation (11.23) shows that the spectrum of light scattered in a reacting fluid mixture would contain information about the kinetic constants k_1 and k_2. However, it seems that it is difficult to obtain the kinetic rate coefficients from actual experiments, see Berne and Pecora (1976) or Martens (2002). We shall comment more on this at the end of Sect. 11.4.2. Equation (11.23) agrees with the results presented in Sect. 8.3.1 of the book by Gardiner (1985), where the same chemical kinetics is considered.

For our purpose here, the most important physical feature is that Eq. (11.23) implies that the concentration fluctuations are spatially short ranged. Specifically, if we calculate from $S(\omega, q)$ the static structure factor $S(q)$ by integrating Eq. (11.23) over the frequency in accordance with Eq. (3.24), we obtain:

$$S(\mathbf{q}) = \frac{n_p \, k_2}{k_1 + k_2} = a_{\text{eq}}. \tag{11.26}$$

The intensity of the fluctuations is independent of the wave number q, meaning that the equal-time autocorrelation function $(t = t')$ of concentration fluctuations will be spatially short ranged. Indeed, by applying to Eq. (11.26) a double inverse Fourier transformation over $\mathbb{R}^3$ and taking into account Eq. (11.22), we obtain:

$$\langle \delta a(\mathbf{r}, t) \cdot \delta a(\mathbf{r}', t) \rangle = a_{\text{eq}} \, \delta(\mathbf{r} - \mathbf{r}'), \tag{11.27}$$

where the spatially short-ranged character of the concentration fluctuations is evident. Consequently, at a given time t, the concentration fluctuations at a given position in the equilibrium reacting mixture are completely uncorrelated with the fluctuations at any other position. We also note that the intensity of the concentration fluctuations does not contain any information about the chemical reactions in the system. Hence, information about the kinetics of the reaction can only be obtained from dynamic experiments.

We have elucidated in this section that the concentration fluctuations in an equilibrium reacting dilute solution are short ranged for the simple chemical reaction (11.7). However, this result is general and applies to any equilibrium chemical reaction, even for concentrated (not dilute) species. In the case of concentrated solutions, the prefactors multiplying the delta functions in (11.27) will contain concentration derivatives of the chemical potentials. A first approach to this more general case has been considered, for instance, by Grossmann (1976).

11.4.2 Nonequilibrium concentration fluctuations in reaction-diffusion

In contrast to equilibrium, concentration fluctuations become long ranged when the chemical reactions are driven out of equilibrium. We shall illustrate this feature by considering, as in the previous section, a solution with two active solutes A and B involved in two chemical reactions with kinetic coefficients k_1 and k_2. However, to drive the system to a nonequilibrium steady state we assume that the inverse reaction proceeds through a mechanism different from that of the direct reaction, namely:

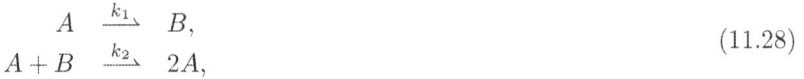

$$
\begin{aligned}
A &\xrightarrow{k_1} B, \\
A + B &\xrightarrow{k_2} 2A,
\end{aligned}
\tag{11.28}
$$

where the first reaction destroys one type-A molecule and creates one type-B molecule, while the second reaction performs the opposite. This particular kinetic scheme (11.28) has been the subject of some recent studies in the literature, where it is known as the WOH (van Wijland, Oerding, Hilhorst) model (Kree et al., 1989; van Wijland et al., 1998; de Freitas et al., 2000; Janssen, 2001; Fulco et al., 2001). It was originally introduced to study the effects of pollution in critical population models, but it has also been widely interpreted as a set of coupled chemical reactions, as we do here. The evolution equations for the concentrations $a(\mathbf{r}, t)$ and $b(\mathbf{r}, t)$ can be written in this case as (van Wijland et al., 1998):

$$
\begin{aligned}
\frac{\partial a}{\partial t} &= D \, \nabla^2 a - k_1 a + k_2 ab + F_0(\mathbf{r}, t), \\
\frac{\partial b}{\partial t} &= D \, \nabla^2 b + k_1 a - k_2 ab + F_1(\mathbf{r}, t),
\end{aligned}
\tag{11.29}
$$

where, to expedite the discussion, we have already added a two-component random noise term $\mathbf{F}(\mathbf{r}, t)$. As in the previous section, it contains contributions from diffusion and from the chemical reaction. Furthermore, without loss of generality for our purpose and to simplify the calculations, we have assumed in Eq. (11.29) that $D_A = D_B = D$.

Because of the assumption of equal diffusion coefficients, Eq. (11.29) implies that the total concentration $n_p = a + b$ of the chemically active species is locally conserved in time since, when averaged over fluctuations, the random noise terms vanish: $\langle F_0(\mathbf{r}, t) \rangle = 0$ and $\langle F_1(\mathbf{r}, t) \rangle = 0$. We therefore use n_p as a parameter to describe the homogeneous steady-state solutions of Eq. (11.29). A simple calculation (de Freitas et al., 2000) shows that there exist two such solutions. The first solution is simply: $a_{\mathrm{ss}} = 0$ and $b_{\mathrm{ss}} = n_p$. However, we focus here on the second solution, for which the concentration of A is not zero:

$$
\begin{aligned}
a_{\mathrm{ss}} &= n_p - \frac{k_1}{k_2}, \\
b_{\mathrm{ss}} &= \frac{k_1}{k_2}.
\end{aligned}
\tag{11.30}
$$

A linear stability study (van Wijland et al., 1998; de Freitas et al., 2000) shows that the first homogeneous stationary solution ($a_{\mathrm{ss}} = 0$) is stable for $n_p \leq n_{p,\mathrm{c}}$, while the second solution, Eq. (11.30), is stable for $n_p > n_{p,\mathrm{c}}$, where $n_{p,\mathrm{c}} = k_1/k_2$. In the following we shall be interested in studying small fluctuations around the solution with $a_{ss} \neq 0$; we thus assume that $n_p > n_{p,c}$. The state described by Eq. (11.30), although homogeneous and stationary, it is not a thermodynamic equilibrium state.

Most of the papers on the WOH model presented in the literature have been concerned with a "nonequilibrium phase transition" from the stationary solution with $a_{ss} = 0$ to the solution (11.30) that appears as $n_p \to n_{p,c}$. As near critical points in equilibrium phase transitions, critical fluctuations appear when the two solutions exchange their stabilities, which have been investigated by renormalization-group techniques (van Wijland et al., 1998). Calculations of corresponding critical exponents have received particular attention (de Freitas et al., 2000), and have caused some debate (Janssen, 2001) about the universality class of the WOH model. However, here we have a different goal and we shall not consider critical fluctuations. We shall assume that $n_{p,c} \ll n_p$ so that the solution (11.30) is stable and far from the critical concentration. The fluctuations around (11.30) are then sufficiently small to be adequately described by a linearized mean-field theory supplemented with appropriate random noise terms, as already assumed in Eq. (11.29).

To calculate the autocorrelation function of the concentration fluctuations, we substitute $a(\mathbf{r}, t) = \delta a(\mathbf{r}, t) + a_{ss}$ and $b(\mathbf{r}, t) = \delta b(\mathbf{r}, t) + b_{ss}$ into Eqs. (11.29) and neglect terms of second order in the fluctuations. As in Sect. 11.4.1, we apply a Fourier transformation in space and time and obtain the following system of algebraic equations for the Fourier transforms of the fluctuations:

$$\begin{pmatrix} i\,\omega + Dq^2 & k_1 - k_2 n_p \\ 0 & i\,\omega + Dq^2 - k_1 + k_2 n_p \end{pmatrix} \begin{pmatrix} \delta a(\omega, \mathbf{q}) \\ \delta b(\omega, \mathbf{q}) \end{pmatrix} = \mathbf{F}(\omega, \mathbf{q}). \quad (11.31)$$

We can readily obtain an expression for the concentration fluctuations by inverting Eq. (11.31). As before, for the calculation of the structure factor we need the autocorrelation functions between the components of the Fourier transform of the random noise term $\mathbf{F}(\omega, \mathbf{q})$. We obtain these by the same method used in Sect. 11.4.1, which is based on the chemical master equation (with diffusion). Thus, we first consider the chemical master equation corresponding to the kinetic scheme (11.28), next we divide the system in small volume elements ΔV, but each one containing enough molecules so as to justify a local system-size expansion (Gardiner, 1985). Finally, we take into account the effects of diffusion (assuming again an ideal infinitely diluted mixture). By this chain of arguments, instead of (11.21), we obtain for the correlation matrix:

$$\mathsf{C}(q) = \begin{bmatrix} 2Dq^2 a_{ss} + k_1 a_{ss} + k_2 a_{ss} b_{ss} & -(k_1 a_{ss} + k_2 a_{ss} b_{ss}) \\ -(k_1 a_{ss} + k_2 a_{ss} b_{ss}) & 2Dq^2 b_{ss} + k_1 a_{ss} + k_2 a_{ss} b_{ss} \end{bmatrix}, \quad (11.32)$$

where a_{ss} and b_{ss} are given by Eqs. (11.30). Again, neither Boltzmann's constant nor temperature appear in the expression of the correlation matrix.

In the previous section, we saw that the correlation matrix (11.21) obtained from the master equation when the chemical reactions are in equilibrium gives the same result that can be deduced from the thermodynamic FDT by using the appropriate expression for the entropy production (see Chapter 3). When the chemical reactions are in a nonequilibrium steady state, it seems also possible to deduce Eq. (11.32) from purely thermodynamic considerations. For example, Vlad and Ross (1994a,b,c) have considered nonequilibrium fluctuations in chemical-reacting systems based on an "excess free energy" introduced by Ross et al. (1992). Alternatively and probably equivalently, a method by Pagonabarraga et al. (1997), who introduced a (mesoscopic) variable to describe intermediate reaction states, also seems to be able to reproduce (11.32) from thermodynamic considerations.

To obtain the structure factor we shall continue to assume that $S(\omega, q)$ is related to the autocorrelation function of the concentration fluctuations by the same relation (11.22) employed in the previous section when the active species A and B were in chemical equilibrium. Then, by inverting Eq. (11.31) and using Eqs. (11.32) and (11.22), we obtain for the nonequilibrium structure factor $S(\omega, q)$:

$$S(\omega, q) = \frac{a_{ss}}{2Dq^2 + k_2 n_p - k_1} \left\{ \frac{2Dq^2(2Dq^2 + k_2 n_p)}{\omega^2 + D^2 q^4} \right.$$
$$\left. + \frac{k_1(Dq^2 + k_2 n_p - k_1)}{\omega^2 + (Dq^2 + k_2 n_p - k_1)^2} \right\}. \quad (11.33)$$

As usual, integration of $S(\omega, q)$ over the frequency ω, yields the static structure factor:

$$S(q) = a_{ss} \left\{ 1 + \frac{2k_1}{2Dq^2 + k_2 n_p - k_1} \right\}. \quad (11.34)$$

We note that outside equilibrium $S(q)$ exhibits an explicit dependence on the wave number q, implying that the equal-time autocorrelation function of the concentration fluctuations is spatially long ranged. The asymptotic behavior of the nonequilibrium structure factor in Eq. (11.34) is proportional to q^{-2} for large q. This behavior is to be compared with a q^{-2} proportionality found for an isothermal one-component fluid under shear or with a q^{-4} proportionality found for the nonequilibrium fluctuations in the RB problem.

By applying a double inverse Fourier transformation in $\mathbb{R}^3$ to Eq. (11.34) and taking into account Eq. (11.22), we find for the real-space dependence of the concentration fluctuations:

$$\langle \delta a(\mathbf{r}, t) \cdot \delta a(\mathbf{r}', t) \rangle = a_{\mathrm{ss}}$$

$$\times \left\{ \delta(\mathbf{r} - \mathbf{r}') + \frac{k_1}{4\pi|\mathbf{r} - \mathbf{r}'|} \exp\left(-|\mathbf{r} - \mathbf{r}'| \sqrt{\frac{k_2 n_p - k_1}{2D}} \right) \right\}. \quad (11.35)$$

Comparison with Eq. (11.27) shows that expression (11.35) for the concentration fluctuations can be decomposed into a short-ranged part, which is the same as if species A were in equilibrium at the concentration a_{ss}, plus a long-ranged-part, which is due to the fact that chemical reactions are out of equilibrium. We conclude that, when chemical reactions are in equilibrium, equal-time concentration fluctuations are spatially short ranged in accordance with Eq. (11.27). However, when chemical reactions are outside equilibrium, equal-time concentration fluctuations are spatially long ranged in accordance with Eq. (11.35). This was the main goal of our analysis of fluctuations in chemical-reacting systems.

It is also noteworthy that the long-ranged nonequilibrium contribution in Eq. (11.35) diverges at $\mathbf{r} = \mathbf{r}'$. This fact has no special physical meaning, but is a consequence of having represented the local correlations as proportional to delta functions. For the same reason the contribution in Eq.(11.35) from the equilibrium fluctuations also diverges at $\mathbf{r} = \mathbf{r}'$. In spite of these divergences, a calculation of any physically relevant quantity, such as the fluctuations in the number of particles A contained in a volume element ΔV, yields meaningful results, as discussed in Sect. 8.3.2 of the book by Gardiner (1985). Any such physically meaningful quantities are obtained by double integration over $\mathbf{r}$ and $\mathbf{r}'$, yielding well-defined results. The dependence of the nonequilibrium contribution on $|\mathbf{r} - \mathbf{r}'|^{-1}$ indicates that the result obtained for any physical quantity will depend on the size ΔV of the volume where it is calculated.

As discussed in Chapter 10, the structure factor of a fluid can be experimentally measured by light scattering or by shadowgraph techniques. However, such measurements are difficult and, in spite of several attempts (Berne and Pecora, 1976), conclusive light-scattering measurements for chemically reacting fluid systems do not seem to be available, not even for the simpler case of chemical reactions in equilibrium. Furthermore, when the chemical reactions are in a steady nonequilibrium state, the experimental results that would be obtained by light scattering will depend on the size of the scattering volume (see above). Because of these complications the long-ranged nature of nonequilibrium concentration fluctuations in reaction-diffusion systems has not yet been verified experimentally. However, the appearance of chemical patterns (Turing patterns) above a threshold in certain chemical reactions can be viewed as a confirmation of the long-ranged nature of these fluctuations, just as the appearance of convection patterns in the RB problem originate from long-ranged nonequilibrium fluctuations below the threshold for convection, as discussed with detail in Chapter 8.

Chapter 12

Epilogue

This book has dealt with thermally excited hydrodynamic fluctuations in fluids and fluid mixtures that are either in a thermodynamic equilibrium state or in a stationary quiescent non-convective non-turbulent nonequilibrium state. There is a qualitative difference between thermal fluctuations in fluids in equilibrium and in nonequilibrium states. With the exception of states near a critical point, temperature and/or concentration fluctuations in fluids in thermodynamic equilibrium are generally uncorrelated at hydrodynamic length scales. Hence, the corresponding real-space hydrodynamic correlation functions in equilibrium are proportional to a delta function of the distance variable r, as explained in Sects. 3.3.2 and 11.4.1. On the other hand, in nonequilibrium states, due to generic scale invariance, the hydrodynamic correlation functions exhibit an algebraic dependence on the distance variable r, so that the nonequilibrium fluctuations become generically long ranged. In the presence of a temperature and/or concentration gradient, the correlations even extend over the entire system, so that buoyancy and finite-size effects need to be included to specify the wave-number dependence of the nonequilibrium structure factors, as explained in Chapters 6-9.

Nonequilibrium thermodynamics is based on a local-equilibrium assumption which implies that at each point in space and time the local thermodynamic properties are related by the same thermodynamic relations as for a fluid in equilibrium. While this assumption is valid for the thermodynamic properties themselves, it turns out that the local-equilibrium assumption no longer holds for hydrodynamic correlation functions. This failure is obvious for fluids undergoing thermal convection or turbulence, but it also fails in quiescent nonequilibrium states far away from any hydrodynamic instability.

Near the critical point of fluids long-ranged dynamic correlations are caused by a coupling between the hydrodynamic modes resulting from nonlinear terms in the hydrodynamic equations. However, in nonequilibrium

269

fluids a coupling between hydrodynamic modes is caused by the presence of an externally imposed gradient that induces a dissipative flux. These couplings usually arise from advection (in the case of the RB problem, the term $(\mathbf{v} \cdot \boldsymbol{\nabla})T$ in the heat equation), so that the origin of the coupling may be, initially, referred to as nonlinear. Hence, the importance of coupling in the theory of nonequilibrium fluctuations was first recognized by researchers interested in nonlinear mode-coupling in kinetic theory (Kirkpatrick et al., 1982b). However, the presence of a gradient causes advection to produce in nonequilibrium states terms that are linear in the fluctuating fields. This is the reason why many mode-coupling effects in nonequilibrium fluids can be treated within the framework of linearized fluctuating hydrodynamics. The spatial long-ranged nature of nonequilibrium fluctuations has a simple physical origin. For instance, in the case of the RB problem, fluctuations in the vertical component of the fluid velocity (parallel to the gradient) are mixing regions with different values of the local temperature, thus causing an enhancement of temperature fluctuations as extensively discussed in Chapter 4.

To extend the theory of fluctuating hydrodynamics to fluctuations in nonequilibrium fluids we have assumed that the noise correlation functions are still given by a local application of the FDT. In this book we have formulated the FDT in terms of the dissipative fluxes. Dissipative fluxes are the manifestation at a macroscopic scale of the random molecular motions ("the kind of motion we call heat"), so that they have been considered in this book as intrinsically stochastic variables, or processes when the dynamics is considered. The characterization of the dissipative fluxes as stochastic processes (for an equilibrium state) has been discussed in Sect. 3.1. In this context, the FDT is indeed a "theorem", that follows from the Einstein hypothesis. Many authors have discussed the FDT in terms of the (equal-time) correlation functions among the fluctuating fields. Of course, when the system under consideration is in a global equilibrium state, it does not matter whether the FDT is presented in terms of dissipative fluxes or in terms of fluctuating fields, since correlation functions among dissipative fluxes or fluctuating variables are both spatially short-ranged. However, as amply discussed in this book, when the system is in a nonequilibrium state, the random part of the dissipative fluxes continues to be spatially short ranged, while the fluctuating part of the thermodynamic fields actually becomes spatially long ranged. For this reason, one has to be careful when discussing apparent "violations" of the FDT in nonequilibrium states. In view of the results for nonequilibrium states, we suggest that a "thermodynamic" formulation of the FDT in terms of dissipative fluxes would be preferable. We believe that such a formulation would be more fundamental than in terms of fluctuating fields, the latter being only valid for global equilibrium states. These ideas regarding the FDT can be justified on the basis of kinetic theory (Chapter 3), and on the basis of accurate experiments

(Chapter 10).

In recent years attempts have been made to extend the theory of fluctuations to systems that are very far from equilibrium. Fluctuation theorems for nonequilibrium steady states of various kinds have been formulated by Gallavotti and Cohen (1995a,b), by Evans and Searles (2002) and others. A closely related set of results which allows nonequilibrium processes to be used to obtain equilibrium properties has been developed by Jarzynski (1997, 2001, 2002). The relationship between the results of Jarzynski and the other approaches has been discussed by Crooks (2000).

While these new developments have opened the possibility of dealing with fluctuations far away from equilibrium, we have demonstrated that there still exist a variety of nonequilibrium fluctuations that can be dealt with in terms of an extension of Landau and Lifshitz (1958, 1959) fluctuating hydrodynamics. We have illustrated this specifically with a detailed analysis of fluctuations in fluids and fluid mixtures subjected to a stationary temperature gradient. In this book we only considered temperature and concentration fluctuations in newtonian molecular fluids. With the exception of some theoretical studies of nonequilibrium fluctuations in liquid crystals, discussed in Sect. 11.3, there exists little information on nonequilibrium fluctuations induced by a temperature gradient in complex fluids.

In the case of liquid crystals nonequilibrium fluctuations below the onset of EHC are particularly interesting. As reviewed in Sect. 11.3, the current theory has been developed in the context of an amplitude equation valid only for fluctuations very close to the electro-convective threshold. Hence, a suitable topic for further research would be to develop a more complete theory for nonequilibrium fluctuations in a liquid crystal below the onset of EHC, similar to the one presented in this volume for the RB problem. Another topic on nonequilibrium fluctuations that probably deserves more theoretical work is that of the contribution of fluctuations to heat or mass transfer, which was very briefly mentioned in Sect. 7.6.

In Sect. 8.6 we saw that the mode-coupling effects incorporated in fluctuating hydrodynamics yield an amplitudes of the wave-number-dependent nonequilibrium fluctuations near the onset of Rayleigh-Bénard convection that differ from the those found from traditional hydrodynamic instability analysis. It would be interesting to also investigate mode-coupling effects in fluids in laminar flow prior to the transition to turbulence. In addition, the recent experimental realization of Turing patterns (Castets et al., 1990) makes it highly desirable to develop a theory of nonequilibrium fluctuations for reaction-diffusion problems with more realistic chemical kinetics than the one considered in Sect. 11.4. Finally, among the topics worthy of further investigation, we may mention the recent developments in the theory of nonlinear stochastic differential equations, particularly in relation to the KPZ (Kandar et al., 1986; Katzav and Schwartz, 2004) or Burgers (1974) equations. An effort will be required to integrate these developments with the theory presented in this book.

In conclusion, we hope that this book provides the basic principles of using fluctuating hydrodynamics in dealing with a variety of nonequilibrium fluctuations in fluids.

Bibliography

Ackerson, B. J., Hanley, H. J. M., 1980. Rayleigh scattering from a methane-ethane mixture. J. Chem. Phys. **73**, 3568.

Ahlers, G., 1980. Effect of departures from the Oberbeck-Boussinesq approximation on the heat transport of horizontal convecting fluid layers. J. Fluid Mech. **98, Part I**, 137.

Ahlers, G., 1991. Experiments with pattern-forming systems. Physica D **51**, 421.

Ahlers, G., Cross, M. C., Hohenberg, P. C., Safran, S., 1981. The amplitude equation near the convective threshold: application to time-dependent heating experiments. J. Fluid Mech. **110**, 297.

Alder, B. J., Wainwright, T., 1970. Decay of velocity autocorrelation functions. Phys. Rev. A **1**, 18.

Allain, C., Cummins, H. Z., Lallemand, P., 1978. Critical slowing down near the Rayleigh-Bénard convective instability. J. Physique (Paris) Lettres **39**, L473.

Alms, G. R., Gierke, T. D., Patterson, G. D., 1977. Observation and analysis of the depolarized Rayleigh doublet in isotropic MBBA and measurement of the de Gennes viscosity coefficients. J. Chem. Phys. **67**, 5779.

Anisimov, M. A., Agayan, V. A., Povodyrev, A. A., Sengers, J. V., Gorodetskii, E. E., 1998. Two-exponential decay of dynamic light scattering in near-critical fluid mixtures. Phys. Rev. E **57**, 1946.

Baras, F., Malek Mansour, M., Pearson, J. E., 1996. Microscopic simulation of chemical bistability in homogeneous systems. J. Chem. Phys. **105**, 8257.

Barrat, J.-L., Hansen, J.-P., 2003. Basic Concepts for Simple and Complex Liquids. Cambridge University Press, Cambridge.

Batchelor, G. K., 1967. An Introduction to Fluid Dynamics. Cambridge Univ. Press, Cambridge.

Bedeaux, D., Mazur, P., 1974. Renormalization of the diffusion coeffcient in a fluctuating field. Physica **73**, 431.

Bedeaux, D., Oppenheim, I., 1978. Hydrodynamic response and free surface modes for two immiscible fluids. Physica **90A**, 39.

Behringer, R., Ahlers, G., 1977. Heat transport and critical slowing down near the Rayleigh-Bénard instability in cylindrical containers. Phys. Lett. **62A**, 329.

Belitz, D., Kirkpatrick, T. R., Votja, T., 2005. Influence of generic scale invariance on classical and quantum phase transitions. Rev. Mod. Phys. **77**, 579.

Berne, B. J., Frisch, H. L., 1967. Light scattering as a probe of fast reaction kinetics. J. Chem. Phys. **47**, 3675.

Berne, B. J., Pecora, R., 1976. Dynamic Light Scattering. Wiley, New York. Dover edition, 2000.

Berry, M. V., Bodenschatz, E., 1999. Caustics, multiply reconstructed by Talbot interference. J. Modern Optics **46**, 349.

Beysens, D., 1983. Fluctuations anisotropy arising from velocity or temperature gradients. Phyica **118A**, 250.

Beysens, D., Garrabos, Y., Zalczer, G., 1980. Experimental evidence for Brillouin asymmetry induced by a temperature gradient. Phys. Rev. Lett. **45**, 403.

Bird, R. B., Armstrong, R. C., Hassager, O., 1986. Dynamics of Polymeric Liquids. Volume 1. Fluid Mechanics, 2nd Ed. Wiley, New York.

Bird, R. B., Stewart, W. E., Lightfoot, E. N., 2002. Transport Phenomena, 2nd Ed. Wiley, New York.

Bisang, U., Ahlers, G., 1998. Thermal fluctuations, subcritical bifurcation, and nucleation of localized states in electroconvection. Phys. Rev. Lett. **80**, 3061.

Bixon, M., Zwanzig, R., 1969. Boltzmann-Langevin equation and hydrodynamic fluctuations. Phys. Rev. **187**, 267.

Blum, L., Salsbury, Z. W., 1969. Light scattering from chemically reactive systems. II. Case with diffusion. J. Chem. Phys. **50**, 1654.

Bodenschatz, E., de Bruyn, J. R., Ahlers, G., Cannell, D. S., 1991. Transitions between patterns in thermal convection. Phys. Rev. Lett. **67**, 3078.

Bodenschatz, E., Pesch, W., Ahlers, G., 2000. Recent developments in Rayleigh-Bénard convection. Annu. Rev. Fluid Mech. **32**, 709.

Bodenschatz, E., Zimmermann, W., Kramer, L., 1988. On electrically driven pattern-forming instabilities in planar nematics. J. Physique (Paris) **49**, 1875.

Bogoliubov, N. N., 1946. J. Phys. (USSR) **10**, 256, 265.

Bogoliubov, N. N., 1962. Problems of a dynamical theory in statistical physics. In: de Boer, J., Uhlenbeck, G. E. (Eds.), Studies in Statistical Mechanics I, Part A. North-Holland, Amsterdam, pp. 5–118.

Boon, J. P., Allain, C., Lallemand, P., 1979. Propagating thermal modes in a fluid under thermal constraint. Phys. Rev. Lett. **43**, 199.

Boon, J. P., Yip, S., 1980. Molecular Hydrodynamics. McGraw-Hill, New York. Dover edition, 1991.

Born, M., Wolf, E., 1989. Principles of Optics, 6th Ed. Pergamon, Oxford.

Brenner, H., 2005a. Kinematics of volume transport. Physica A **349**, 11.

Brenner, H., 2005b. Navier-Stokes revisited. Physica A **349**, 60.

Breuer, H.-P., Petruccione, F., 1994. A master equation approach to fluctuating hydrodynamics: Heat conduction. Phys. Lett. A **185**, 385.

Brogioli, D., Vailati, A., 2001. Diffusive mass transfer by nonequilibrium fluctuations: Fick's law revisited. Phys. Rev. E **63**, 012105.

Brogioli, D., Vailati, A., Giglio, M., 2000a. Giant fluctuations in diffusion processes. J. Phys.: Condens. Matter **12**, A39.

Brogioli, D., Vailati, A., Giglio, M., 2000b. Universal behavior of nonequilibrium fluctuations in free diffusion processes. Phys. Rev. E **61**, R1.

Brogioli, D., Vailati, A., Giglio, M., 2003. A schlieren method for ultra-low-angle light scattering measurements. Europhys. Lett. **63**, 220.

Brush, S. G., 1972. Kinetic Theory, Vol. 3. Pergamon, New York.

Burgers, J. M., 1974. The Nonlinear Diffusion Equation: Asymptotic Solutions and Statistical Problems. D. Reidel Pub. Co, Dordrecht-Holland.

Busse, F. H., 1967. Stability of finite amplitude cellular convection and its relation to an extremum principle. J. Fluid Mech. **30**, 625.

Camacho, J. F., Hijar, H., Rodríguez, R. F., 2005. Hydrodynamic correlation functions for a nematic liquid crystal in a stationary state. Physica A **348**, 252.

Castets, V., Dulos, E., Boissonade, J., de Kepper, P., 1990. Experimental evidence of a sustained standing Turing-type nonequilibrium chemical pattern. Phys. Rev. Lett. **64**, 2953.

Chandrasekhar, S., 1961. Hydrodynamic and Hydromagnetic Stability. Oxford Univ. Press, Oxford. Dover edition, 1981.

Chasnov, J. R., Lee, K. L., 2001. Turbulent penetrative convection with an internal heat source. Fluid Dyn. Res. **28**, 397.

Chistyi, I. L., 1977. Tr. Fiz. Inst. Akad. Nauk SSSR **102**, 129.

Choudhuri, A. R., 1998. The Physics of Fluids and Plasmas: An Introduction for Astrophysicists. Cambridge Univ. Press, Cambridge.

Chu, 1991. Laser Light Scattering: Basic Principles and Practice, 2nd Ed. Academic Press, Boston.

Chung, D. S., Lee, K. Y. C., Mazur, E., 1988. Light scattering from nonequilibrium interfaces. Int. J. Thermophys. **9**, 729.

Chung, D. S., Lee, K. Y. C., Mazur, E., 1990. Spectral asymmetry in the light scattered from a nonequilibrium liquid interface. Phys. Lett. A **145**, 348.

Cicuta, P., Vailati, A., Giglio, M., 2000. Equilibrium and nonequilibrium fluctuations at the interface between two fluid phases. Phys. Rev. E **62**, 4920.

Cohen, C., Sutherland, J. W. H., Deutch, J. M., 1971. Hydrodynamic correlation functions for binary mixtures. Phys. Chem. Liquids **2**, 213.

Cohen, E. R., Giacomo, P., 1987. Symbols units, nomenclature and fundamental constants in physics. Physica **146A**, 1.

Courant, R., Hilbert, D., 1953. Methods of Mathematical Physics. Wiley, New York. Wiley Classics Library edition, 1996.

Crooks, G. E., 2000. Path-ensemble averages in systems driven far from equilibrium. Phys. Rev. E **61**, 2361.

Cross, M. C., 1980. Derivation of the amplitude equation at the Rayleigh-Bénard instability. Phys. Fluids **23**, 1727.

Cross, M. C., Hohenberg, P. C., 1993. Pattern formation outside of equilibrium. Rev. Mod. Phys. **65**, 851.

Cross, M. C., Kim, K., 1988. Linear instability and the codimension-2 region in binary fluid convection between rigid impermeable boundaries. Phys. Rev. A **37**, 3909.

Cugliandolo, L. F., Kurchan, J., Peliti, L., 1997. Energy flow, partial equilibration, and effective temperatures in systems with slow dynamics. Phys. Rev. E **55**, 3898.

de Bruyn, J. R., Bodenschatz, E., Morris, S. W., Trainoff, S. P., Hu, Y., Cannell, D. S., Ahlers, G., 1996. Apparatus for the study of Rayleigh-Bénard convection in gases under pressure. Rev. Sci. Instrum. **67**, 2043.

de Freitas, J. E., Lucena, L. S., da Silva, L. R., Hilhorst, H. J., 2000. Critical behavior of a two-species reaction-diffusion problem. Phys. Rev. E **61**, 6330.

de Gennes, P. G., Prost, J., 1993. The Physics of Liquid Crystals. Clarendon Press, Oxford.

de Giorgio, V., 1978. Dynamics of convective instabilities in a horizontal liquid layer. Phys. Rev. Lett. **41**, 1293.

de Groot, S. R., Mazur, P., 1962. Non-Equilibrium Thermodynamics. North-Holland, Amsterdam. Dover edition, 1984.

Demirel, Y., 2002. Nonequilibrium Thermodynamics. Elsevier, Amsterdam.

Denn, M. M., 2004. Fifty years of non-newtonian fluid dynamics. AIChE J. **50**, 2335.

Dennin, M., Treiber, M., Kramer, L., Ahlers, G., Cannell, D. S., 1996. Origin of travelling rolls in electroconvection of nematic liquid crystals. Phys. Rev. Lett. **76**, 319.

Derrida, B., Lebowitz, J. L., Speer, E. R., 2001. Free energy functional for nonequilibrium systems: an exactly solvable case. Phys. Rev. Lett. **87**, 150601.

Derrida, B., Lebowitz, J. L., Speer, E. R., 2002. Exact free energy functional for a driven diffusive open stationary nonequilibrium system. Phys. Rev. Lett. **89**, 030601.

Domínguez Lerma, M. A., Ahlers, G., Cannell, D. S., 1984. Marginal stability curve and linear growth rate for rotating Couette-Taylor flow and Rayleigh-Bénard convection. Phys. Fluids **27**, 856.

Dorfman, J. R., 1975. Kinetic and hydrodynamic theory of time correlation functions. In: Cohen, E. G. D. (Ed.), Fundamental Problems in Statistical Mechanics, Vol. 3. North-Holland, Amsterdam, pp. 277–330.

Dorfman, J. R., Cohen, E. G. D., 1967. Difficulties in the kinetic theory of gases. J. Math. Phys. **8**, 282.

Dorfman, J. R., Cohen, E. G. D., 1970. Velocity-correlation functions in two and three dimensions. Phys. Rev. Lett. **25**, 1257.

Dorfman, J. R., Cohen, E. G. D., 1975. Velocity-correlation functions in two and three dimensions. II. Higher density. Phys. Rev. A **12**, 292.

Dorfman, J. R., Kirkpatrick, T. R., Sengers, J. V., 1994. Generic long-range correlations in molecular fluids. Annu. Rev. Phys. Chem. **45**, 213.

Dubois, M., Bergé, P., 1971. Experimental study of Rayleigh scattering related to concentration fluctuations in binary solutions: Evidence of a departure from ideality. Phys. Rev. Lett. **26**, 121.

Einstein, A., 1910. Theorie der Opaleszenz von homogenen Flüssigkeiten und Flüssigkeitsgemischen in der Nähe des kritischen Zustandes. Annalen der Physik **33**, 1275.

Ernst, M. H., 2005. Universal power law tails of time correlation functions. Phys. Rev. E **71**, 030101.

Ernst, M. H., Hauge, E. H., van Leeuwen, J. M. J., 1971. Asymptotic time behavior of correlation functions. 1. Kinetic terms. Phys. Rev. A **4**, 2055.

Ernst, M. H., Hauge, E. H., van Leeuwen, J. M. J., 1976a. Asymptotic time behavior of correlation functions. 2. Kinetic and potential terms. J. Stat. Phys. **15**, 7.

Ernst, M. H., Hauge, E. H., van Leeuwen, J. M. J., 1976b. Asymptotic time behavior of correlation functions. 3. Local equilibrium and mode-coupling theory. J. Stat. Phys. **15**, 23.

Evans, D. J., Searles, D. J., 2002. The fluctuation theorem. Adv. Phys. **51**, 1529.

Fabelinskii, I. L., 1965. Molecular Scattering of Light. Nauka, Moscow.

Fabelinskii, I. L., 1994. Spectra of molecular ligth scattering and some of their applications. Physics-Uspekhi **37**, 821.

Fisher, M. E., 1964. Correlation functions and the critical region of simple fluids. J. Math. Phys. **5**, 944.

Fitts, D. D., 1962. Nonequilibrium Thermodynamics. McGraw-Hill, New York.

Fixman, M., 1967. Transport coefficients in the gas critical region. J. Chem. Phys. **47**, 2808.

Fleury, P. A., Boon, J. P., 1969. Brillouin scattering in simple liquids: Argon and neon. Phys. Rev. **186**, 244.

Foch, J., 1971. Stochastic equations for fluid mixtures. Phys. Fluids **14**, 893.

Forster, D., 1975. Hydrodynamic Fluctuations, Broken Symmetry and Correlation Functions. Vol. 47 of Frontiers in Physics. W.A. Benjamin, Reading, MS. Advanced Book Classics (Westview Press) edition, 1995.

Fox, R. F., 1982. Testing theories of non-equilibrium processes with light-scattering techniques. J. Phys. Chem. **86**, 2812.

Fox, R. F., 1984. Theoretical analysis of long-time-tail observations by light scattering off polystyrene spheres. Phys. Rev. A **30**, 2590.

Fox, R. F., Uhlenbeck, G. E., 1970a. Contributions to non-equilibrium thermodynamics. I. Theory of hydrodynamical fluctuations. Phys. Fluids **13**, 1893.

Fox, R. F., Uhlenbeck, G. E., 1970b. Contributions to non-equilibrium thermodynamics. II. Fluctuation theory for Boltzmann equation. Phys. Fluids **13**, 2881.

Fröba, A. P., , Leipertz, A., 2003. Accurate determination of liquid viscosity and surface tension using surface light scattering (SLS): Toluene under saturation conditions between 260 and 380 K. Int. J. Thermophys. **24**, 895.

Fröba, A. P., Will, S., Leipertz, A., 2000. Diffusion modes of an equimolar methane-ethane mixture from dynamic light scattering. Int. J. Thermophys. **21**, 603.

Fulco, U. L., Messias, D. N., Lyra, M. L., 2001. Critical behavior of a one-dimensional diffusive epidemic process. Phys. Rev. E **63**, 066118.

Gallavotti, G., Cohen, E. G. D., 1995a. Dynamic ensembles in nonequilibrium statistical mechanics. Phys. Rev. Lett. **74**, 2694.

Gallavotti, G., Cohen, E. G. D., 1995b. Dynamical ensembles in stationary states. J. Stat. Phys. **80**, 931.

Garcia, A. L., Malek Mansour, M., Lie, G. C., Clementi, E., 1987. Numerical integration of the fluctuating hydrodynamics equations. J. Stat. Phys. **47**, 209.

García-Colín, L. S., 1995. Extended irreversible thermodynamics - an unfinished task. Mol. Phys. **86**, 697.

García-Colín, L. S., Velasco, R. M., 1982. Modified Navier-Stokes model for nonequilibrium stationary states. Phys. Rev. A **26**, 2187.

García Ojalvo, J., Hernández Machado, A., Sancho, J. M., 1993. Effects of external noise on the Swift-Hohenberg equation. Phys. Rev. Lett. **71**, 1542.

García Ojalvo, J., Sancho, J. M., 1999. Noise in Spatially Extended Systems. Springer, New York.

Gardiner, C. W., 1985. Handbook of Stochastic Methods, 2nd Ed. Springer, Berlin.

Gardiner, C. W., Chaturvedi, S., 1977. The Poisson representation. I. A new technique for chemical master equation. J. Stat. Phys. **17**, 429.

Garrido, P. L., Lebowitz, J. L., Maes, C., Spohn, H., 1990. Long-range correlations for conservative dynamics. Phys. Rev. A **42**, 1954.

Giglio, M., Vendramini, A., 1977. Buoyancy-driven instability in a dilute solution of macromolecules. Phys. Rev. Lett. **39**, 1014.

Gillespie, D. T., 2000. The chemical Langevin equation. J. Chem. Phys. **113**, 297.

Gillespie, D. T., 2001. Approximate accelerated stochastic simulation of chemically reacting systems. J. Chem. Phys. **115**, 1716.

Glansdorff, P., Prigogine, I., 1971. Structure, Stability and Fluctuations. Wiley Interscience, London.

Gradstein, I. S., Ryzhik, I. M., 1994. Table of Integrals, Series, and Products, 5th Ed. Academic Press, San Diego.

Graham, R., 1974. Hydrodynamic fluctuations near the convection instability. Phys. Rev. A **10**, 1762.

Graham, R., Pleiner, H., 1975. Mode-coupling theory of the heat convection threshold. Phys. Fluids **18**, 130.

Grant, M., Desai, R. C., 1983. Fluctuating hydrodynamics and capillary waves. Phys. Rev. A. **27**, 2577.

Gray, D. D., Giorgini, A., 1976. The validity of the Boussinesq approximation for liquids and gases. Int. J. Heat Mass Transfer **19**, 545.

Grinstein, G., 1991. Generic scale-invariance in classical nonequilibrium systems. J. Appl. Phys. **69**, 5441.

Grossmann, S., 1976. Langevin forces in chemically reacting multicomponent fluids. J. Chem. Phys. **65**, 2007.

Haase, R., 1969. Thermodynamics of Irreversible Processes. Addison-Wesley, Reading MA. Dover edition, 1990.

Hansen, J. P., McDonald, I. R., 1986. Theory of Simple liquids, 2nd Ed. Academic Press, London.

Hattori, K., Sakai, K., Takagi, K., 1996. Brillouin scattering under a temperature gradient. Physica B **220**, 553.

Hattori, K., Sakai, K., Takagi, K., 1998. Low-angle Brillouin scattering under a temperature gradient. J. Phys.: Condens. Matter **10**, 3333.

Hohenberg, P. C., Halperin, B. L., 1977. Theory of dynamic critical phenomena. Rev. Mod. Phys. **49**, 435.

Hohenberg, P. C., Swift, J. B., 1992. Effects of additive noise at the onset of Rayleigh-Bénard convection. Phys. Rev. A **46**, 4773.

Hollinger, S., Lücke, M., 1998. Influence of the Soret effect on convection of binary fluids. Phys. Rev. E **57**, 4238.

Hollinger, S., Lücke, M., Müller, H. W., 1998. Model for convection in binary liquids. Phys. Rev. E **57**, 4250.

Hort, W., Linz, S. J., M. Lücke, 1992. Onset of convection in binary gas mixtures: Role of the Dufour effect. Phys. Rev. A **45**, 3737.

Janssen, H. K., 2001. Comment on "Critical behavior of a two-species reaction-diffusion problem". Phys. Rev. E **64**, 050101.

Jarzynski, C., 1997. Nonequilibrium entropy for free energy differences. Phys. Rev. Lett. **78**, 2690.

Jarzynski, C., 2001. How does a system respond when driven away from thermal equilibrium? PNAS **98**, 3636.

Jarzynski, C., 2002. What is the microscopic response of a system driven far away from equilibrium? In: Garbaczewski, P., Olkiewicz, R. (Eds.), Dynamics of Dissipation. Vol. 597 of Lecture Notes in Physics. Springer, Berlin, pp. 63–82.

Jeffreys, H., 1926. The stability of a layer of fluid heated from below. Phil. Mag. (7)2, 833.

Jou, D., Casas-Vázquez, J., Lebon, G., 1993. Extended Irreversible Thermodynamics. Springer, Heidelberg.

Jou, D., Criado-Sancho, M., Casas-Vázquez, J., 2005. Nonequilibrium temperature and fluctuation-dissipation in flowing gases. Physica A 358, 49.

Kadanoff, L. P., Swift, J., 1968. Transport coefficients near the liquid-gas critical point. Phys. Rev. 166, 89.

Kandar, M., Parisi, G., Zhang, Y., 1986. Dynamic scaling of growing interfaces. Phys. Rev. Lett. 56, 889 892.

Kato, T., 1980. Perturbation theory for linear operators, 2nd Ed. Springer, New York.

Katzav, E., Schwartz, M., 2004. Numerical evidence for stretched exponential relaxations in the Kandar-Parisi-Zhang equation. Phys. Rev. E 69, 052603.

Kawasaki, K., 1970. Kinetic equations and time correlation functions of critical fluctuations. Ann. Phys. (NY) 61, 1.

Kawasaki, K., 1976. Mode coupling and critical dynamics. In: Domb, C., Green, M. S. (Eds.), Phase Transitions and Critical Phenomena, Vol 5A. Academic, New York, p. 165.

Keizer, J., 1978. A theory of spontaneous fluctuations in viscous fluids far from equilibrium. Phys. Fluids 21, 198.

Kiefte, H., Clouter, M. J., Penney, R., 1984. Experimental confirmation of nonequilibrium steady-state theory: Brillouin scattering in a temperature gradient. Phys. Rev. B 30, 4017.

Kirkpatrick, T. R., Belitz, D., Sengers, J. V., 2002. Long-time tails, weak localization, and classical and quantum critical behavior. J. Stat. Phys. 109, 373.

Kirkpatrick, T. R., Cohen, E. G. D., 1983. Kinetic theory of fluctuations near a convective instability. J. Stat. Phys. 33, 639.

Kirkpatrick, T. R., Cohen, E. G. D., Dorfman, J. R., 1979. Kinetic-theory of light-scattering from a fluid not in equilibrium. Phys. Rev. Lett. 42, 862.

Kirkpatrick, T. R., Cohen, E. G. D., Dorfman, J. R., 1980. Hydrodynamic theory of light scattering from a fluid in a nonequilibrium steady state. Phys. Rev. Lett. **44**, 472.

Kirkpatrick, T. R., Cohen, E. G. D., Dorfman, J. R., 1982a. Fluctuations in a nonequilibrium steady state; Basic equations. Phys. Rev. A **26**, 950.

Kirkpatrick, T. R., Cohen, E. G. D., Dorfman, J. R., 1982b. Light scattering by a fluid in a nonequilibrium steady state. II. Large gradients. Phys. Rev. A **26**, 995.

Ko, L. F., Cohen, E. G. D., 1987. Propagating viscous modes in a Taylor-Couette system. Phys. Lett. A **125**, 231.

Köhler, W., Müller, B., 1995. Soret and mass diffusion coefficients of toluene/n-hexane mixtures. J. Chem. Phys. **103**, 4367.

Köhler, W., Wiegand, S., 2002. Thermal Nonequilibrium Phenomena in Fluid Mixtures. Vol. 584 of Lecture Notes in Physics. Springer, Berlin.

Krall, N. A., Trivelpiece, A. W., 1973. Principles of Plasma Physics. McGraw-Hill, New York.

Kramer, L., Pesh, W., 1995. Convection instabilities in nematic liquid crystals. Annu. Rev. Fluid Mech. **27**, 515.

Kree, R., Schaub, B., Schmittmann, B., 1989. Effects of pollution on critical population dynamics. Phys. Rev. A **39**, 2214.

Kubo, R., Toda, M., Hatshitsume, N., 1991. Statistical Physics II. Nonequilibrium Statistical Mechanics, 2nd Ed. Springer, Berlin.

La Porta, A., Surko, C. M., 1998. Convective instability in a fluid mixture heated from above. Phys. Rev. Lett. **80**, 3759.

Lallemand, P., Allain, C., 1980. Propagative thermal excitations in a stratified fluid layer. J. Physique (Paris) **41**, 1.

Landau, L. D., Lifshitz, E. M., 1958. Statistical Physics. Part I. Pergamon, London.

Landau, L. D., Lifshitz, E. M., 1959. Fluid Mechanics. Pergamon, London. 2nd revised English version, 1987.

Landsberg, G. S., Shubin, A. A., 1939. Zh. Eksp. Teor. Fiz. **9**, 1309.

Law, B. M., Gammon, R. W., Sengers, J. V., 1988. Light-scattering observations of long-range correlations in a nonequilibrium liquid. Phys. Rev. Lett. **60**, 1554.

Law, B. M., Nieuwoudt, J. C., 1989. Noncritical liquid mixtures far from equilibrium: The Rayleigh line. Phys. Rev. A **40**, 3880.

Law, B. M., Segrè, P. N., Gammon, R. W., Sengers, J. V., 1990. Light-scattering measurements of entropy and viscous fluctuations in a liquid far from equilibrium. Phys. Rev. A **41**, 816.

Law, B. M., Sengers, J. V., 1989. Fluctuations in fluids out of thermal equilibrium. J. Stat. Phys. **57**, 531.

Legros, J. C., Platten, J. K., Poty, P. G., 1972. Stability of a two-component fluid layer heated from below. Phys. Fluids **15**, 1383.

Leipertz, A., 1988. Transport properties of transparent liquids by photon-correlation spectroscopy. Int. J. Thermophys. **9**, 897.

Lekkerkerker, H. N. W., Laidlaw, W. G., 1977. Pretransitional effects near the convective instability in binary mixtures. J. Physique (Paris) **38**, 1.

Leontovich, M. A., 1935. Dokl. Akad. Nauk SSSR **1**, 97.

Levich, V. G., 1962. Physicochemical Hydrodynamics. Prentice Hall, Englewood Cliffs, NJ.

Lhost, O., Linz, S. J., Müller, H. W., 1991. Onset of convection in binary liquid mixtures: improved Galerkin approximations. J. Physique (Paris) II **1**, 279.

Li, W. B., Segrè, P. N., Gammon, R. W., Sengers, J. V., 1994a. Small-angle Rayleigh scattering from nonequilibrium fluctuations in liquids and liquid mixtures. Physica A **204**, 399.

Li, W. B., Segrè, P. N., Sengers, J. V., Gammon, R. W., 1994b. Nonequilibrium fluctuations in liquids and liquids mixtures subjected to a stationary temperature gradient. J. Phys.: Condens. Matter **6**, A119.

Li, W. B., Sengers, J. V., Gammon, R. W., Segrè, P. N., 1995. Measurement of transport properties of liquids with equilibrium and nonequilibrium Rayleigh scattering. Int. J. Thermophys. **16**, 23.

Li, W. B., Zhang, K. J., Sengers, J. V., Gammon, R. W., J. M. Ortiz de Zárate, 1998. Concentration fluctuations in a polymer solution under a temperature gradient. Phys. Rev. Lett. **81**, 5580.

Li, W. B., Zhang, K. J., Sengers, J. V., Gammon, R. W., J. M. Ortiz de Zárate, 2000. Light scattering from nonequilibrium concentration fluctuations in a polymer solution. J. Chem. Phys. **112**, 9139.

Liepmann, H. W., Roshko, A., 1957. Elements of Gasdynamics, 2nd Ed. Wiley, New York. Dover edition, 2002.

Lifshitz, E. M., Pitaevskii, 1986. Statistical Physics. Part II. Pergamon, London.

Lin, J. L., Taylor, W. L., Rutherford, W. M., 1991. Secondary coefficients. In: Wakeham, W. A., Nagashima, A., Sengers, J. V. (Eds.), Measurement of the Transport Properties of Fluids. Blackwell Scientific, Oxford, pp. 323–387.

Liu, I. S., 2002. Continuum Mechanics. Springer Verlag, Heidelberg.

López de Haro, M., del Río, J. A., Vázquez, F., 2002. Light-scattering spectrum of a viscoelastic fluid subjected to an external temperature gradient. Revista Mexicana de Física **48 - Supl. 1**, 230.

Lord Rayleigh, 1916. On convective currents in a horizontal layer of fluid when the higher temperature is on the under side. Phil. Mag **(6)32**, 529.

Lorenz, E., 1964. The problem of deducing the climate from the governing equations. Tellus **16**, 1.

Lücke, M., Barten, W., Büchel, P., Fütterer, F., St. Hollinger, Ch. Jung, 1998. Pattern formation in binary fluid convection and in systems with throughflow. In: Busse, F. H., Müller, S. C. (Eds.), Evolution of Spontaneous Structures in Dissipative Continuous Systems. Vol. 55m of Lecture Notes in Physics. Springer, Berlin, pp. 128–196.

Lutsko, J. F., Dufty, J. W., 1985. Hydrodynamic fluctuations at large shear rate. Phys. Rev. A **32**, 3040.

Lutsko, J. F., Dufty, J. W., 2002. Long-ranged correlations in sheared fluids. Phys. Rev. E **66**, 041206.

Machta, J., Oppenheim, I., Procaccia, I., 1980. Statistical mechanics of stationary states. V. Fluctuations in systems with shear flow. Phys. Rev. A **22**, 2809.

Malek Mansour, M., García, A. L., Turner, J. W., Mareschal, M., 1988. On the scattering function of simple fluids in finite systems. J. Stat. Phys. **52**, 295.

Malek Mansour, M., Turner, J. W., Garcia, A. L., 1987. Correlation functions for simple fluids in a finite system under nonequilibrium constrains. J. Stat. Phys. **48**, 1157.

Mandelstam, L. I., 1934. Dokl. Akad. Nauk SSSR **2**, 219.

Manneville, P., 1990. Dissipative Structures and Weak Turbulence. Academic Press, San Diego.

March, N. H., Tosi, M. P., 1976. Atomic Dynamics in Liquids. Macmillan Press, London. Dover edition, 1991.

Martens, C. C., 2002. Qualitative dynamics of generalized Langevin equations and the theory of chemical reaction rates. J. Chem. Phys. **116**, 2516.

Mashiyama, K. T., Mori, H., 1978. Origin of Landau-Lifshitz hydrodynamic fluctuations in non-equilibrium systems and a new method for deducing the Boltzmann-equation. J. Stat. Phys. **18**, 385.

Mazenko, G. F., Ramaswamy, S., Toner, J., 1983. Breakdown of conventional hydrodynamics for smectic-A, smectic-B, and cholesteric liquid crystals. Phys. Rev. A **28**, 1618.

Mazur, E., Chung, D. S., 1987. Light scattering from the liquid-vapor interface. Physica A **147**, 387.

Meyer, C. W., Ahlers, G., Cannell, D. S., 1991. Stochastic influences on pattern formation in Rayleigh-Bénard convection: Ramping experiments. Phys. Rev. A **44**, 2514.

Michels, A., Sengers, J. V., van der Gulik, P. S., 1962. The thermal conductivity of carbon dioxide in the critical region. II. Measurements and conclusions. Physica A **28**, 1216.

Mihaljan, J. M., 1962. A rigorous exposition of the Boussinesq approximation applicable to a thin layer of fluid. Astrophys. J. **136**, 1126.

Mills, I., Cvitas, T., Homann, K., Kallay, N., Kuchitsu, K., 1988. Quantities, Units and Symbols in Physical Chemistry. Blackwell, Oxford.

Mountain, R. D., 1966. Spectral distribution of scattered light in a simple fluid. Rev. Mod. Phys. **38**, 205.

Napolitano, L. G., 1978. Thermodynamics and dynamics of pure interfaces. Acta Astronaut. **5**, 655.

Nepomnyashchy, A. A., Velarde, M. G., Colinet, P., 2002. Interfacial Phenomena and Convection. CRC Press, Boca Ratón, FL.

Niederländer, J., Lücke, M., Kamps, M., 1991. Weakly nonlinear convection: Galerkin model, numerical simulation, and amplitude equation. Z. Phys. B **82**, 135.

Nieuwoudt, J. C., Law, B. M., 1990. Theory of light scattering by a non-equilibrium binary mixture. Phys. Rev. A **42**, 2003.

Normand, C., Pomeau, Y., Velarde, M. G., 1977. Convective instability: A physicist's approach. Rev. Mod. Phys. **49**, 581.

Oh, J., Ahlers, G., 2003. Thermal-noise effect on the transition to Rayleigh-Bénard convection. Phys. Rev. Lett. **91**, 094501.

Oh, J., Ortiz de Zárate, J. M., Sengers, J. V., Ahlers, G., 2004. Dynamics of fluctuations in a fluid below the onset of Rayleigh-Bénard convection. Phys. Rev. E **69**, 021106.

Onsager, L., 1931a. Reciprocal relations in irreversible processes. I. Phys. Rev. **37**, 405 426.

Onsager, L., 1931b. Reciprocal relations in irreversible processes. II. Phys. Rev. **38**, 2265 2279.

Onuki, A., 1997. Phase transitions of fluids under shear flow. J. Phys.: Condens. Matter **9**, 6119.

Onuki, A., 2002. Phase Transition Dynamics. Cambridge University Press, Cambridge.

Oono, Y., 1976. The meaning of δz^2 of Glandsdorff and Prigogine. Phys. Lett. **57A**, 207.

Ortiz de Zárate, J. M., Muñoz Redondo, L., 2001. Finite-size effects with rigid boundaries on nonequilibrium fluctuations in a liquid. Eur. Phys. J. B **21**, 135.

Ortiz de Zárate, J. M., Peluso, F., Sengers, J. V., 2004. Nonequilibrium fluctuations in the Rayleigh-Bénard problem for binary fluid mixtures. Eur. Phys. J. E **15**, 319.

Ortiz de Zárate, J. M., Pérez Cordón, R., Sengers, J. V., 2001. Finite-size effects on fluctuations in a fluid out of thermal equilibrium. Physica A **291**, 113.

Ortiz de Zárate, J. M., Sengers, J. V., 2001. Fluctuations in fluids in thermal nonequilibrium states below the convective Rayleigh-Bénard instability. Physica A **300**, 25.

Ortiz de Zárate, J. M., Sengers, J. V., 2002. Boundary effects on the non-equilibrium structure factor of fluids below the Rayleigh-Bénard instability. Phys. Rev. E **66**, 036305.

Ortiz de Zárate, J. M., Sengers, J. V., 2004. On the physical origin of long-ranged fluctuations in fluids in thermal nonequilibrium states. J. Stat. Phys. **115**, 1341.

Osenda, O., Briozzo, C. B., Caceres, M. O., 1997. Stochastic Lorenz model for periodically driven Rayleigh-Benard convection. Phys. Rev. E **55**, R3824.

Osenda, O., Briozzo, C. B., Caceres, M. O., 1998. Noise and pattern formation in periodically driven Rayleigh-Benard convection. Phys. Rev. E **57**, 412.

Otnes, K., Riste, T., 1980. Observation by neutron correlation spectroscopy of a nonlinear soft mode at the Rayleigh-Bénard instability in para-azoxyanisole. Phys. Rev. Lett. **37**, 1490.

Öttinger, H. C., 2005. Beyond Equilibrium Thermodynamics. Wiley, New York.

Pagonabarraga, I., Pérez-Madrid, A., Rubí, J. M., 1997. Fluctuating hydrodynamics approach to chemical reactions. Physica A **237**, 205.

Pedersen, A. M., Riste, T., 1980. Real time observations by neutron scattering of fluctuations near the Rayleigh-Bénard instability. Z. Phys. B **37**, 171.

Pellew, A., Southwell, R. V., 1940. On maintained convection in a fluid heated from below. Proc. R. Soc. A **176**, 312.

Pérez Cordón, R., Velarde, M. G., 1975. On the (nonlinear) foundations of Boussinesq approximation applicable to a thin layer of fluid. J. Physique (Paris) **36**, 591.

Pérez-Madrid, A., Reguera, D., Rubí, J. M., 2003. Origin of the violation of the fluctuation-dissipation theorem in systems with activated dynamics. Physica A **329**, 357.

Platten, J. K., Chavepeyer, G., 1975. Finite amplitude instability in the two-component Bénard problem. In: Prigogine, I., Rice, S. (Eds.), Advances in Chemical Physics. Vol. 32. Wiley, New York, pp. 281–322.

Platten, J. K., Legros, J. C., 1984. Convection in Liquids. Springer, Berlin.

Pleiner, H., Brand, H., 1983a. Light scattering in nematic liquid crystals in a nonequilibrium steady state. Phys. Rev. A **27**, 1177.

Pleiner, H., Brand, H., 1983b. Light scattering in nematic liquid crystals in the presence of shear flow. J. Physique (Paris) Lettres **44**, L23.

Pomeau, Y., Résibois, P., 1975. Time dependent correlation functions and mode-mode coupling theories. Phys. Reports **19**, 63.

Procaccia, I., Ronis, D., Collins, M. A., Ross, J., Oppenheim, I., 1979. Statistical mechanics of stationary states. I. Formal theory. Phys. Rev. A **19**, 1290 1306.

Proccacia, I., Ronis, D., Oppenheim, I., 1979. Light scattering from nonequilibrium stationary states: The implication of broken time-reversal symmetry. Phys. Rev. Lett. **42**, 287.

Qiu, X., Ahlers, G., 2005. Dynamics of fluctuations below a stationary bifurcation to electroconvection in the planar nematic liquid crystal N4. Phys. Rev. Lett. **94**, 087802.

Quentin, Rehberg, I., 1995. Direct measurement of hydrodynamic fluctuations in a binary mixture. Phys. Rev. Lett. **74**, 1578 1581.

Rehberg, I., Hörner, F., Chiran, L., Richter, H., Winkler, B. L., 1991a. Measuring the intensity of director fluctuations below the onset of electroconvection. Phys. Rev. A **44**, R7885.

Rehberg, I., Rasenat, S., de la Torre Juárez, M., Schöpf, W., Hörner, F., Ahlers, G., Brand, H. R., 1991b. Thermally induced hydrodynamic fluctuations below the onset of electroconvection. Phys. Rev. Lett. **67**, 596.

Robinson, R. A., Stokes, R. H., 2002. Electrolyte Solutions: Second Revised Edition. Dover, New York.

Rodríguez, R. F., Camacho, J. F., 2002a. Nonequilibrium effects on the light-scattering spectrum of a nematic driven by a pressure gradient. Rev. Mex. Física **48, Supl. 1**, 144.

Rodríguez, R. F., Camacho, J. F., 2002b. Nonequilibrium thermal light-scattering from nematic liquid crystals. In: Macias, A., Díaz, E., Uribe, F. (Eds.), Statistical Physics and Beyond. Vol. B of Recent Developments in Mathematical and Experimental Physics. Kluwer, New York, pp. 209–224.

Ronis, D., Procaccia, I., 1982. Nonlinear resonant coupling between shear and heat fluctuations in fluids far from equilibrium. Phys. Rev. A **26**, 1812.

Ronis, D., Procaccia, I., Machta, J., 1980. Statistical mechanics of stationary states. VI. Hydrodynamic fluctuation theory far from equilibrium. Phys. Rev. A **22**, 714.

Ronis, D., Procaccia, I., Oppenheim, I., 1979. Statistical mechanics of stationary states. III. Fluctuations in dense fluids with applications to light scattering. Phys. Rev. A **19**, 1324.

Ross, J., Hunt, K. L. C., Hunt, P. M., 1992. Thermodynamic and stochastic theory for nonequilibrium systems with multiple reactive intermediates: The concept and role of excess work. J. Chem. Phys. **96**, 618.

Rowlinson, J. S., Widom, B., 1982. Molecular Theory of Capillarity. Clarendon, Oxford.

Rubí, J. M., 1984. Fluctuations around equilibrium. In: Casas-Vázquez, J., Jou, D., Lebon, G. (Eds.), Recent Developments in Nonequilibrium Thermodynamics. Vol. 199 of Lecture Notes in Physics. Springer, Berlin, pp. 233–266.

Rubí, J. M., Mazur, P., 2000. Nonequilibrium thermodynamics and hydrodynamic fluctuations. Physica A **276**, 477.

Sahoo, D., Sood, A. K., 1983a. Comments on: "Light scattering from a fluid in a nonequilibrium steady state: Plane Couette flow". Phys. Lett. **93A**, 476.

Sahoo, D., Sood, A. K., 1983b. Light scattering from a fluid in a nonequilibrium steady state: Plane Couette flow. Phys. Lett. **93A**, 476.

Sahoo, D., Sood, A. K., 1984. Possibility of modification of the Rayleigh line in a nonequilibrium fluid with a constant shear velocity gradient. Phys. Rev. A **30**, 2802.

Santamaría-Holek, I., Pérez-Madrid, A., Rubí, J. M., 2004. Local quasi-equilibrium description of slow relaxation systems. J. Chem. Phys. **120**, 2818.

Sawada, Y., 1978. Acoustic observation of critical slowing down for the convectional instability by configuration inversion. Phys. Lett. A **65**, 5.

Schechter, R. S., Prigogine, I., Hamm, J. R., 1972. Thermal diffusion and convective instability. Phys. Fluids **15**, 379.

Schechter, R. S., Velarde, M. G., Platten, J. K., 1974. The two-component Bénard problem. In: Prigogine, I., Rice, S. A. (Eds.), Advances in Chemical Physics, Vol. 26. Wiley, New York, pp. 265–301.

Scherer, M. A., Ahlers, G., Hörner, F., Rehberg, I., 2000. Deviations from linear theory for fluctuations below the supercritical primary bifurcation to electroconvection. Phys. Rev. Lett. **85**, 3754.

Schmitt, S., Lücke, M., 1991. Amplitude equation for modulated Rayleigh-Bénard convection. Phys. Rev. A **44**, 4986.

Schmitz, R., 1988. Fluctuations in nonequilibrium fluids. Phys. Reports **171**, 1.

Schmitz, R., 1994. Fluctuations in a nonequilibrium colloidal suspension. Physica A **206**, 25.

Schmitz, R., Cohen, E. G. D., 1985a. Fluctuations in a fluid under a stationary heat flux. I. General theory. J. Stat. Phys. **39**, 285.

Schmitz, R., Cohen, E. G. D., 1985b. Fluctuations in a fluid under a stationary heat flux. II. Slow part of the correlation matrix. J. Stat. Phys. **40**, 431.

Schmitz, R., Cohen, E. G. D., 1987. Brillouin scattering from fluids subject to large thermal gradients. Phys. Rev. A **35**, 2602.

Schöpf, W., Zimmermann, W., 1993. Convection in binary fluids: Amplitude equations, codimension-2 bifurcation, and thermal fluctuations. Phys. Rev. E **47**, 1739.

Segrè, P. N., Gammon, R. W., Sengers, J. V., 1993a. Light-scattering measurements of nonequilibrium fluctuations in a liquid mixture. Phys. Rev. E **47**, 1026.

Segrè, P. N., Gammon, R. W., Sengers, J. V., Law, B. M., 1992. Rayleigh scattering in a liquid far from thermal equilibrium. Phys. Rev. A **45**, 714.

Segrè, P. N., Schmitz, R., Sengers, J. V., 1993b. Fluctuations in inhomogeneous and nonequilibrium fluids under the influence of gravity. Physica A **195**, 31.

Segrè, P. N., Sengers, J. V., 1993. Nonequilibrium fluctuations in liquid mixtures under the influence of gravity. Physica A **198**, 46.

Sengers, J. V., 1966. Behavior of viscosity and thermal conductivity of fluids near the critical point. In: Green, M. S., Sengers, J. V. (Eds.), Critical Phenomena. Vol. 273 of National Bureau of Standards Miscellaneous Publications. U.S. Government Printing Office, Washington, DC, pp. 165–178.

Sengers, J. V., 1985. Transport properties of fluids near critical points. Int. J. Thermophys. **6**, 203.

Sengers, J. V., Gammon, R. W., Ortiz de Zárate, J. M., 2000. Thermal-diffusion driven concentration fluctuations in a polymer solution. In: Dadmun, M., Noid, D., Sumpter, B., Melnichenko, Y. (Eds.), Computational Studies, Nanotechnology and Solution Thermodynamics. Kluwer, New York, pp. 37–44.

Sengers, J. V., Keyes, P. H., 1971. Scaling of the thermal conductivity near the gas-liquid critical point. Phys. Rev. Lett. **26**, 70.

Sengers, J. V., Ortiz de Zárate, J. M., 2001. Finite-size effects on Soret-induced nonequilibrium concentration fluctuations in binary liquids. Revista Mexicana de Física **48 Supl. 1**, 14.

Sengers, J. V., Ortiz de Zárate, J. M., 2002. Nonequilibrium concentration fluctuations in binary liquid systems induced by the Soret effect. In: Köhler, W., Wiegand, S. (Eds.), Thermal Nonequilibrium Phenomena in Fluid Mixtures. Vol. 584 of Lecture Notes in Physics. Springer, Berlin, pp. 121–145.

Shapiro, A. H., 1953. The Dynamics and Thermodynamics of Compressible Fluid Flow. Wiley, New York.

Skelland, A. H. P., 1967. Non-Newtonian Flow and Heat Transfer. Wiley, New York.

Soret, C., 1880. Influence de la température sur la distribution des sels dans leurs solutions. C.R. Acad. Sci., Paris **91**, 289.

Spiegel, E. A., Veronis, G., 1960. On the Boussinesq approximation for a compressible fluid. Astrophys. J. **131**, 442.

Staliunas, K., 2001. Spatial and temporal spectra of noise driven stripe patterns. Phys. Rev. E **64**, 066129.

Steele, W. A., 1969. Time-correlation functions. In: Hanley, H. J. M. (Ed.), Transport Phenomena in Fluids. Marcel Dekker, New York, pp. 209–312.

Stephen, M. J., Straley, J. P., 1974. Physics of liquid crystals. Rev. Mod. Phys. **46**, 617.

Suave, R. N., de Castro, A. R. B., 1996. Asymmetry in the Brillouin spectra of organic fluids exposed to a temperature gradient. Phys. Rev. B **53**, 5330.

Swift, J. B., Hohenberg, P. C., 1977. Hydrodynamic fluctuations at the convective instability. Phys. Rev. A **15**, 319.

Trainoff, S. P., 1999. Rayleigh-Bénard convection in the presence of a weak lateral flow. Ph. D. thesis, University of California, Santa Barbara.

Trainoff, S. P., Cannell, D. S., 2002. Physical optics treatment of the shadowgraph. Phys. Fluids **14**, 1340.

Treiber, M., 1996. Analytic expressions for the stochastic amplitude equation for Taylor-Couette flow. Phys. Rev. E **53**, 577.

Treiber, M., Kramer, L., 1994. Stochastic envelope equations for nonequilibrium transitions and application to thermal fluctuations in electroconvection in nematic liquid cristals. Phys. Rev. E **49**, 3184.

Tremblay, A. M. S., 1984. Theories of fluctuations in nonequilibrium systems. In: Casas Vázquez, J., Jou, D., Lebon, G. (Eds.), Recent developments in Nonequilibrium Thermodynamics. Vol. 199 of Lecture Notes in Physics. Springer, Berlin, pp. 267–317.

Tremblay, A. S., Arai, M., Siggia, E. D., 1981. Fluctuations about simple nonequilibrium steady states. Phys. Rev. A **23**, 1451.

Tritton, D. J., 1988. Physical Fluid Dynamics, 2nd Ed. Oxford University Press, Oxford.

Turner, J. S., 1985. Multicomponent convection. Annu. Rev. Fluid Mech. **17**, 11.

Tyrrell, H. J. V., 1961. Diffusion and Heat Flow in Liquids. Butterworths, London.

Vailati, A., Giglio, M., 1996. q Divergence of nonequilibrium fluctuations and its gravity-induced frustration in a temperature stressed liquid mixture. Phys. Rev. Lett. **77**, 1484.

Vailati, A., Giglio, M., 1997a. Giant fluctuations in a free diffusion process. Nature **390**, 262.

Vailati, A., Giglio, M., 1997b. Very low-angle static light scattering from steady-state and time-dependent nonequilibrium fluctuations. Prog. Colloid. Polym. Sci. **104**, 76.

Vailati, A., Giglio, M., 1998. Nonequilibrium fluctuations in time-dependent diffusion processes. Phys. Rev. E **58**, 4361.

Val'kov, A. Y., Romanov, V. P., Shalaginov, A. N., 1994. Fluctuations and ligth scattering in liquid crystals. Physics-Uspekhi **37**, 139.

van Beijeren, H., Cohen, E. G. D., 1988a. Comment on "Initial stages of pattern formation in Rayleigh-Bénard convection". Phys. Rev. Lett. **60**, 1208.

van Beijeren, H., Cohen, E. G. D., 1988b. The effects of thermal noise in a Rayleigh-Bénard cell near its first convective instability. J. Stat. Phys. **53**, 77.

van der Zwan, G., Bedeaux, D., Mazur, P., 1981. Light scattering from a fluid with a stationary temperature gradient. Physica A **107**, 491.

van Kampen, N. G., 1976. The equilibrium distribution of a chemical mixture. Phys. Lett. **59A**, 333.

van Kampen, N. G., 1982. Stochastic Processes in Physics and Chemistry. North-Holland, Amsterdam.

van Wijland, F., Oerding, K., Hilhorst, H. J., 1998. Wilson renormalization of a reaction-diffusion process. Physica A **251**, 179.

Vázquez, F., López de Haro, M., 2001. Fluctuating hydrodynamics and irreversible thermodynamics. J. Non-Equilib. Thermodyn. **26**, 279.

Velarde, M. G., Schechter, R. S., 1972. Thermal diffusion and convective instability. II. An analysis of the convected fluxes. Phys. Fluids **15**, 1707.

Velasco, R. M., García-Colín, L. S., 1991. Viscoheat coupling in a binary mixture. J. Phys. A: Math. Gen. **24**, 1007.

Villar, J. M. G., Rubí, J. M., 2001. Thermodynamics "beyond" local equilibrium. PNAS **98**, 11081.

Vlad, M. O., Ross, J., 1994a. Fluctuation dissipation relations for chemical systems far from equilibrium. J. Chem. Phys. **100**, 7268.

Vlad, M. O., Ross, J., 1994b. Random paths and fluctuation dissipation dynamics for one-variable chemical systems far from equilibrium. J. Chem. Phys. **100**, 7279.

Vlad, M. O., Ross, J., 1994c. Thermodynamic approach to nonequilibrium chemical fluctuations. J. Chem. Phys. **100**, 7295.

Vladimirskii, V. V., 1943. Dokl. Akad. Nauk SSSR **38**, 229.

Wada, H., 2004. Shear-induced quench of long-ranged correlations in a liquid mixture. Phys. Rev. E **69**, 031202.

Wada, H., Sasa, S., 2003. Anomalous pressure in fluctuating shear flow. Phys. Rev. E **67**, 065302(R).

Weeks, J. D., 1977. Structure and thermodynamics of the liquid-vapor interface. J. Chem. Phys. **67**, 3106.

Wegdam, G. H., Keulen, N. M., Michielsen, J. C. F., 1985. Small-angle Rayleigh scattering in CCl_4 subjected to a temperature gradient. Phys. Rev. Lett. **55**, 630.

Westried, J., Pomeau, Y., Dubois, M., Normand, C., Bergé, P., 1978. Critical effects in Rayleigh-Bénard convection. J. Physique (Paris) **39**, 725.

Widom, A., 1971. Velocity fluctuations of a hard-core brownian particle. Phys. Rev. A **3**, 1394.

Winkler, B. L., Decker, W., Richter, H., Rehberg, I., 1992. Measuring the growth rate of electroconvection by means of thermal noise. Physica D **61**, 284.

Wu, G., Fiebig, M., Leipertz, A., 1988. Thermal diffusivity of transparent liquids by photon correlation spectroscopy II. Measurements in binary mixtures with a small difference in the refractive index of both pure components. Int. J. Heat Mass Transfer **31**, 1471.

Wu, M., Ahlers, G., Cannell, D. S., 1995. Thermally induced fluctuations below the onset of Rayleigh-Bénard convection. Phys. Rev. Lett. **75**, 1743.

Zaitsev, V. M., Shliomis, M. I., 1971. Hydrodynamic fluctuations near the convection threshold. Sov. Phys. JETP **32**, 866.

Zhang, K. J., Briggs, M. E., Gammon, R. W., Sengers, J. V., 1996. Optical measurement of the Soret coefficient and the diffusion coefficient of liquid mixtures. J. Chem. Phys. **104**, 6881.

Zwanzig, R., 1965. Time-correlation functions and transport coefficients in statistical mechanics. Ann. Rev. Phys. Chem. **16**, 67.

Zwanzig, R., 2001. A chemical Langevin equation with non-gaussian noise. J. Phys. Chem. B **105**, 6472.

Zwanzig, R., Bixon, M., 1970. Hydrodynamic theory of the velocity correlation function. Phys. Rev. A **2**, 2005.

Zwanzig, R., Nordholm, K. S. J., Mitchell, W. C., 1972. Memory effects in irreversible thermodynamics: Corrected derivation of transport equations. Phys. Rev. A **5**, 2680.

List of symbols and corresponding SI units

Subscripts and superscripts

Latin lowercase subscripts i, j,... generally run over the cartesian coordinates x, y and z. Subscript k generally runs over the components of a mixture. Latin uppercase subscripts N, M,... run from 1 to ∞ and usually distinguishes among hydrodynamic modes. Greek subscripts and superscripts α, β,... run over the different dissipative fluxes (and corresponding conjugate forces) characterizing a giving nonequililibrium thermodynamic problem. Quantities with a tilde, like $\tilde{q}$, $\tilde{q}_{\parallel}$,... always represent dimensionless variables.

Boldface symbols

$\mathbf{f}(\mathbf{r}, t)$ Density of volumetric forces (N m^{-3}), page 11

$\mathbf{J}_k(\mathbf{r}, t)$ Diffusion flow of component k (kg m^{-2}s^{-1}), page 10

$\mathbf{q}$ Wave vector of the fluctuations (m^{-1}), page 49

$\mathbf{q}_{\parallel}$ Component of the wave vector of the fluctuations in the horizontal xy-plane (m^{-1}), page 71

$\mathbf{Q}(\mathbf{r}, t)$ Heat flow (J m^{-2}s^{-1}), page 18

$\mathbf{Q}'(\mathbf{r}, t)$ Dissipative energy flux for a binary mixture (J m^{-2} s^{-1}), Eq. (2.64), page 29

$\mathbf{v}(\mathbf{r}, t)$ Center of mass local velocity (m s^{-1}), page 7

$\mathbf{v}_k(\mathbf{r}, t)$ Local velocity of the k component of a mixture (m s^{-1}), page 10

Caligraphic symbols (dimensionless)

$\mathcal{I}$ Shadowgraph signal, page 225

$\mathcal{R}$ Rayleigh factor ratio, page 107

$\mathcal{R}_T$ Ratio of random stress to random heat correlation intensities, page 195

$\mathcal{R}_s$ Ratio of nonequilibrium enhancement due to mode-coupling, to enhancement due to inhomogeneously correlated noise, page 88

Greek letters

α_p Thermal expansion coefficient (K^{-1}), page 7

β Dimensionless solutal expansion coefficient, page 91

$\dot{\gamma}$ Shear rate (s^{-1}), page 249

γ Dimensionless adiabatic index, page 48

Γ Decay rate (s^{-1}), various subindexes are used for different problems, page 54

$\hat{\Gamma}_s$ Sound attenuation coefficient ($m^2\ s^{-1}$), page 50

$\tilde{\Gamma}$ Dimensionless decay rate, page 125

$\delta\psi$ Divergence of velocity fluctuations (s^{-1}), page 49

$\delta\xi$ Random degree of advancement of a chemical reaction ($m^{-3}\ s^{-1}$), page 259

ΔV Volume element (m^3), page 9

ϵ Dimensionless relative distance to the convective threshold, page 137

ϵ_B Dimensionless nonequilibrium asymmetry of the Brillouin component of the scattering spectrum, page 79

ϵ_D Dimensionless Dufour-effect ratio, page 33

η Shear viscosity ($kg\ m^{-1}\ s^{-1}$), page 26

η_v Bulk viscosity ($kg\ m^{-1}\ s^{-1}$), page 26

$\varkappa_T$ Isothermal compressibility ($N^{-1}\ m^2$), page 7

λ Thermal conductivity ($J\ m^{-1}\ s^{-1}\ K^{-1}$), page 25

λ_0 Wavelength of light in vacuo (m), page 215

$\mu = \mu_1 - \mu_2$ Difference in chemical potentials per unit mass between the two components of a binary mixture (J kg^{-1}), page 28

μ_k Chemical potential per unit mass of the k component of a mixture (J kg^{-1}), page 20

∇T_0 Stationary temperature gradient (K m^{-1}), page 66

∇c_0 Stationary concentration gradient (m^{-1}), page 89

ν Kinematic viscosity (m^2 s^{-1}), page 66

ω Frequency of the fluctuations (s^{-1}), page 49

Φ Phase accumulated by light as crosses a shadowgraph medium (dimensionless), page 224

$\Pi(\mathbf{r}, t)$ Deviatoric stress tensor (N m^{-2}), page 15

Ψ Dissipation function (J kg^{-1} s^{-1}), page 21

ψ Dimensionless separation ratio, page 92

$\psi(\mathbf{r}, t)$ Total specific potential energy (J kg^{-1}), page 17

$\psi_k(\mathbf{r})$ Specific potential energy corresponding to k-component of a mixture (J kg^{-1}), page 16

$\rho(\mathbf{r}, t)$ Mass density (kg m^{-3}), page 7

$\rho_k(\mathbf{r}, t)$ Mass density of component k in a mixture (kg m^{-3}), page 10

σ Surface tension (N m^{-1}), page 35

τ_0 Growth rate coefficient (of the SH approximation), page 137

θ Scattering angle, page 215

ξ_0^2 Dimensionless parameter of the SH approximation, page 173

ζ Dimensionless variable, page 187

Latin letters

a_T Thermal diffusivity (m^2 s^{-1}), page 27

$\tilde{c}$ Dimensionless normalization of the SH hydrodynamic mode, page 173

$c(\mathbf{r}, t)$ Single concentration (mass fraction) needed to describe the composition of a binary mixture (dimensionless), page 29

c_s Adiabatic sound speed (m s^{-1}), page 49

c_k Concentration of the k-component of a mixture in mass fraction (dimensionless), page 20

c_p Specific heat at constant pressure (J kg^{-1} K^{-1}), page 27

c_V Specific heat at constant volume (J kg^{-1}K^{-1}), page 27

D Mass diffusion coefficient (m^2 s^{-1}), page 30

D_V Longitudinal kinematic viscosity (m^2 s^{-1}), page 49

g Gravitational acceleration constant (m s^{-2}), page 11

I_0 Intensity of incident light in a scattering or shadowgraph medium (W m^{-2}), page 215

$I_\mathrm{s}(\omega)$ Spectrum of light scattered by a fluid (W m^{-2} s), page 215

k_p Dimensionless barodiffusion ratio (for a binary mixture), page 31

k_T Dimensionless thermal diffusion ratio (for a binary mixture), page 31

k_B Boltzmann's constant (J K^{-1}), page 42

L Thickness: distance between confinement plates of a fluid layer (m), page 63

m_0 Mass of a single molecule in a one-component fluid (kg), page 51

n Index of refraction (dimensionless), page 99

$p(\mathbf{r}, t)$ Local hydrostatic pressure (N m^{-2} or kg m^{-1} s^{-2}), page 7

q Wave number (m^{-1}), page 50

$q_\parallel$ Magnitude of the component of the wave vector $\mathbf{q}_\parallel$ in the horizontal plane (m^{-1}), page 71

q_0 Wave number of light in vacuo (m^{-1}), page 215

q_i Wave number of light in an incident medium (m^{-1}), page 215

R Distance from the scattering volume to the detector (m), page 215

$S(\mathbf{q})$ Static structure factor (dimensionless for an one-component fluid), *c.f.* Eq. (3.24), page 52

$S(\omega, \mathbf{q})$ Dynamic structure factor (s, for a one-component fluid), page 51

S_T Soret coefficient (K^{-1}), page 31

$s(\mathbf{r}, t)$ Local specific entropy (J kg^{-1} K^{-1}), page 21

$\breve{T}(q_{\parallel})$ Shadowgraph optical transfer function (dimensionless), page 227

$T(\mathbf{r}, t)$ Local temperature (K), page 7

$u(\mathbf{r}, t)$ Internal specific energy (J kg^{-1}), page 18

Sans serif symbols (matrices combining several units unless indicated)

$\mathsf{C}(q)$ Noise correlation matrix, page 51

$\mathsf{G}^{-1}(\omega, q)$ Inverse of the linear response function (or matrix), page 49

$\mathsf{M}(\mathbf{r}, t)$ Dissipation matrix, or matrix of phenomenological coefficients, page 23

$\mathsf{P}(\mathbf{r}, t)$ Pressure tensor (N m^{-2}), page 12

Dimensionless numbers

Le Lewis number, *c.f.* Eq. (5.6), page 94

Pr Prandtl number, *c.f.* Eq. (4.7), page 66

Ra Rayleigh number, *c.f.* Eq. (4.8), page 66

Sc Schmidt number, *c.f.* Eq. (5.6), page 94

List of abbreviations

Subject index

www.ingramcontent.com/pod-product-compliance
Lightning Source LLC
Chambersburg PA
CBHW060331220326
41598CB00023B/2677